Studies in Ecological Economics

Volume 8

Series Editors
R. Kerry Turner, School of Environmental Sciences, University of East Anglia, Norwich, UK
Robert Costanza, JG Crawford Building 132, Australian National University, Acton, ACT, Australia
Joshua Farley, University of Vermont, Burlington, VT, USA

The series "Ecological Economics" publishes peer reviewed monographs and edited volumes that offer fresh perspectives and overviews on a range of topics in Ecological Economics. Special attention will be given to integrative research methods and techniques and to pluralistic approaches. Fundamental and critical discussions are invited of: theoretical assumptions, ethical starting points, behavioural models, (co)evolutionary change, integration concepts, the ecosystem approach and ecosystem services environmental macroeconomics, industrial ecology, spatial dimensions, thermodynamics and production functions, policy goals and instruments, international policy dimensions, alternative welfare measures, valuation and benefits transfer. Particular emphasis will be placed on the interaction between valuation, modelling and evaluation in a multidisciplinary setting; the link between ecology, biodiversity, ecosystem services, economics politics and environmental management; the incorporation of physical flows in economic models; and the interface between development, poverty, technology and sustainability. In addition, applied and policy oriented research is welcomed, addressing specific resources, substances, materials, regions, sectors, countries or environmental problems. International comparative studies are also encouraged. The ultimate aim of this series is to present a rigorous but broad perspective on contemporary and future environmental policy questions.

The series welcomes new proposals! To submit a proposal, please fill in a "book proposal form" available on the series page, or contact juliana.pitanguy@springer.com.

Sergio Villamayor-Tomas • Roldan Muradian
Editors

The Barcelona School of Ecological Economics and Political Ecology

A Companion in Honour of Joan Martinez-Alier

Editors
Sergio Villamayor-Tomas
ICTA
Autonomous University of Barcelona
Barcelona, Spain

Roldan Muradian
Faculty of Economics
Universidade Federal Fluminense
Niterói, Rio de Janeiro, Brazil

ISSN 1389-6954   ISSN 2542-9531   (electronic)
Studies in Ecological Economics
ISBN 978-3-031-22568-0   ISBN 978-3-031-22566-6   (eBook)
https://doi.org/10.1007/978-3-031-22566-6

This work contributes to the 'María de Maeztu Unit of Excellence' (CEX2019-000940-M). This work has been financed by the Fundació Autònoma Solidària, Universitat Autònoma de Barcelona

© The Editor(s) (if applicable) and The Author(s) 2023. This is an open access publication.

Open Access This book is licensed under the terms of the Creative Commons Attribution 4.0 International License (http://creativecommons.org/licenses/by/4.0/), which permits use, sharing, adaptation, distribution and reproduction in any medium or format, as long as you give appropriate credit to the original author(s) and the source, provide a link to the Creative Commons license and indicate if changes were made.

The images or other third party material in this book are included in the book's Creative Commons license, unless indicated otherwise in a credit line to the material. If material is not included in the book's Creative Commons license and your intended use is not permitted by statutory regulation or exceeds the permitted use, you will need to obtain permission directly from the copyright holder.

The use of general descriptive names, registered names, trademarks, service marks, etc. in this publication does not imply, even in the absence of a specific statement, that such names are exempt from the relevant protective laws and regulations and therefore free for general use.

The publisher, the authors, and the editors are safe to assume that the advice and information in this book are believed to be true and accurate at the date of publication. Neither the publisher nor the authors or the editors give a warranty, expressed or implied, with respect to the material contained herein or for any errors or omissions that may have been made. The publisher remains neutral with regard to jurisdictional claims in published maps and institutional affiliations.

This Springer imprint is published by the registered company Springer Nature Switzerland AG
The registered company address is: Gewerbestrasse 11, 6330 Cham, Switzerland

Prefaces

A Grateful Appreciation of Joan Martinez-Alier

Herman Daly
University of Maryland, College Park, MD, USA
e-mail: hdaly@umd.edu

Joan (Juan) Martinez-Alier has been a pioneer in ecological economics. In addition to being one of the founding editors of the journal *Ecological Economics*, he contributed enormously to its theoretical development, as well as to uncovering and explaining its foundations in the history of economic thought. As if that were not enough, he served as president of the International Society for Ecological Economics, as professor and mentor of many European students in this new field at the University of Barcelona and as a social activist for ecological justice all over the world.

Back in the mid-1960s, I read something he had written that impressed me, and I wrote him about it. Later while attending a conference in Barcelona, I managed to meet him briefly. Our friendship grew at early conferences on ecological economics in Stockholm and Barcelona and in Costa Rica. On a trip to the USA, he visited me at Louisiana State University, and later, I managed to get him as a consultant at the World Bank for a month, although neither his efforts nor mine had much influence on that institution. I regret that the Atlantic Ocean prevented more frequent and closer contact between us. Nevertheless, collegial friendship with Joan has been a great personal benefit to me, as it has also been to many others. His many intellectual and moral contributions are a gift to all ecological economists, and to the world. *Felicidades y gracias, Joan!*

Memories Concerning the Career of Joan Martinez-Alier

Laurence Whitehead
Nuffield College, Oxford, UK

> Andaluces de Jaén.
> Aceituneros altivos,
> decidme en el alma, quién,
> quien levanto los olivos?
> No los levanto la nada,
> ni el dinero, ni el señor,
> sino la tierra callada
> el trabajo y el sudor.
> Unidos al agua pura
> y a los planetas unidos,
> los tres dieron la hermosura
> de los troncos retorcidos.
> Cuantos siglos de aceituna,
> los pies y las manos presas,
> sol a sol y luna a luna,
> pesan sobre vuestros huesos?
> Jaén, levántate, brava,
> sobre tus piedras lunares,
> no vayas a ser esclava
> con todos tus Olivares.
> Andaluces de Jaén (Miguel Hernandez).

My mother tried (unsuccessfully) to become a volunteer nurse on the Republican side in the Spanish Civil War and prompted me at an early age to read George Orwell's *Homage to Catalonia*. In my fifteenth summer, before the tourist boom, she took me for a week to Tossa del Mar, partly in the hope that I would learn a little Spanish. I remember on the beach meeting another teenager, a couple of years senior to me, and enquiring (inarticulately) what he thought of Franco. At first he looked all round, then led me to a quiet spot. "Odio a Franco." I tried to understand why, but the barrier separating us was more than linguistic. No-one I knew in London was as intense as this. "After the war he killed my father, he killed my two older brothers. The only reason I survived was because I was so little."

Such antecedents shaped my reactions when, 8 years later, I first met Joan Martinez-Alier – at the Latin American Centre in Oxford University, where I was beginning my academic career. My spoken Spanish was still shaky, but I had learnt to read quite well. Joan had just published his first book with *Ruedo Iberico*, an ethnography of olive growing in southern Spain, and he needed someone to help him translate into grammatical English. After a few months I became remarkably proficient in the specialized vocabulary of olive planting, hoeing and harvesting, even though I had never knowingly clapped eyes on an olive grove, and was still barely capable of a social conversation in Spanish on more familiar matters. What did penetrate my consciousness was the spirit of the Miguel Hernandez poem quoted above.

Little by little I became acquainted with more of Joan's work and outlook, and over the ensuing half century, our paths have crossed from time to time, although only rarely and fleetingly. Since the great hinterland of his life is closed to me, all that I can contribute here are the following fragmentary and episodic recollections. They concern his impact on me, rather than the overall arc of his existence.

Reading about his *oliveros* reinforced one of my early professional biases. Whatever grand theory or social science orthodoxy happened to absorb my attention at any particular time, Joan (like my mother in this respect) always reminded me of the need to listen to ordinary people, to value close observation of real conditions, to resist the intellectual's temptation to substitute a fancy generalization for the granular texture of actual human practices. I was more of a welfare statist than Joan (always the anarchist at heart). But Joan's encouragement of patient ethnographic enquiry, informed by a genuine openness to humanity's strange diversity of "really existing" conditions, beliefs and practices, has proved a vital antidote to the "ivory tower" distortions of a lifetime spent in Oxford Colleges.

Oxford's Latin American Centre has always needed some capacity to monitor and explain the course of the Cuban Revolution, and (unlike in Spain) there is almost no-one in British academia who has the opportunity to pursue that thread of enquiry. On a modest scale that task has normally fallen to me, and in the early days, Joan played a creative role in strengthening that project. The core problem has always been that the Fidelista lobby can tolerate no critical commentary, while the massed weight of US (and therefore English language) scholarship on Cuba is suffused with anti-fidelista and Miami-centric bias to such an extent that little in-between can get a hearing. But a few of us could listen to both sides without being captured by either orthodoxy, and the mandate of the Latin American Centre was to pursue the evidence that both camps distorted and disvalued. In particular, my problem has always been that although I could easily access both Havana and Miami, I lacked the face-to-face contact needed to reliably bypass the clashing ideologies. In this regard, at an early stage in my career, I recall making some approximate and unfounded guesses about peasant life in communist Cuba. Joan quietly but firmly informed me of the sad and sorry realities, at least as he had observed them firsthand in one small community in the neglected eastern interior. Since then I have I have been more careful to remember what I did not know about the island.

My early lived experiences of Latin America came from Bolivia (later also Chile and Argentina) rather than the Caribbean. In due course, Joan also undertook fieldwork in the Andes, and again he saw from the bottom-up realities that I was mainly aware from the top (the capital cities) down. In particular, I recall that he taught in the Universidad de Huancayo (in Central Peru), around the time that it was becoming the epicentre of what would soon become known as *Senderismo Luminoso*. From my privileged vantage point it seemed credible to believe that these "Maoist" intellectuals were indeed true believers in the doctrines then being exported from Beijing, but again Joan provided a sobering corrective. I recall learning two points in particular. His students came to his courses suffering from malnutrition; and they did not need to believe that a Maoist seizure of state power was really attainable. It was worthwhile to join the revolutionary struggle so that at least those with enough to eat would find out what it was like to suffer, even if the prospect of emancipation was just a fantasy serving the ambitions of their leaders. Again, Joan's anarchism proved closer to reality than my welfare statism. (I should have known better from the Khmer Rouge.)

A decade or so later, when I edited the *Journal of Latin American Studies*, I got the chance to publish Joan's ideas on "the environmentalism of the poor". This was a highly controversial piece, which I believe *Past and Present* declined on the ground that it romanticized the attitudes of the peasantry towards the natural environment. My impression was (and remains) that this was an ideal rather than a strictly empirical argument, and as such it provided a much-needed corrective to the then prevailing cult of the market.

However that may be, Joan then progressed to a deeper level of social science theorization. Other contributors can say more about his work in the theory of ecological economics. My comment here only reflects a distant personal impression. When I first went to Cuba I recall enquiring in one of their economics departments about how they could calculate national product in the absence of a price mechanism. "We are Marxists, so we use the labour theory of value" came the reply. It seemed that the more labour they put into production, the greater the value it must generate. Trained as a Keynesian I took price signals as the markers of social value, although like the Cubans I distrusted exploitative property relations (and I overrated the physical outputs of commodities, while downgrading the importance of financial services). By the time Joan was developing his challenge to economic orthodoxy both the Cuban and the Keynesian approaches were being displaced by what can be briefly summed up as "neoliberalism". All those who thought that something essential was missing from the new calculus of society's measurement of economic optimization needed to develop a workable method of precise aggregation other than that underpinning the prevailing orthodoxy. It was a bold and truly radical idea to measure the worth of production in terms of its energy content and environmental impact. Since neither the labour theory of value, nor the physical commodities perspective, could meet this challenge that left a void that ecological economics could endeavour to fill.

My two most recent encounters with Joan were both occasioned by his return visits to Oxford. He played a leading role in an informal 1-day seminar that I organized in Nuffield place. At that time what most struck me was the impact his ideas were having outside their initial audiences in southern Europe and in Latin America. Again on his latest visit (only last year), it was clear that ecological historians and contemporary radical policymakers in Asia, as well Europe and the Americas, were developing his insights. I also know that he continues to play a lively (and provocative) role in Catalan academia and politics. Perhaps my original informant from Tossa del Mar has rallied to his banner? And maybe there are still Andaluces de Jaen who honour Miguel Hernandez and who cultivate their olive groves on Joan's principles of ecological soundness.

Conversations with a Catalan Polymath

Ramachandra Guha
Krea University, Sri City, Andhra Pradesh, India

I first heard of Juan/Joan Martinez-Alier sometime in the second half of 1987. I had just returned to India, after a spell as a visiting lecture at Yale University, prior to which I had written a dissertation in Calcutta on the history and prehistory of the Chipko movement. Now, living and working in Bangalore, I met a man called Paul Kurian, who had studied at the Jawaharlal Nehru University (JNU) in New Delhi. JNU was a hotbed of student radicalism, sort of an Indian Berkeley. Like most males who had passed through that university, Paul wore a beard and carried a *jhola*; however, unlike them, he thought for himself. The left-wing student leaders at JNU worshipped at the altar of Stalin and Mao. Paul told his classmates that if they indeed wanted to combine intellectualism and Marxism, then the chap whose writings they should study was named Leon Trotsky.

After JNU, Paul went off to work with the Sandinistas in Nicaragua and with Solidarity in Poland. Somewhere along the way he picked up a book called *Ecological Economics*, a history of how environmental ideas had been suppressed by mainstream economists, whether Marxist or Neo-classical or Keynesian.

Paul Kurian was fascinated by the book and determined to bring its author to India. He gifted me a faded photocopy of *Ecological Economics*, and I read and marked it up very closely. I was deeply impressed by its learning and its scholarhip but had doubts about some of its conclusions.

Three-and-a-half decades later, I still have with me that bound photocopy of Juan Martinez-Alier's *Ecological Economics*. It has travelled with me as I have moved jobs and cities, always finding a place in my bookshelves. Through the 1990s, when I was an active researcher in environmental studies, I dipped into it often, but in recent years I have not had a chance to look at it. Since many readers of this festschrift know the book well, let me share some of the marginal comments I made on it when I first read it three-and-a-half decades ago.

Early on, on page 5, I underlined this phrase with approval: 'There has been a long-standing divorce between Marxism and ecology'. Then, a few pages later, I endorsed this methodological credo: 'I have tried to avoid writing this book in the form of an a priori exercise in legitimation of a putative disciple of ecological economics which would have been in statu nascendi for about 100 years. I have also tried not to alter the other authors' ideas in order to turn them into background support for today's left-wing "ecologism"'. Leafing further, I find that among the phrases I had highlighted was this one, from page 43: 'However, a decrease in the price of oil does not mean that oil reserves in the world have increased'.

Many of the thinkers profiled by Martinez-Alier were new to me, and they wrote in languages I had no access to. My copy of *Ecological Economics* has many passages from these writers underlined for possible future use in my own work. I was learning an enormous amount as I went along, but, being young and argumentative, and Indian besides, I wanted to express my disagreements as well. Thus, when the nineteenth-century anarchist Sergei Podolinsky (one of the book's heroes) is quoted as saying that 'Every nation suffers from foreign rule', I appended the comment in pencil: 'Marx on British India' (a reference to the fact that in a newspaper article from the 1850s, Marx had argued that colonialism sometimes had *both* destructive and regenerative effects). Reading the section on Patrick Geddes, I find that, while underlining many phrases and sentences that I found striking, I also remarked: 'Doesn't seem to have read his Indian town plans'.

Martinez-Alier had shown how the chemist Frederick Soddy thought 'science had proved at least as much a curse as a blessing to mankind', adding: 'The anarchists, in their few remaining strongholds, did not read books or at least did not read Soddy and believed fervently in technical progress'. This elicited from me the comment: 'Only in the West – cf Gandhi in the 1920s!' (for that gentle Indian anarchist certainly believed that science had proved as much a curse as a blessing to humankind). Then, when the author mourned the lack of interest that Marx and Engels had in the work of proto-ecologists like Podolinsky, I appended the remark: 'Technological optimism is one of the main stumbling blocks with regard to a rapprochement between Marxism and ecologism'.

Martinez-Alier's magnificent, scholarly and utterly non-teleological history of ecological ideas in economic thought ended with a 'Political Epilogue'. The first comment I made in this epilogue was complimentary. Against the author's observation: 'A new field of knowledge must be constructed not only intellectually but socially. At least in the short-term, it is of little use publishing if you do not fit into an academic or political group; "parish or publish"' – I had written: 'Nice!' But as I carried on reading I became more combative. On the basis of his research into the past, and the ecological devastation he was seeing in the present, Martinez-Alier remarked: 'I am puzzled by the fact that that left-wing ecologism has grown in the 1970s, and is still growing, not so much in the Third World among part of the youth of some of the most over-developed countries'. He asked: 'Why are there not strong ecological movements in India, in Africa?' To this I answered, on the margin: 'There are – see the CSE [Centre for Science and Environment] reports'. Then, when Martinez-Alier wrote (on the basis of reading books by parochial British authors) that 'there are almost no ecological social movements with roots in the Third World', I responded: 'Rubbish'.

Notwithstanding these disputes and disagreements, I agreed with my friend Paul Kurian that this was a work of defining importance, and we had to find a way to get the author to Bangalore. Now Paul was both a renegade Marxist and an impecunious one. However, he had a brother, Siddhartha, who ran a well-funded NGO in Bangalore. Paul prevailed upon Siddhartha to organize a conference on the Indian environmental movement. The programme was drafted by Paul and myself; we invited scholars and activists from around the country, as well as one foreigner, the author of the aforementioned *Ecological Economics*, Professor Juan Martinez Alier of the Autonomous University of Barcelona. Among the Indian participants were the country's leading ecologist, Professor Madhav Gadgil of the Indian Institute of

Prefaces xi

Science, and one of our most experienced environmental activists, Ashish Kothari or Kalpavriksh.

The conference made possible by the Kurian brothers was organized in Bangalore in August 1988. The venue was an ecumenical Christian centre on St. Marks Road, named Aashirvad, where the out-of-station participants were also staying. The morning the meeting began, I went early, to have breakfast with the speakers. I was asked by Paul to sit with the chief (and only foreign) guest. Juan Martinez Alier was of medium height and wore rimless spectacles. He walked slowly and talked softly. In that first conversation he introduced himself as 'a lapsed Marxist'. It was a brilliant description, and I soon adopted it as my own.

After the conference ended, Juan had a free day before returning to Barcelona, so my wife Sujata and I took him to see the sights outside Bangalore – the Hoysala temples of Belur and Halebid, and the Jain shrine on a hill, Shravanbelagola. Although much older than my wife and I, Juan steadily strode up the hillside, while we unfit Indians panted and stumbled behind him.

At the time, Juan was 50, while I was 30 – an age difference that loomed much larger then than it does now. Through that long day, driving from temple to temple, I was able to flesh out the story of his life – his upbringing in Franco's Spain, his education in Oxford and the years of exile away from his homeland, his first studies of agrarian sociology in Andalusia, his travels in Latin America, his return to Spain after the return of democracy to the country, his greenward turn after reading the works of the British chemist Frederick Soddy and the maverick Ukranian socialist Sergei Podolinsky.

Juan Martinez-Alier was both a polyglot and a polymath. His first languages were Catalan and Spanish, yet he spoke French, German and English well enough to make puns and jokes in them. Hitler's slogan *Blut und Boden*, he liked to say, had in practice become *Blud und Autobahnen*. (And he had a more than adequate knowledge of Portuguese and Italian as well.) To me, who had one-and-a-half languages (fluent English plus conversational Hindustani), this was at once deeply impressive and deeply humiliating. Meanwhile, seeing him effortlessly traverse the disciplines, an economist who became an anthropologist before moving to history and ecology, encouraged me not to be embarrassed about my own (more hesitant and more limited) intellectual transgressions. Juan had extensive first-hand knowledge of North America, Latin America, and Europe; and this trip to India was the beginning of a long immersion in the ecology and politics of a fourth continent, Asia.

I had been greatly influenced by a youthful reading of George Orwell's *Homage to Catalonia*, through it (and some other encounters) rejecting the dogmatic Marxism of my teachers in Calcutta. When Juan was growing up, the book was banned in his homeland, so he only read it when he went to Oxford to study. Later, on a visit to Paris, he met some Spaniards who had worked alongside Orwell in the anti-Stalinist resistance group called POUM. He hung out with them in the émigré bookshop in the Latin Quarter called Ruedo Iberico, nourishing for future use the democratic spirit that Franco had sought to extinguish within Spain itself.

When we set out that morning for our tour of the Mysore countryside, Juan Martizez Alier and I were merely scholarly acquaintances. We returned to the city at night, as friends – for life. Some months after he had returned to Barcelona, I wrote to him that I was contemplating a long essay on the ecological thought of Lewis Mumford, whose books and fugitive essays I had read at Yale. I had, with Juan's assistance, identified three kinds of environmental ideologues, whom I called 'Scientific Industrialists', 'Agrarians' and 'Wilderness Thinkers', respectively. I now told Juan that my study of Mumford showed that he, almost uniquely, did not fit into any of these categories. In fact, he incorporated and transcended all three.

Juan wrote back:

> Do not forget to trace [the Scottish polymath] Patrick Geddes' influence on Lewis Mumford, and to explain also which were the origins of Mumford's anarchism. Geddes belongs to some extent to your 'Scientific Industrialists', in fact he does not, because what you mean is rather 'Ecological Managerialism' (even perhaps 'Socio-Ecological Engineering'), into which the American Techocrats of the 1930s, and also [the forester Gifford] Pinchot, would fit. Geddes was more of an organicist, but not an Agrarian either, and not a Wilderness mystic. There is a current of ecological and Urban Planners and Regional Planners (Geddes, Ebenezer Howard, Ballod-Atlanticus, later Mumford). Geddes was influenced by [the anarchist geographers] Kropotkin and Reclus, but also by [the sociologist Frederic] Le Play.

Juan's letter continued:

> If you look up [Friedrich] Hayek's 'Counterrevolution of Science' (1952) you will see that he classifies Mumford with Otto Neurath, [Frederick] Soddy, [Lancelot] Hogben, Geddes as 'social engineers', too concerned with the study of energy flow, all of them enemies of the Market, all of them descendants of [Henri] Saint Simon, all of them potentially totalitarian utopianists. You have to deal with this issue, how an anarchist utopianist as Mumford, could be classified as an enemy of freedom. ... It would also be interesting to see his position on the Spanish Civil War (I remember vaguely a connection with Luisa Berneri, who wrote a book on utopias, and whose brother Camilo was an Italian anarchist killed in Barcelona in May [19]37, in the communist-anarchist fights). It would be interesting also to see how he weathered the McCarthy period.

This was a letter of advice, instruction and encouragement, the sort of letter I could never have got from one of my mentors at Yale, or from anyone anywhere else in the world. The breathtaking lack of parochialism was at once cultural, geographical and intellectual.

In 1991 I was invited to spend a term at St. Anthony's College, Oxford. I wrote to Juan, who hopped across from Barcelona to visit me. He knew the town well, since he had done his doctorate at St. Anthony's. I found that all his old friends in Oxford called him Whoo-an, pronouncing the J as Y; whereas I, who knew him only as someone who had renounced Marxism for Ecology as well as Spain for Catalonia, called him Jooan.[1]

---

[1] A few years later, my friend was to formally change the spelling of his name, becoming Joan, as in his fellow Catalan, the painter Miro. This has caused much confusion; since some of his early books have him as Juan. Did he, some scholars now wonder, have a sex change? Like the economist Deirdre, once Donald, McCloskey? Or are these two different people, husband and wife or perhaps father and daughter? When citing his works myself, I solve the problem by referring to him as 'J. Martinez Alier'.

One day, while walking to Juan/Joan's old College, we bumped into Jairus Banaji, a brilliant Indian Marxist who had been an undergraduate at Oxford and a graduate student at JNU before working with trade unions in Bombay. Now, in his forties, he had returned to Oxford to write a doctoral thesis on the olive oil economy of ancient Rome. When I introduced Juan, Jairus more or less prostrated himself. 'The author of *Landlords and Labourers in Southern Spain*!', he exclaimed. 'That's the finest modern treatment of the dialectic between the formal and the real subsumption of labour. On my shelf, your book lies between [Karl Marx's] *Capital* and [V. I.] Lenin's *The Development of Capitalism in Russia*.'

Fortunately, Juan did not spoil the moment by telling Jairus what he had already told me; that he was by now a *lapsed* Marxist.

Through the 1990s, Juan and I met every other year, in places around the world. Under the auspices of the Social Sciences Research Council, we assembled a working group on 'The Environmentalism of the Poor', whose members included the Indian feminist economist Bina Agarwal, the Mexican agro-ecologist Victor Toledo and the American political scientist Eric Hershberg. One meeting I remember with particular poignancy was organized by me in New Delhi in January 1993. It was an anxious time for me, both politically and personally; the Babri Masjid had just been demolished, catalysing Hindu-Muslim riots across northern India, and Sujata was going through a difficult pregnancy. In between our work sessions at the India International Centre Juan and I went for walks in Lodi Gardens, our conversations calming my nerves and soothing my anxieties.

Another meeting I remember for its display of Juan's sense of humour. Both of us had been invited by the Institute for Socio-Ecological Research in Frankfurt to speak at a conference on 'Sustainability and the Social Sciences'. After the conference ended, the participants were taken for a tour of the city, with the hosts proudly pointing out local initiatives to promote sustainability. One such was an incinerator that burnt the city's wastes to produce energy. As the guide told us in excited detail about this cutting-edge technology, Juan laconically remarked, 'At least they only burn human wastes in incinerators nowadays'.

Juan's wit was mostly mischievous, and I later told him that this caustic exception was perhaps some sort of Freudian rebellion against his romantic companions. His first wife, the anthropologist Verona Stolcke, was German; while his second wife Martha Giralt taught German. Although he had many German friends (including his collaborator on *Ecological Economics*, Klaus Schlüpmann), he had some sort of special feeling against the country, which may have had historic roots – Hiter's support to Franco – but also perhaps aesthetic reasons, the fact that compared to the Englishmen he had lived with Germans had a rather heavy-handed sense of humour.

This Frankfurt meeting was held, if memory serves, in 1997. The next year Sujata and I left our kids with her parents and went to Spain for a working holiday. Juan had arranged for me to give talks in his university in Barcelona, and in Granada and Jaén too, with a visit to Cordoba added on. Juan took us on an architectural tour of his native city, the museum on Picasso's juvenilia and Gaudi's buildings among its highlights, and for a drive into the Catalan countryside as well. Then we went to

Granada, where Juan's friend, the environmental historian Manuel Gonsalez dé Molina, was likewise a splendid host. We walked around the old city, where we encountered a group of Catholic nuns, sourced from Africa since the locals were becoming ever more irreligious and spent an afternoon in the Alhambra.

From Granada we drove to Cordoba, where our host was the sociologist Eduardo Sevilla. After allowing us to see the Mesquita and other sites on our own, in the evening Eduardo took us for a tour of the old town. The sociologist had, like his friend Juan, spent a lot of time in Latin America, where he had developed a liking for the drink known as maté. So, as the darkness fell and the lamps came on, we sat in the square named after Cervantes, the three of us sipping the drink Eduardo had brought along. A quarter-of-a-century later, I retain vivid memories of that enchanted evening, as our host, large of body and larger of heart, filled our cups as we talked.

As a designer with a serious interest in architecture, Sujata was enthralled by what she saw in Catalonia and in Andalusia. She was very fond of Juan and enjoyed his sense of humour almost as much as I did. It remains the nicest trip I ever made with my wife. The only problem was the food. It was difficult, really difficult, to be a vegetarian in Spain, although I told Juan that he might console himself that in Germany Sujata had found it even harder.

There were other meetings in Spain organized by Juan where I went alone. One was held in the Andalusian town of Baeza, intense discussions in a church-turned-conference centre with a visit to the lovely little town of Ibizza thrown in. Another meeting was held in Andorra, where on an off-day I went for a long solitary walk, the oak forests and the streams reminding me of my boyhood in the Himalayan foothills. On the drive back to Barcelona, Juan told me of the complicated history of the principality, suggesting that its peaceful recent past might be a model for a solution in Kashmir, which could likewise become an autonomous dominion in which the two large countries bordering it had an avuncular rather than avaricious interest.

Some years later, Juan came to stay with us in Bangalore, with a draft manuscript and his son Ricard in town. During the day, Richard went to school with our children while Sujata and I and Juan worked. In the evenings we chatted and gossiped. Juan was particularly taken with our dog, a gentle black Labrador whom Sujata had named Foucalt (pronounced 'Fuko'), as a joke of her own, aimed at the pomo poco stuff then all the rage in Indian academia. It was a happy as well as productive time, for the manuscript Juan was revising was published as the book we know as *The Environmentalism of the Poor*.

By now, I had myself moved away from environmental research. However, while Juan could no longer try out new ideas on me, I could certainly try out mine on him. I was now writing on the history of Indian democracy, and my friend's formidable knowledge of European and Latin American history helped me place my findings in some sort of comparative context. He no longer had any need to send me drafts of his manuscripts, whereas I had an instrumental interest in continuing to send him drafts of mine. I remember with particular gratitude the help he gave me in writing the introduction and epilogue to an anthology of Indian political thought.

In our years working together, Joan and I had developed a shared distaste for an environmentalist who travelled around the world (always in the first-class cabin) while preaching the virtues of village economics and small-scale agriculture. We thought this person hypocritical as well as wrong-headed, for relentlessly demonizing modern science and valourizing ancient Hindu 'wisdom' as the solution to our environmental challenges today. In November 2009, after Juan had sent me mails from various locations in Western Europe, the United States and Latin America and then fallen silent, I wrote to him: 'You back in Barcelona? Your carbon footprint is approaching [name redacted]...'. I then added: 'By the way, my air travel and hence impact on the earth has drastically reduced once I abandoned environmental for political history – why, I wonder?'

To this taunt Joan responded: 'Am planning to live 95 years, and spend the last 15, or 10, in a village, growing my own food (if any), or perhaps in a sailing boat, so that the CO_2 average comes down noticeably'. I responded: 'Keep those years for writing your memoirs, which I will render into the Queen's English'.

Nine years later we had an email exchange that was infinitely more portentous. In September 2017, the campaigning journalist Gauri Lankesh, who also lived in Bangalore and whom I had known, was murdered by Hindu fundamentalists. The day after she was killed I received a mail from Joan. Since his first visit to India in 1987, my friend had returned often, working with and inspiring young ecological economists, and travelling through the countryside mapping environmental conflicts. In the 30 years since we had first met he had come to know my country rather well.

Now, on hearing of the assassination of a writer in my city, Joan thought I might be at risk too. He knew I often attacked Hindu fundamentalism in my newspaper articles, where I also made clear my distaste for the ideology and personality of Prime Minister Narendra Modi. So he wrote suggesting that I go out India for a spell, perhaps to London, 'a 3 months working holiday', or else in Barcelona, where he lived. 'Living outside for a while would not stop your presence in internet and newspapers in India', he remarked.

Joan's advice stemmed from personal affection, and from an acute sense of history. He had himself grown up in Franco's Spain and then spent many years overseas. His first books were published in emigré editions in Paris. He had gone back to his homeland only after democracy was restored to it.

I wrote back to Joan, saying that I was in London, en route to the States, to see my daughter (then studying at Harvard) and to speak at a conference. I would be back home in 2 weeks, I said, adding: 'And then I will stay put there. For the prospect of an enforced and extended exile fills me with dread. My karmabhumi [place of work] is India; there I will live and die.'

To my profession of patriotism, Juan answered:

> If they killed Gandhi, getting rid of the biographer might seem a minor job for some of these crazy people, in Bangalore or elswhere in India, protected by powerful people. Your roots in India cannot be doubted at this point of your life, you will be 60 next year. This is where you have lived, have written your books, have influenced and do influence public opinion and will die and be cremated – no doubt with some appropriate rituals.
>
> However, death should come as late as possible (as also in my case), provided that our brains are in working order, as they obviously are. This is one of your duties, to remain alive

(for your family but also for your country). If there was a yellow fever or cholera epidemic in Bangalore, you would take some precautions. I am glad you are in London, and then in Harvard, good libraries to spend some months there and correct the proofs of the Gandhi book and write your columns.... A working holiday. Nobody talks of exile, this is not the word. It's rather coming and going.

'You have seen harsh and horrible times in your own country, when you were young', I wrote back to Joan: 'I am seeing them in my country when I am in late middle age. Modi too shall pass, as Franco did.'

Joan now sent a mail with the subject line: 'One thing you could do', the text reading: 'In your travels and lecture tours in America etc. and in Europe you could get a group of younger people from India (mostly Indian graduate students and young professionals) to set up a civil society organization representing the values that you defend, based on the Constitution of India and its founding mothers and fathers. Something opposing Hindutva at the ideological level, with a name that represents this. There are hundreds of thousands, millions of Indians outside India. Many of them manual workers, shopkeepers... also.'

This suggestion emanated no doubt from Joan's own life history. It was important to sustain, outside India, the democratic values of the Indian Republic, so that – as had happened in Spain – these values could one day reclaim the land and its Constitution from those who sought to destroy it.

I'd like to end this essay where I began, with the book I read just before I first met my friend and companion of three decades and counting. In the 'Political Epilogue' to *Ecological Economics*, Juan (not yet Joan) Martinez-Alier asked 'the question on the plausibility of international ecological neo-narodnism as an ideology for the dispossessed of the earth'. He continued: 'It is doubtful, however, whether ideas originating in the First World are fit for consumption in the Third World. Who will be the intermediaries, what distortions will take place in transit?' To these questions I had, back in 1987, posed a question of my own on the photocopied page: 'What about ideas originating in the Third World?'

After Juan/Joan came to India later that year, he began asking that question too. The meeting in Bangalore acquainted the visiting Catalan scholar with a vigorous environmental debate in India, itself inspired by popular struggles such as the Chipko movement in the Himalaya and the fisherfolk's movement in Kerala. Unlike the 'full-stomach' environmentalism of the West, these livelihood struggles represented an emerging environmentalism of the poor.

Through the 1990s and beyond Joan came often to India, being inspired by scholars and activists and teaching them a great deal in return. He was also spending a lot of time in Latin America, particularly in Ecuador, studying shrimp farming on the coast and indigenous knowledge systems in the Andes. Within his own continent he embraced the European project enthusiastically, making close connections in France and Germany in particular, even as he was becoming more fervently committed to the creation of an independent Catalan state. And he was visiting and speaking in the great universities of North America too.

Back in 1987, Joan hoped for the emergence of 'international ecological neo-narodnism as an ideology for the dispossessed of the earth'. But, he wondered, 'whether ideas originating in the First World are fit for consumption in the Third World. Who will be the intermediaries, what distortions will take place in transit?' Thirty-five years later, we can say that many such intermediaries have since emerged, among whom certainly the foremost is the author of *Ecological Economics*. He has interpreted Catalonia to Spain, Spain to Europe, Europe to North America, Latin America to India and India to Latin America, the Third World to the First World and vice versa. He has not been alone in this task, of course, but amongst all of us he has conveyed the most wisdom, as well as been responsible for the fewest distortions.

[Ramachandra Guha's books include *The Unquiet Woods* (1989), *Environmentalism: A Global History* (2000), *India after Gandhi* (2007) and *Gandhi: The Years That Changed the World* (2018). He lives in Bengaluru, which was formerly known as Bangalore.]

Joan Martinez-Alier and the Crisis of Civilization, Knowledge, and the Human Species

Víctor M. Toledo
Universidad Nacional Autooma de Mexico, Mexico City, Mexico
e-mail: vtoledo@cieco.unam.mx

ONE. Time has passed. It has almost run out. We will soon be or not be assessed by historians of environmental and critical thinking. Of course, so long as humankind can transcend what is the second most dangerous phase in human history (paleontologists agree that the first phase occurred when *Homo sapiens* were reduced to a minimum and were trapped on the coast of South Africa due to a freezing climate). After all, our species is the only survivor of the ten species constituting our genus. Time has passed and what we have watched like a distant horror movie has drawn close without our even noticing. We are now enveloped in it. From mere fanatical movie spectators, over a few decades we have become actors and actresses in the drama. The crisis of the human species is above all a crisis of civilization. It is the crisis of a modern, industrial, capitalist, technocratic, patriarchal, and anti-ecological world. However, it is also a crisis of knowledge since we are experiencing an epistemological turning point. Western and Eurocentric thinking has been breaking apart, and the cracks have reached not only the defenders of the system but also its critics. This epistemological crisis, a profound reframing of science's main theories and methods, constitutes an extensive scientific transformation in the sense set forth by Thomas Kuhn in his book *The Structure of Scientific Revolutions* (1996). This work is a reflection about the most important science written about in the twentieth century, with more than 110,000 citations (Google Scholar). For all of the aforementioned, we are experiencing the dawning of a new scientific paradigm, a new way of looking at the world through a scientific lens. This requires adopting a holistic or comprehensive approach (which is interdisciplinary, multidisciplinary, transdisciplinary, etc.) that conducts analyses that conjoin social and natural phenomena. It is a question of substituting "normal science," the science that was gradually imposed and duplicated to the point of leaving a fragmented vision of the world and an over-specialized and alienated science. Today, we are facing the challenge of offering relevant and above all viable analyses in order to overcome the crisis confronting the human species (see González-Márquez & Toledo, 2020). Authors such as Fritjof Capra (*The Turning Point*), Edgar Morin (complex thought), Arthur Koestler (*The Ghost in the Machine*), Silvio Funtowicz and Jerome Ravetz (post-normal science), Enrique Leff (environmental epistemology), and others have contributed to this analysis. Besides, this deep epistemological transformation has made way for a set of what I refer to as "hybrid disciplines," which are reactions to or attempts that emerge from the main fields of social science in order to explore their own objects of study in relation to ecological or natural processes (Fig. 1). Lastly, there exist hundreds of often weak or even naïve proposals to construct a general theory for the study of society and nature. To my way of thinking, only two merit inclusion: the Theory of Socio-Ecological Resilience and the Theory of Social Metabolism.

Fig. 1 Hybrid disciplines. During recent decades, we have noted the emergence of twenty new fields of study that seek to integrate ecological principles to a particular social or applied science. In many cases, there are many more papers, books, congresses, societies, and university courses and degrees emerging from hybrid disciplines than from the "mother disciplines." (Source: González-Márquez & Toledo, 2020)

To the former, another unimaginable innovation should be added – the recognition and acceptance of the existence of other forms of non-scientific knowledge of a pre-modern origin, which Claude Lévi-Strauss (1966) called the "science of the concrete" (Paleolithic and Neolithic), which today survives among 7000 indigenous peoples worldwide. If science has existed for some 300 years, as revealed by key milestones such as the foundation of the Royal Society in England, in 1662, and the French Royal Academy of Sciences, in 1666, what has been dubbed traditional ecological knowledge (TEK) has existed since the origin of the human species around 300,000 years ago. This has given way to another hybrid discipline, a new field of study, Ethnoecology (Toledo, 2013; Toledo & Barrera-Bassols, 2008) which has elicited what is referred to as a "dialogue of forms of knowledge" between scientists and members of local communities – an intercultural dialogue.

TWO. In this context, briefly described above, Joan Martinez-Alier and a whole generation of critical and radical thinkers have generated ideas. We are a diverse and heterogeneous generation that embraces concepts such as de-colonization, sustainability, degrowth, post-development, and post-modernity, among others. Given his prolific innovative ideas, the enormous number of students, collaborators, and colleagues with whom he has interacted, and the projects he has spawned (journals, scientific societies and observatories, as well as databases), Martinez-Alier has become an extraordinary figure within this symphony of radical voices. To the above, should be added an unusual characteristic for European authors, who are normally self-absorbed. Martinez-Alier has travelled more than an airplane pilot, visiting almost all the Latin American countries and various regions of India, besides his own European continent. Officially educated in Economic History, he is known to transgress the split of knowledge into fields of study, which places him as the undisputed founder of two hybrid disciplines: ecological economics and political ecology. Since it would not only be senseless but also impossible to fully survey Martinez-Alier's vast work, I will focus on four themes emerging from my knowledge about the academic career of Joan, who I met by the first time during the First Conference of the International Society for Ecological Economics (ISEE) in 1990, in Washington, DC. In spite of not having coauthored a single article, chapter, or book with Joan, nor belonging to the Barcelona School, my writings spring from a certain "natural attunement" of our visions, which have become enhanced throughout three decades, a period in which we have participated together in courses, conferences, congresses, and travels, and of course, my intellectual, fraternal, and family-like relationship with him.

THREE. I discovered Spain in 1992, when I attended three unorthodox academic events and met Eduardo Sevilla (from Córdoba), Manuel González de Molina (from Granada) and Joan Martínez Alier (from Barcelona), with whom I have frequently exchanged ideas throughout my scientific trajectory. They are my three main Iberian *interlocutors*. Together with them and several other close Latin American and Spanish colleagues, we have constructed alternatives to the civilization crisis in at least four hybrid disciplines: agroecology, political ecology, ecological economics, and environmental history. With them, I found an academic attunement that enabled me to overcome intellectual solitude. They were not only perceiving the world through a lens that articulated ecology and Marxism, but also from an agrarian perspective in which the core stakeholders are the peasants of the world. That date coincided with the launching of a journal entitled *Ecología Política* (*Political Ecology*) in Barcelona, in 1991, created by Martinez-Alier. I had the honor to open the first issue with an article about Mexican peasantry. This journal has published 60 issues so far and has been joined by other similar journals in the United Kingdom (*The Ecologist*), the USA (*Capitalism, Nature, Socialism*), France (*Écologie Politique*), Italy (*Capitalismo, Natura, Socialismo*), Greece (*Society and Nature*), and India (*Down to Earth*). Joan enthusiastically promoted the interchange of essays between these journals, which came to form a kind of editorial consortium. *Ecología Política* has been a decisive publication that has stimulated and influenced thousands of scholars, activists, and social organizations in the Spanish-speaking world.

FOUR. Another of Martinez-Alier's contributions is his pioneering role in the theoretical construction and application of the idea of *Social Metabolism* (see Martinez-Alier, 2004, 2009). All hybrid disciplines point to the need to formulate a general theory conjoining society and nature. So far, social metabolism has been the most promising theory. The origin of this concept dates to the nineteenth century and relates to none other than Darwin and Marx. The nineteenth century was the era of the British Empire and London was not only the largest city in the world (with a population of 6.7 million inhabitants), but also the world capital of finances, commerce, politics, and intellectual creation. Numerous scholars and researchers from the most advanced intellectual circles of that time lived in London, including two giants of thought: Charles Darwin (1809–1892) and Karl Marx (1818–1883). Darwin established his home in a town near London in 1842 after returning from his exploration voyage to the southern hemisphere. Marx arrived in London 7 years later as part of his journey through several European cities. He remained there until his death. Although they lived within sixteen miles of each other, Darwin and Marx never met in person. Their contact was limited to correspondence initiated by Marx, who so admired Darwin's work that he felt inclined to dedicate the second volume of *Capital: A Critique of Political Economy* to him. Both Marx and Engels acquired and immediately read Darwin's *On the Origin of Species*, published in November, 1859.

Marx had not only extensively read the works of naturalists of his time, but he had also perused the work of a key author from the Netherlands, Jacob Moleschott (1822–1893), who was widely recognized in the European natural science circles and wrote *Der Kreislauf des Leben* (*The Circuit of Life*). From this work, Marx derived the key concept that allowed him to build his critical theory of capitalism: *Stoffwechsel*, translated as metabolism or organic exchange.

The concept of social metabolism remained virtually dormant during decades until Marina Fisher-Kowalski formally relaunched the concept in a chapter of the book *Handbook of Environmental Sociology* published in 1997 presenting it as a conceptual star that serves to analyze energy and material flows. Another two precursors were Alfred Schmidt and his "The Concept of Nature in Marx" (1971), the English version of which I came across in a book store at Harvard University in 1973. Undoubtedly, it was Joan who promoted the study of social metabolism, in

Barcelona, and who paved the way for three clearly defined tendencies: that which he promoted in several Latin American countries through various theses on the interchange of energy and materials; that which Mario Giampietro led and developed; and that which was developed by economists and geographers from Catalonia through the spatial studies led by Enric Tello, Joan Marull, and other authors.

Over the last decades, the concept of social metabolism has gained prestige as a theoretical instrument for required analyses, to such an extent that there are now dozens of researchers, thousands of articles, and several books that have adopted and used this concept. However, there is a wide variety of definitions and interpretations, as well as different methodologies surrounding this concept, which hinders the consolidation of a unified field of new knowledge. In *The Social Metabolism* (2014) coauthored by M. González de Molina and I, we have attempted to provide a theory of social metabolism that actually takes up the socio-ecological character of this concept in its full complexity. The theoretical and methodological innovations of this work include: (a) the rigorous definition of a basic model for the process of social metabolism and its components (beyond the "black box"); (b) the distinction of two main types of metabolic processes: tangible and intangible; (c) a detailed discussion regarding the concept of nature appropriation; (d) an analysis of social metabolism at different scales (spatial dimension); (e) a historical analysis of social metabolism (temporal dimension and socio-ecological change); (f) overcoming the merely "systemic" or "cybernetic" nature of approaches; (g) providing agency and visibility to collective action, as well as to the consequences and explanations of the above; and (h) integration of an ethical and political dimension into this theory.

FIVE. *The Environmentalism of the Poor* (2004), published in English, Spanish, and Portuguese, is Martinez-Alier's most widely consulted and cited work. Environmentalism emerged half a century ago in the central and most industrialized countries (Europe, the USA, Japan, and others) and emphasized on urban and industrial problems regarding waste and residues. Antinuclear struggles, which were unable to stop the proliferation of the energy industry, characterized this first stage of world environmentalism. Nevertheless, the expansion of extractive activities unleashed an ever-increasing number of environmental conflicts in rural areas throughout the world to such an extent that in a few decades the center of environmental injustice spread to rural regions of the peripheral countries. The expansion of agricultural monocropping and cattle-raising, the over-exploitation of forests and fishing, mining, the extraction of fossil fuels, dam-building, the multiplication of tourist hubs, expansion of highways, and more recently the new wind and solar energy projects gave way to innumerable conflicts. These conflicts were raised among local populations (communities, municipalities, and regions) forced to defend their territories from attacks by private and government projects of all kinds. The environmental struggle spread, therefore, to marginalized sectors in rural areas: small-scale farmers, fishermen, shepherds, indigenous and Afro-descendant peoples, seasonal workers, and foragers. Joan's book focuses on characterizing this second type of environmentalism. He was ahead of what today represents the main socio-environmental battles worldwide. It should be noted that two decades later, this theoretical work was followed by a pioneering cyber project: *The Global Atlas of Environmental Justice* (EJAtlas) devoted to documenting socio-environmental conflicts throughout the world. Making these events visible to the public worldwide has been an indisputable achievement. The EJAtlas that today records almost 3500 conflicts constitutes an extremely valuable instrument of analysis and condemnation (including the filing of lawsuits). Its example has been duplicated on a national scale in countries like Mexico, and there exists a Latin American Observatory of Environmental Conflicts (http://www.olca.cl).

SIX. I conclude this brief review by resorting to three enduring personal and family reminiscences. The first is when, in 1996, he invited my family to his home in Barcelona, during a 6-month stay there. He prepared a strange and delicious *quinoa paella*, and our respective wives began an ongoing deep friendship, which exists until today. The second memory is the "anti-wedding" they held when Joan and Marta were obliged to marry by the United States government in order to be able to conduct research at Yale University, a most enjoyable but bizarre event. The third was a trip taken by our families in Mexico, from Xalapa to Oaxaca and from there to Morelia. The van we used unbelievably was able to transport the eight of us, our respective baggage and myriad cooking utensils, table games, and bedding. These are life-celebrating memories.

Barcelona, Spain   Sergio Villamayor-Tomas

Niterói, Rio de Janeiro, Brazil   Roldan Muradian

References

Fischer-Kowalski, M. (1997). Society's metabolism: On the childhood and adolescence of a rising conceptual star. In M. Redclift & G. Woodgate (Eds.), *The international handbook of environmental sociology* (pp. 119–137). Edward Elgar.
González-Márquez, I., & Toledo, V. M. (2020). Sustainability science: A paradigm in crisis? *Sustainability, 12*, 2802. https://doi.org/10.3390/su12072802
González de Molina, M., & Toledo, V. M. (2014). *The social metabolism*. Springer. 355 pp.
Kuhn, T. S. (1996). *The structure of scientific revolutions*. The University of Chicago Press.
Levi-Strauss, C. (1966). *The savage mind*. The University of Chicago Press. 310 pp.
Martínez-Alier, J. (2004). Marx, energy and social metabolism. *Encyclopedia of Energy, 3*, 825–834.
Martínez-Alier, J. (2009). Social metabolism, ecological distribution conflicts and languages of valuation. *Capitalism Nature Socialism, 20*(1), 58–87.
Schmidt, A. (2014). *The concept of nature in Marx*. New Left Books-Verso.
Toledo, V. M. (2013). Indigenous peoples and biodiversity. In S. A. Levin (Ed.), *Encyclopedia of biodiversity* (2nd ed., pp. 269–278). Elsevier.
Toledo, V. M., & Barrera-Bassols, N. (2008). *La Memoria biocultural*. Icaria Editorial.

Contents

Part I Introduction

1 **Justification and Scope of the Book**........................... 3
Roldan Muradian and Sergio Villamayor-Tomas

2 **A Barcelona School of Ecological Economics and Political Ecology**.. 9
Joan Martínez-Alier

3 **The Barcelona School of Ecological Economics and Political Ecology: Building Bridges Between Moving Shores**....................................... 17
Sergio Villamayor-Tomas, Brototi Roy, and Roldan Muradian

Part II Epistemological Foundations

4 **Metaphysical Midwifery and the Living Legacy of Nicholas Georgescu-Roegen**............................... 37
Katharine N. Farrell

5 **Languages of Valuation** 47
Christos Zografos

6 **Post-development: From the Critique of Development to a Pluriverse of Alternatives** 59
Federico Demaria, Ashish Kothari, Ariel Salleh, Arturo Escobar, and Alberto Acosta

7 **Indigenous and Local Knowledge Contributions to Social-Ecological Systems' Management** 71
Victoria Reyes-García

8 **Degrowth and the Barcelona School** 83
Giorgos Kallis

Part III Social Metabolism

9 **Agrarian Metabolism and Socio-ecological Transitions to Agroecology Landscapes** 93
Enric Tello and Manuel González de Molina

10 **Multi-scale Integrated Analysis of Societal and Ecosystem Metabolism** 109
Mario Giampietro

11 **Materials Flow Analysis in Latin America** 123
Mario Alejandro Pérez-Rincón

12 **Biophysical Approaches to Food System Analysis in Latin America** 137
Jesus Ramos-Martin and Fander Falconí

13 **Ecologically Unequal Exchange: The Renewed Interpretation of Latin American Debates by the Barcelona School**............. 147
Beatriz Macchione Saes

14 **Flow/Fund Theory and Rural Livelihoods** 157
Jose Carlos Silva-Macher

15 **Deceitful Decoupling: Misconceptions of a Persistent Myth**........ 165
Alevgul H. Sorman

Part IV Environmental Justice Conflicts and Alternatives

16 **Does the Social Metabolism Drive Environmental Conflicts?** 181
Arnim Scheidel

17 **Critical Mapping for Researching and Acting Upon Environmental Conflicts – The Case of the EJAtlas** 195
Daniela Del Bene and Sofia Ávila

18 **The EJAtlas: An Unexpected Pedagogical Tool to Teach and Learn About Environmental Social Sciences** 211
Mariana Walter, Lena Weber, and Leah Temper

19 **Commons Regimes at the Crossroads: Environmental Justice Movements and Commoning** 219
Sergio Villamayor-Tomas, Gustavo García-López, and Giacomo D'Alisa

20 **(In)Justice in Urban Greening and Green Gentrification**.......... 235
Isabelle Anguelovski

21 **From the Soil to the Soul: Fragments of a Theory of Economic Conflicts**... 249
Julien-François Gerber

Part V Science and Self-Reflected Activism

22 Activism Mobilizing Science Revisited 261
Marta Conde and Martí Orta-Martínez

23 Iberian Anarchism in Environmental History 271
Santiago Gorostiza

**24 The Barcelona School of Ecological Economics
and Social Movements for Alternative Livelihoods** 283
Claudio Cattaneo

**25 The Ups and Downs of Feminist Activist Research:
Positional Reflections** 293
Sara Mingorria, Rosa Binimelis, Iliana Monterroso,
and Federica Ravera

**26 From the Environmentalism of the Poor and the Indigenous
Toward Decolonial Environmental Justice** 305
Brototi Roy and Ksenija Hanaček

Part VI Public Policy Applications

27 Agrobiodiversity in Mexican Environmental Policy 319
Nancy Arizpe and Dario Escobar-Moreno

**28 Conventional Climate Change Economics:
A Way to Define the Optimal Policy?** 327
Jordi Roca and Emilio Padilla

**29 Contribution of Global Cities to Climate Change
Mitigation Overrated** 335
Jeroen C. J. M. van den Bergh

30 Reconciling Waste Management and Ecological Economics 347
Ignasi Puig Ventosa

**31 Work and Needs in a Finite Planet:
Reflections from Ecological Economics** 357
Erik Gómez-Baggethun

32 The Environmentalism of the Paid 367
Esteve Corbera and Santiago Izquierdo-Tort

33 Collective Action in Ecuadorian Amazonia 383
Fander Falconí and Julio Oleas

Index 395

Part I
Introduction

Chapter 1
Justification and Scope of the Book

Roldan Muradian and Sergio Villamayor-Tomas

Academic schools are important institutions for consolidating and disseminating ideas. They can be defined as diffused communities held together by a collectively constructed body of knowledge, a shared worldview and a network of social relations. Identifying and delimiting academic schools might be a hard endeavor, since they are embedded in a fluid "state of spirit," which is context- and time-specific, and not always easy to systematize and communicate. However, we think that the notion (or metaphor) of "school" still makes sense to characterize the bonding elements that hold together and give coherence to the diverse ideas, debates and approaches represented in this book.

This book is the first self-reflective and systematic attempt to delineate the scope and boundaries of what we have called the "Barcelona school of ecological economics and political ecology." As stated above, any academic school is constituted by the intersection between a social network and a knowledge dimension. The former refers to the actors and organizations involved and their relations, while the latter has to do with the epistemic principles and methodological approaches that constitute content-wise bonding elements. As the book shows, the thematic scope of the Barcelona school is very broad, including contributions in social metabolism analysis, environmental valuation, ecological knowledge systems, environmental justice, management of the commons, (agro)biodiversity, climate and urban policies and degrowth. We think, however, that some core elements (that we call the pillars of the school) create the sense of a consistent body of knowledge that permeates all these subjects. The book is structured along these pillars.

R. Muradian (✉)
Faculty of Economics, Universidade Federal Fluminense, Niterói, Rio de Janeiro, Brazil

S. Villamayor-Tomas
UAB, Barcelona, Spain

© The Author(s) 2023
S. Villamayor-Tomas, R. Muradian (eds.), *The Barcelona School of Ecological Economics and Political Ecology*, Studies in Ecological Economics 8,
https://doi.org/10.1007/978-3-031-22566-6_1

We want to stress that this compilation is not fully comprehensive. Many possible contributors, who would consider themselves as part of the school, unfortunately, have not been included in this volume, due to diverse reasons (lack of time, different sorts of agenda mismatches or just editorial mistakes). We apologize for any unintended absence. Nevertheless, our intention was to be as inclusive as possible, aiming to reflect the complexity and diversity of approaches that can be gathered together under the metaphorical umbrella of the "Barcelona school". We have used a broad criteria for inviting and including authors in this compilation, such as scholars that have spent some time working or studying in Barcelona and who were influenced by the academic contributions of J. Martinez-Alier. In the book, there is a high proportion of researchers who have completed their PhD at the Institute of Environmental Science and Technology (ICTA) of the Autonomous University of Barcelona (UAB), current post-docs or permanent researchers from this organization. The educational program at ICTA has played definitively a key role in the consolidation and dissemination of core approaches and ideas that represent the bonding elements of the Barcelona School. However, not all contributors have been passed directly by the UAB, though likely most of them have been influenced by the ideas developed there. The authors of the more than 30 chapters composing this book represent several generations of scholars that have contributed to the development of the school over the past 20 years. Though not all the authors that potentially belong to the school are present, we think that the number and composition of authors are representative of the social network that forms the basis of the school.

With regard to the content-wise bonding elements, we have identified three pillars that can be considered as the foundations of the school: (i) an emphasis on the biophysical dimension of the economic system, (ii) an interest in the political and historical aspects underlying the environmental performance of contemporary capitalism, and (iii) the study of alternative ways of knowing, valuing and organizing social life in order to achieve a fairer and more sustainable relationship with the environment, including the engagement with or support of activism.

The book aims to be a reading companion for students as well as both young and experienced scholars interested in becoming acquainted with the complex interactions between human societies and their natural environments, from a critical and socially engaged perspective. This compendium of contributions also has the objective to pay homage to the academic trajectory and work of Joan Martínez-Alier, who has played a decisive role in not only helping to set the conceptual foundations of the different branches of the school but also teaching and supervising students, as well as establishing learning bridges between the academia and civil society.

The book is divided into 6 parts. The second part deals with the analytical foundations of the school (key shared concepts and approaches). Parts three, four and five cover the main transversal threads (pillars) of the school: social metabolism analysis, environmental justice conflicts and activism mobilizing science. The sixth part shows a series of cutting-edge examples of the contributions of the school in policy analysis. Additionally, the book includes an introduction to the history, thematic evolution and social network that characterize the school as a research community, an overview from Joan Martinez-Alier, and words of recognition from other prominent social scientists to Martinez-Alier's important contributions to the fields of ecological economics and political ecology.

Part two (epistemological foundations) sets the theoretical and historical basis of the Barcelona school. The part introduces key concepts and approaches for better understanding the remaining parts of the book. The first chapter, written by K. Farrell, covers the connections between the school's longstanding contributions to social metabolism studies and the work of pioneering scholars such as Nicholas Georgescu-Roegen. Georgescu-Roegen, who introduced core ideas of modern ecological economics, stressed two fundamental themes that are part of the overarching vision of the school: the biophysical constraints to the economic system and the need to define moral goals for guiding the organization of the social life and human-environmental interactions. A key issue for the second theme has been the recognition of the plurality of values that different societies attach to nature and social life. The second and third chapters of this part do justice to this epistemological stepping stone of the school. In the second chapter, C. Zografos addresses key theoretical and methodological tenets of the school's approach to environmental valuation, including paying attention to the diversity of languages of valuation. Different social groups might hold not only contrasting worldviews but also express those views using languages that often are not easily commensurable. In the third chapter, F. Demaria et al. systematize criticisms to the idea of sustainable development. By means of contesting the idea of limitless economic growth and linear human progress, the school comes close to the postulates of "post-development" approaches, which acknowledge the plurality of notions of what constitutes a "good life."

A critical discussion about what constitute legitimate knowledge and research is another epistemological foundation of the school. This discussion is covered in the last two chapters of the part. In the fourth chapter, V. Reyes-García makes a comprehensive review of the literature about non-Western forms of knowing and valuing nature. A concern for understanding and incorporating into decision-making non-Western ways of knowing is a key part of a research agenda that acknowledges the diversity of ways humans relate with the natural environment and one of the two sides of the activism mobilizing science approach of the school. The other side is covered in the final chapter of this part. Building on the experience of the "research and de-growth" movement, the chapter by G. Kallis discusses the challenges and opportunities of promoting new policy agendas for triggering socio-ecological transitions and the practice of academically informed activism for promoting changes both in individual practices and public policies.

The third part (on social metabolism) focuses on one of the foundational themes in ecological economics research, to which the Barcelona school has made seminal contributions: social metabolism. Namely, the material/energy profile of the economic process. Social metabolism analysis has a rather long history that can be traced back to early agrarian studies. The first chapter of this part, written by E. Tello and M. González de Molina, recounts these historical roots. In the second chapter of this part, M. Giampietro's deals with more recent developments in this field. More specifically, it draws an overview of MUSIASEM, a comprehensive framework for understanding, classifying and assessing the diversity of material and energy flows that take place in contemporary societies at multiple scales and sectors. Another important vertex of the school's contributions has to do with the application of material flow accounting at the national scale and over time, in order to measure the unequal ecological trade that certain economies have historically been subject to, as illustrated in the third, fourth and fifth chapters of the third part of the book. In the third chapter, M. Pérez-Rincón makes a comprehensive literature review of material flow analysis in Latin America and the Caribbean. The fourth chapter, by J. Ramos-Martín and F. Falconí, focuses on the biophysical analysis of food systems. In the fifth chapter of this part, B. Saes discusses the concept of ecological unequal trade and analyses the advancement in the theory during the last three decades. The notion of social metabolism can connect ecological distribution conflicts and the limits to growth discourses. As illustrated in the sixth chapter of this part, by J. Silva-Macher, trade relations have determined to a large extent land use transformation in resource-exporting countries, often driving socio-environmental conflicts. As pointed out in the final chapter of this part, by A. Sorman, social metabolism thinking can also help to unveil the myths and realities behind the possibility to decouple growth from the material throughput of economic systems.

The fourth part (on environmental justice conflicts and alternatives) deals with the second pillar of the Barcelona school. Namely, it has to do with the empirical and conceptual work on environmental justice. Several of these contributions are derived from the Environmental Justice Atlas project. A key working hypothesis of this project is that the rise of material and energy throughput results in more ecological distribution conflicts. Building on this postulate, A. Scheidel's chapter examines the need for integrating biophysical and social dynamics for achieving a more nuanced and comprehensive understanding of socio-environmental conflicts. The second and third chapters of this part also belong to this research program. In the second chapter, D. Del Bene and S. Ávila introduce the Global Atlas of Environmental Justice (EJAtlas) as one of the examples of research projects of the Barcelona school and elaborate on how it has been developed as a novel tool for spatial, comparative and statistical political ecology. In the third chapter, M. Water et al. examine the role of the EJAtlas as a tool for "public political ecology" by describing how it is used in formal teaching. These efforts are currently one of the most visible examples of the longstanding commitment of the school to strengthen the relationship between science and activism.

The environmental justice program is in continuous evolution, including exchanges with other approaches, and this is reflected in the last three chapters of the part. In the fourth chapter, S. Villamayor-Tomas, G. García-López and G. D'Alisa connect with the commons scholarship by analyzing the interface between environmental justice movements and community-based natural resource management. In the fifth chapter of this part, I. Angelovsky connects with the urban environmental justice tradition by discussing the work around the "greening paradox" of urban environmental policies that disregard social inclusion concerns. In the final chapter of the part, J.F. Gerber reminds us of the intricate connections between ecological conflicts and a variety of other social conflicts, from a historical and degrowth perspective. Each of these chapters proposes new frontiers for the Barcelona school, which involves re-politicizing older debates, with new perspectives and methods.

The fifth part (on science and self-reflected activism) addresses the third pillar of the Barcelona school, which is related to the postmodern take of science as a practice that needs to be understood in its own historical and political context and whose role needs to be continuously revised through self-reflection. Much of this understanding has to do with one of the previously mentioned epistemological foundations of the school: the importance of plurality of knowledge systems (e.g., chapter by V. Reyes-García) and of the political projects behind sustainability research (e.g., chapter by Kallis). In this spirit, the chapters of this part combine academic language with storytelling and self-reflection. In the first chapter of this part, M. Conde and M. Orta discuss the importance given within the school to "activism mobilizing science." They review key aspects of this approach in the scientific community and then illustrate it with three examples in different contexts. The importance given to "science within a political context" and self-reflection is tackled in the following three chapters. In the second chapter, S. Gorostiza systematizes the historical roots of the Barcelona school's research agenda, which can be traced back to anarcho-syndicalism and its influence on agrarian studies in the early twentieth century. In the third chapter, C. Cattaneo self-reflects on the commitment of the school with "theory in the making" and the connections between science and the lives we practice. In the fourth chapter of this part, S. Mingorria et al. discuss the strong connections between research and personal choices through a self-reflective account of the growing commitment in the school to gender studies and the role of women in science. The last chapter of this part, written by. B. Roy and K. Hanaček, discusses the limits of activist-academic interaction and the co-production of knowledge. It does so by advancing decolonial methods and theory within the school's research in environmental justice.

The last part of the book (on public policy applications) illustrates a variety of applications of the concepts and methodological approaches reviewed in the previous part to analyze the development and performance of public policies. This part addresses the variety of problemsheds and cultural contexts to which the school's approaches have been applied. A common thread that links all chapters within this part is the dismantling of policy myths, i.e., a reassessment of policy programs that tend to be praised for their (potential) success. The part opens with a chapter by N. Arizpe and D. Escobar, which critically assesses top-down policy approaches for agrobiodiversity conservation and highlights the importance of bottom-up processes led by local communities and environmental justice movements. Next, J. Roca and E. Padilla systematize and challenge the main economic premises on which mainstream climate change policies have been traditionally elaborated, including the common assumptions of economic models, the arbitrariness of discount rates, and issues dealing with (un)certainty and commensurability. The third chapter, by J. van den Bergh, also addresses the policy challenge of climate change by questioning widespread optimism about the contribution of cities to global reduction of greenhouse gas (ghg) emissions. Then I.Puig, in the fourth chapter, discusses the promise of the circular economy through a detailed analysis of waste management policies at European and municipal levels. In the fifth chapter of this part, E. Gómez-Baggethun critically assesses the merits and pitfalls of universal basic income policies in the context of current debates about the future of paid work. The sixth chapter of the part, by F. Falconi and J. Oleas, discusses the development and difficulties of implementing innovative policy propositions aiming to reduce biodiversity loss in the Ecuadorian Amazon. Lastly, E. Corbera and S. Izquierdo-Tort undertake a discussion about the broad implications of payments for ecosystem services, a policy instrument that has received considerable attention among academic and policy circles during the past two decades, in terms of motivation, behavior and social mobilization for environmental protection.

We hope that the readers will enjoy going through this compilation of diverse, interesting and relevant contributions of the Barcelona school. Even though the compilation is not exhaustive, we think it is representative of the main ideas, methods and approaches that have been developed by the school. This book is the result of a large effort by contributors, editors and the publishing company. We want to thank all the persons involved for their engagement, commitment and patience. We expect that this book will contribute with innovative and interesting ideas and an engaged vision to tackle the big socio-environmental challenges we collectively face nowadays, at a planetary level.

Open Access This chapter is licensed under the terms of the Creative Commons Attribution 4.0 International License (http://creativecommons.org/licenses/by/4.0/), which permits use, sharing, adaptation, distribution and reproduction in any medium or format, as long as you give appropriate credit to the original author(s) and the source, provide a link to the Creative Commons license and indicate if changes were made.

The images or other third party material in this chapter are included in the chapter's Creative Commons license, unless indicated otherwise in a credit line to the material. If material is not included in the chapter's Creative Commons license and your intended use is not permitted by statutory regulation or exceeds the permitted use, you will need to obtain permission directly from the copyright holder.

Chapter 2
A Barcelona School of Ecological Economics and Political Ecology

Joan Martínez-Alier

The first 21 years of my life were spent in Barcelona (all of them under General Franco's regime, since I was born in 1939). The following 14 years I spent in Oxford, Stanford, in Andalusia and again in Oxford (St. Antony's College) until 1973. In between, long stays in Cuba, Peru and Brazil and some periods in Paris, with the publishing house of Ruedo ibérico. At 35 years of age and feeling rather defeated by the lack of "transitional justice" in Spain after Franco's death, I came back to Barcelona, with a chair in the new Universitat Autònoma (UAB) in Economics and Economic History, which I held until I was 70 years of age. I continued my travels in the sabbatical years, to Oxford in 1984–1985, Stanford again in 1988–1989, to Ecuador (the Flacso in Quito) in 1995–1995, to Yale University in 1999–2000 and in the meantime also often to India after my first visit in 1988. My interests and my books followed this trajectory, first some books on agrarian history and land conflicts in Andalusia, Cuba and Peru between 1968 and 1977, then between 1984 and 2022 many books on ecological economics and political ecology.

Influenced by agricultural energetics and ecological anthropology, my first articles on energy and agriculture were published in the late 1970s. It was not until the early 1990s that I could start teaching outside the Faculty of Economics at UAB because a new degree course in Environmental Sciences had opened up in the Faculty of Sciences, where I taught Introduction to Environmental Sciences (with Jaume Terradas) and also Environmental and Resource Economics for another 15 years. In the mid-1990s, Dr. Giuseppe Munda, with a recent doctorate from the University of Amsterdam with Peter Nijkamp and having become an expert on multi-criteria evaluation, joined the UAB (pushed by Silvio Funtowicz to do so), and he suggested in 1997 that we should start a doctoral program in ecological economics. The Faculty of Economics was less open to the idea than the Department of Geography (with David Saurí, who held a doctorate from Clark University in the

---

J. Martínez-Alier (✉)
UAB, Barcelona, Spain

USA). And this is how we began in 1997, with a group of students from different countries and disciplinary backgrounds, including Roldan Muradian, Fander Falconí, Jesús Ramos Martin, Daniela Russi, Begum Özkaynak ... and some visiting professors, among whom was Roger Strand. The chapters of this book trace part of the history of this doctoral program until today. We were all very brave to engage on this path, particularly the students. After a few years, the program was very properly housed at the new ICTA building in the UAB, and the teachers came to include internationally famous scholars such as Jeroen van den Bergh, Mario Giampietro and Giorgos Kallis, all of them ICREA professors.

While ecological economics was not taught with its name at the UAB until the mid-1990s, it dates from 10 years before, the mid-1980s as a field of study with an international society, the International Society for Ecological Economics (ISEE), and a journal, *Ecological Economics*. After some informal meetings in Stockholm and Barcelona, the society held its first meeting in Washington DC in 1990. A seminar of a few dozen people at Wye Island nearby organized by Bob Costanza led to a defining publication, *Ecological Economics: The Science and Management of Sustainability*. The disciplinary origins were varied, with dissident economists that followed from Kenneth Boulding and Nicholas Georgescu-Roegen, and systems ecologists (often trained by Howard T. Odum). Systems ecologists such as Charlie Hall soon contributed to ecological economics with tools like the EROI (energy return on investment). This early story was competently summarized by Røpke (2004).

Among the founders and early presidents of ISEE, there were ecologists like AnnMari Jansson and Bob Costanza and dissident economists like Herman Daly, Dick Norgaard, and myself. Other presidents have included John Proops, Charles Perrings, Peter May, John Gowdy, Bina Agarwal, Marina Fischer-Kowalski, Sabine O'Hara, Clovis Cavalcanti, Joshua Farley, Roldan Muradian and Erik Gomez-Baggethun. The last three are (according to their first university degrees): one economist, one biologist and one ecologist, but they have been ecological economists from an early age. In contrast, the founding members became ecological economists.

There are strong regional societies in Latin America, Europe, the United States, Canada, Australia, New Zealand and India. I am a co-founder of three of them. The journal has had as editors Bob Costanza, Cutler Cleveland, Richard Howarth, Stefan Baumgärtner and, most recently, Begüm Özkaynak, and it has been a fundamental research outlet for the practitioners of ecological economics. Some criticisms have been made, not without reason, about the relative absence of feminist economics in the journal and the relative abundance of mainstream economic articles.

The Beijer Institute at the Royal Swedish Academy of Sciences played a confusing role in the establishment and development of ecological economics in the early 1990s. It continued with Anna-Mari Jansson and her student Carl Folke (today a top author by the number of citations in ecological economics), with a focus on energy and human ecology, but it was in the early 1990s transformed to a Beijer Institute of Ecological Economics, which left out ecological economists in favor of mainstream environmental economists such as Karl-Göran Mäler. Another mainstream environmental economist active in the ISEE at the beginning was David Pearce, who in

1994 was asked to leave the editorial board of the journal. These were not clashes of personality but intellectual conflicts. Pearce promoted "weak sustainability" (all forms of capital – manufactured capital, human capital and "natural" capital – could be measured in the same units and substitute for one another) against the more robust view from ecological economics favoring physical indicators and strong sustainability (a requirement that natural capital be preserved in physical terms so that its functions remain intact).

Robert Ayres had already in 1969 introduced (with Allen Kneese) in an article in *The American Economic Review* the accounting of materials in the economy (Ayres & Kneese, 1969), which later flourished in the Vienna group led by Marina Fischer-Kowalski of studies of the social metabolism measuring the material and energy intensities of the economy (e.g., Fischer-Kowalski & Haberl, 2007). This is ecological economics, overlapping with industrial ecology, urban ecology, and agroecology, which are practiced by many other groups.

The debate and tension between, on the one hand, the economic accounting of environmental damages and of nature's services to humans, and on the other hand, their biophysical assessment, has persisted in ecological economics. Sometimes even those most favorable to a multi-criteria biophysical and social assessment have opted for an economic methodology, such as a modified gross domestic product (GDP), that would produce a single indicator and a single number (e.g., the calculation of the Index of Sustainable Economic Welfare (ISEW) was popular for many years). Sometimes, those who started from human ecology and energetics have gone over to the economic counting of the loss of so-called natural capital thinking that this would impress policymakers.

However, the basic tenets of ecological economics still go so against the grain that efforts to bridge the gap and communicate with mainstream economists and so-called policymakers have sometimes led to contentious compromises. Such tenets are:

(a) The economy is embedded in physical and social realities; it cannot be analyzed as a system of its own. The economists' view of the economy as a circular system (that Georgescu-Roegen called "the merry-go-round") in which producers bring their products to the markets where they are bought by consumers who receive their income for the work or services they provide to producers, is wrong. The industrial economy is clearly not circular, it is entropic. It is increasingly entropic with still an increasing role in absolute terms for fossil fuels.
(b) Externalities are not so much "market failures" as systematic cost-shifting (to use K. William Kapp's term, in 1950, in his book on what we could now call business ecological economics). Firms systematically avoid including environmental liabilities in their accounts.
(c) The damages that the human economy does to Nature (and the contributions that the human economy does sometimes to the reparation and regeneration of Nature) must be counted in a variety of valuation languages. The livelihood values, sacredness, relevance to future generations, and full ecological values cannot be translated into monetary terms. They are not commensurate with

money (as Otto Neurath already discussed in the 1920s in the "socialist calculation debate" against Ludwig Von Mises and F.A. Hayek).
(d) An ecological macroeconomics does not focus on GDP growth but on the social and physical sustainability of the economy. Hence, proposals since 1970 for a "steady state" (originated by Herman Daly) and more recently a vigorous debate on "prosperity without growth" and the need for a period of Degrowth of the rich economy.
(e) Demography is not a field of study outside ecological economics, on the contrary, ecological economists have knowledge and opinions on demography, favoring in general a stop and a certain decrease in the human population. They are familiar with indicators like the human appropriation of net primary productivity (HANPP) and the "ecological footprint". At the same time, they are very much aware of the enormous inequalities in the exosomatic use of energy and materials by humans.
(f) Far from international trade contributing to prosperity, it has contributed to inequality and exhaustion of materials and sources of energy, through "ecologically unequal exchange" that should be measured with physical indicators (Hornborg & Jorgenson, 2010). This creates a link from ecological economics to world systems theory and its concept of frontiers of commodity extraction and waste disposal.

2.1 From Ecological Economics to Political Ecology

At the UAB, ecological economics, perhaps because of my own interests in agrarian conflicts from 60 years ago, has overlapped to some extent with Political Ecology. This is what characterizes the "Barcelona school" in my view. Some senior members of ICTA are mainly concerned with environmental public policies, some with the study of the metabolism of society, some with ecological macroeconomics and "degrowth." There is a variety of interests at ICTA in the environmental social sciences, including a very strong group in ecological anthropology with Victoria Reyes Garcia (who was my student as a first-year undergraduate, some time ago). But the overlap between ecological economics and political ecology (as the study of socio-environmental conflicts) is one of our specialties at ICTA. One first product was the article by Gerber, Veuthey and myself, comparing conflicts on tree plantations in Cameroon and Ecuador (2009). The link between ecological economics and political ecology arises because conflicts (what we call "ecological distribution conflicts") are born from the growth and changes in the social metabolism, which are studied and quantified by ecological economics (and also by industrial ecology). Such conflicts are often "valuation contests"; the social actors of such conflicts express values which cannot be reduced to economic accounting. Political power is used to impose some valuation languages (such as cost-benefit analysis, or monetary compensation for externalities) negating others.

To understand the link between ecological economics and political ecology, let us look first in more detail at the fact that the industrial economy is not circular, it is entropic. (Georgescu-Roegen, 1971). The enormous circularity rift or metabolic gap or "entropy hole" explains the march of the economy to the commodity extraction frontiers and, therefore, the increasing number of environmental conflicts gathered in the EJAtlas that by January 2022 has reached 3600 entries.

It is fashionable to talk about the "circular economy." This could be meant in two senses.

Introductory microeconomics is often taught in terms of what Georgescu-Rogen called "the merry-go-round between consumers and producers," a circular scheme in which producers put goods and services in the market at prices which consumers pay; meanwhile, consumers (as providers of labour, land or other inputs or "factors of production") get money from producers in the form of salaries, rents etc. and they buy, as consumers, the products or services that have been produced. The "merry-go-round" needs energy for running (energy which gets dissipated), and it produces material waste which is not recycled. This is left aside in introductory mainstream economics, or maybe it is introduced much later, in the analysis of the "intergenerational allocation of exhaustible resources" and in the treatment of externalities which are "internalized into the price system."

As ecological critics of mainstream economics since the 1970s and 1980s, we thought that we were slowly convincing the public if not the professional economists that the "merry-go-round" representation of the economy was wrong. The economy is embedded in physical realities. However, to our surprise, the recent novelty is that, from industrial ecology and not only from economics, a circular vision of the economy is also preached. The geologically produced energy and the materials entering the economy are here taken into account, and the waste is very much present, but it is assumed that technical change may close the circle. The waste becomes inputs. The energy (dissipated, of course, because of the Second Law of Thermodynamics) is not a problem because it will come from current sun energy (not fossil fuels, which are exhaustible stocks of photosynthesis from the past). The circular supply chain is supposed to rule physically the economy. We know however that the actual degree of the circularity of the industrial economy is very low, and it is probably decreasing as formerly biomass-based economies complete their transition to an industrial economy based on fossil fuels in India and Africa (Roy & Schaffartzik, 2021).

Georgescu-Roegen in *The Entropy Law and the Economic Process* (1971) and other authors before and after him insisted on the fact that the industrial economy is not circular but entropic. This explains the growth of environmental conflicts at the extraction and waste disposal frontiers. This is lesson number one in a course of ecological economics and political ecology. Of all the materials entering the economy (fossil fuels, building materials, metal ores, biomass), by 2005 only about 6% were recycled (Haas et al., 2015). There is no reason to expect an improvement to have happened since 2005. The low degree of circularity has two main reasons. First, 44% of processed materials were used to provide energy and are thus not available for recycling. Second, socioeconomic stocks were growing at a high rate with net additions to stocks of 17 Gt/year. In the last 120 years, the human population grew five times (from 1.5 to 7.5 billion) while the inputs processed in the global economy (biomass, fossil fuels, building materials, metals) grew approximately thirteen times, from 7.5 to 95 Gt per year (Haas et al., 2015). The economy is becoming less and less circular. The expansion of stocks requires, once in place, a persistent input of materials and energy for their maintenance and operation

Therefore, the industrial economy marches all the time in search of energy and materials towards the commodity extraction frontiers and to the waste disposal frontiers, often inhabited by humans and certainly by other species. (Scheidel et al., 2018, 2020; Hanaček et al., 2022). Hence the growth in the number of Ecological Distribution Conflicts (EDC), and as a response the strength of the environmental justice movements. Sometimes, I have called these movements "the environmentalism of the poor." This does not mean that all poor peasants and indigenous people are environmentalists and behave like environmentalists. It means that in the conflicts over resource extraction and in the conflicts over waste disposal, poor people are often in favour of nature conservation because they live on nature's contributions very directly. That's why Chico Mendes opposed deforestation in Acre in Brazil in 1988 and why in Peru indigenous people in Bagua in 2009 refused to give up their communal lands with their minerals and oil to be placed on the market. Many died. Not too far, in Yanacocha, in Cajamarca, near where Pizarro met Atahualpa, peasants opposed Newmont gold mining because it pollutes the water. Elsewhere, poor people oppose eucalyptus plantations for pulp export. My book *The Environmentalism of the Poor* (2002) collected a few hundred of such ecological distribution conflicts, and my new book *Land, Water, Freedom and Air – The Making of the World Movement for Environmental Justice*, drawing on the EJAtlas, collects many more across the world (Temper et al., 2015, 2018). The protagonists of such conflicts display incommensurable values. When they stop metal mines or coal-fired power plants or object to oil palm plantations, they carry out "degrowth in practice."

References

Ayres, R. U., & Kneese, A. V. (1969). Production, consumption, and externalities. *The American Economic Review, 59*(3), 282–297.

Fischer-Kowalski, M., & Haberl, H. (Eds.). (2007). *Socioecological transitions and global change: Trajectories of social metabolism and land use*. Edward Elgar.

Georgescu-Roegen, N. (1971). *The entropy law and the economic process*. Harvard U.P.

Gerber, J. F., Veuthey, S., & Martinez-Alier, J. (2009). Linking political ecology with ecological economics in tree plantation conflicts in Cameroon and Ecuador. *Ecological Economics, 68*(12), 2885–2889.

Haas, W., Krausmann, F., Wiedenkofer, D., & Heinz, M. (2015). How circular is the global economy? An assessment of material flows, waste production, and recycling in the European Union and the world in 2005. *Journal of Industrial Ecology*. https://doi.org/10.1111/jiec.12244

Hanaček, K., Kröger, M., Scheidel, A., Rojas, F., & Martinez-Alier, J. (2022). In thin ice. The Arctic commodity extraction frontier and environmental conflicts. *Ecological Economics*. https://doi.org/10.1016/j.ecolecon.2021.107247

Hornborg, A., & Jorgenson, A. K. (Eds.). (2010). *International trade and environmental justice: Toward a global political ecology*. Nova Science Publishers.

Kapp, K. W. (1950). *The Social costs of private enterprise*. Harvard University Press.

Røpke, I. (2004). The early history of modern ecological economics. *Ecological Economics, 50*(3–4), 293–314.

Roy, B., & Schaffartzik, A. (2021). Talk renewables, walk coal: The paradox of India's energy transition. *Ecological Economics, 180*.

Scheidel, A., Temper, L., Demaria, F., & Martínez-Alier, J. (2018). Ecological distribution conflicts as forces for sustainability: An overview and conceptual framework. *Sustainability Science, 13*(3), 585–598.

Scheidel, A., Del Bene, D., Juan Liu, G., Navas, G., Mingorría, S., Demaria, F., Avila, S., Roy, B., Ertör, I., Temper, L., & Martínez-Alier, J. (2020). Environmental conflicts and defenders: A global overview. *Global Environmental Change, 63*, 102104. https://www.sciencedirect.com/science/article/pii/S0959378020301424

Temper, L., Del Bene, D., & Martinez-Alier, J. (2015). Mapping the frontiers and front lines of global environmental justice. *Journal of Political Ecology, 22*(1), 255–278.

Temper, L., Demaria, F., Scheidel, A., Del Bene, D., & Martinez-Alier, J. (2018). The Global Environmental Justice Atlas (EJAtlas): Ecological distribution conflicts as forces for sustainability. *Sustainability Science, 13*(3), 573–584.

Open Access This chapter is licensed under the terms of the Creative Commons Attribution 4.0 International License (http://creativecommons.org/licenses/by/4.0/), which permits use, sharing, adaptation, distribution and reproduction in any medium or format, as long as you give appropriate credit to the original author(s) and the source, provide a link to the Creative Commons license and indicate if changes were made.

The images or other third party material in this chapter are included in the chapter's Creative Commons license, unless indicated otherwise in a credit line to the material. If material is not included in the chapter's Creative Commons license and your intended use is not permitted by statutory regulation or exceeds the permitted use, you will need to obtain permission directly from the copyright holder.

Chapter 3
The Barcelona School of Ecological Economics and Political Ecology: Building Bridges Between Moving Shores

Sergio Villamayor-Tomas, Brototi Roy, and Roldan Muradian

3.1 Introduction: The Bonding Elements of the Barcelona School

As stated in the introductory Chap. 1 (by Muradian and Villamayor-Tomas), we have identified the following three aspects as the key bonding elements of the Barcelona school of ecological economics and political ecology:

(i) Paying special attention to the biophysical dimension of the economic system
(ii) An interest in the political and historical aspects underlying the environmental performance of contemporary capitalism
(iii) An emphasis on the study of alternative ways of knowing, valuing and organizing social life

In this introductory section, we briefly explain some key concepts and principles underlying these three bonding elements of the School.

The first element (biophysical dimension of the economic system) revolves around the analysis of social metabolism. This concept refers to the processes of material and energy use, transformation and disposal by societies, associated with self-organization, reproduction and maintenance of internal functions and structures (González de Molina & Toledo, 2014). Social metabolism is rooted in the entropic

---

S. Villamayor-Tomas (✉)
UAB, Barcelona, Spain

B. Roy
ICTA-UAB, Institut de Ciència i Tecnologia Ambientals, Universitat Autònoma de Barcelona, Barcelona, Spain

Department of Environmental Sciences and Policy, Central European University, Wien, Austria

R. Muradian
Faculty of Economics, Universidade Federal Fluminense, Rio de Janeiro, Brazil

© The Author(s) 2023
S. Villamayor-Tomas, R. Muradian (eds.), *The Barcelona School of Ecological Economics and Political Ecology*, Studies in Ecological Economics 8,
https://doi.org/10.1007/978-3-031-22566-6_3

nature of the economic process and its consequences in the form of environmental degradation (see Chap. 4 by K. Farrell in this book). Within ecological economics, the social metabolism lens has been used, among others, to evaluate quantitatively the rate at which communities and societies use resources, comparing, for example, hunter-gatherers, agrarian subsistence communities or industrial societies (Fischer-Kowalski & Haberl, 2007).

The Barcelona School has developed a strong quantitative research agenda on social metabolism, epitomized in the Multi-Scale Integrated Analysis of Societal and Ecosystem Metabolism (MUSIASEM) framework (for a comparison of MUSIASEM to other frameworks see Gerber & Scheidel, 2018). MUSIASEM unfolds into a series of concepts and protocols to translate both quantitative and qualitative data into a common language and analyze metabolic processes at different scales of social aggregation. Applications of MUSIASEM have focused on the metabolic pattern of food, energy, water and their interrelations (see Chap. 10 by M. Giampietro in this book). Other recent applications of social metabolic accounting within the school have featured agri-food systems at the regional scale (Marull et al., 2018; Cattaneo et al., 2018).

The second key interest of the School (political and historical aspects of environmental performance) relies heavily on the notion of ecological distribution conflicts. This concept refers to social conflicts born from the unfair access to natural resources and the unjust distribution of pollution burdens (Martinez-Alier, 2002). These conflicts presuppose some deliberate exercise of cost shifting from polluters, extractive industries or even governments to vulnerable social groups. Cost shifting, a term now well established in the field of ecological economics, has been adopted in the works of Joan Martinez Alier and other colleagues within the school (Muradian & Martinez-Alier, 2001) to refer to those environmental load displacements. This vision questions Coasian expectations that parties infringing environmentally related costs to others would be willing to bargain over compensations for the damages. As put by Martinez-Alier more than 25 years ago, "the issue cannot be resolved by bringing externalities into surrogate markets, but only by social activism against depletion of resources and environmental pollution" (Martines-Alier, 1995, pp. 70–71).

Ecological distribution conflicts are understood as collective claims against environmental injustices. This type of social conflicts often show that environmental degradation is closely linked to processes of marginalization of impoverished and vulnerable communities. This is indeed the basis of the "environmentalism of the poor," a proposition claiming that marginalized communities often (though not always) defend the environment because it is an integral part of their livelihoods. The concept emerged as an alternative to influential environmental discourses that aligned environmentalism with either the need to conserve nature in a pristine state at all cost ("cult of wilderness" discourse) or with the hope that technological progress would decouple economic growth from environmental degradation ("gospel of eco-efficiency" discourse) (Guha & Martinez-Alier, 2013). This concept has given visibility to the environmental concerns of rural communities in the global South, as compared to that of environmental movements in the global North.

The third core aspect of concern for the School (alternative ways of knowing, valuing and organizing social life) rests on a post-modern stand towards Western science. Firstly, this perspective acknowledges that the premises on which science lies, as well as science-policy relations, are embedded in particular historical and cultural backgrounds, and therefore this way of knowing cannot be considered as universal or intrinsically superior in all contexts. Science must be considered as one among several ways of knowing the world. Secondly, science faces inherent limitations in decision contexts characterized by complexity, high uncertainly and high stakes. These limitations can create legitimacy challenges in public decision-making. These two standpoints are expressed in specific theoretical propositions, such as the post-normal science paradigm, as well as in particular research agendas, such as paying attention to the diversity of languages of valuation and to non-Western forms of knowledge.

In this chapter, we explore the connections between these three dimensions. We argue that the theoretical and methodological integration of these areas of concern has been the main contribution of the Barcelona School. By doing so, the School has advanced on creating bridges between ecological economics and political ecology. One of the objectives of the present chapter is precisely to explain the foundations and development of such interaction. In the following section, first we explain the epistemological foundations that in our view inspire the interest around cross-fertilization between ecological economics and political ecology. After, we explore the bridges that have been built between the three dimensions outlined above, including concrete cases of cross-fertilization.

3.2 Epistemological Foundations Inspiring Cross-Fertilization

3.2.1 Ecological Economics as a Place of Convergence and Host of Diversity

The place of inter-disciplinary convergence that has been called Ecological Economics is characterized by the co-existence of heterogeneous analytical approaches and by acknowledging complexity in the way it conceptualizes human-environment interactions. This academic community was the result of a conscious effort to study the biophysical foundation of the economy (Martinez-Alier, 1993). Despite the difficulties it has faced over time, the field has been able to maintain both its identity and intellectual openness, which enables fruitful exchanges with other fields. Debates around the scope of the field and the composition of its epistemic communities are frequent in ecological economics, including calls for epistemological closure (Spash, 2011), for acknowledging sub-fields, such as "institutional ecological economics" (Paavola & Adger, 2005) and "social-ecological economics" (Spash, 2011), or for recognizing academic schools, such as the Vienna Social Ecology School (Fischer-Kowalski & Weisz, 2016; Farrell, 2018) or the Bloomington School (Aligica & Boettke, 2009).

These discussions show that ecological economics has always had moving boundaries, on the intersections between different social sciences but also between them and natural sciences. More specifically, this introductory chapter, and the whole book, aims at recognizing the efforts of Joan Martinez-Alier and other scholars associated with the Barcelona School to connect two related and moving realms of knowledge: ecological economics and political ecology. Indeed, we could state that one of the foundational propositions of the School is that a "fruitful theoretical and methodological frictions" with political ecology and other social sciences is a productive endeavor for ecological economics (see Zimmerer, 2015 for a similar argument with regard to the integration of political ecology and environmental sciences). In sum, adopting ecological economics as a foundational analytical approach has enabled the School to have an open scope and vision, which looks for cross-fertilization among disciplines and methods.

3.2.2 Post-normal Science and the Search for Other Ways of Knowing

The most important contemporary environmental problems usually combine a high level of uncertainty, high stakes (both in terms of entangled values and consequences) and the urgency of the social decisions to be taken. During the 1990s, Silvio Funtowicz and Jerome Ravetz (1993) developed the notion of post-normal science to bring about new epistemological insights about the role and limits of science in supporting social decisions in such context. The core contribution of this set of propositions is the idea of "extended peer community". Due to the high level of complexity and uncertainty that characterize the decision contexts described above, knowledge holders that traditionally have had high power to influence policy design (e.g. scientists), face serious difficulties to provide both accurate and legitimate inputs to public decision-making processes. In order to solve such democratic deficit in the policy-science interface, the post-normal science framework proposed to shift the source of quality and legitimacy of public decisions from expertise to participation. In a democratic setting, by expanding the community of "peers" (people empowered to voice their opinions and judgments) to non-scientists, public decision processes can gain support ("procedural quality"), even though uncertainties and complexities remain in place.

Post-normal science has been very influential in ecological economics, as well as in the Barcelona school. More specifically, the works around social multi-criteria evaluation (SMCE) and traditional ecological knowledge can be justified by the propositions of post-normal science. SMCE offers a way to integrate different value systems when facing a problem of social choice. This could be done with participatory methods, where criteria selection, weighting and aggregation steps are performed with the voice of a broad group of actors (Munda, 2008). Driven by an interest in the incommensurability of values in the context of environmental conflicts (Gerber et al., 2012), scholars from the School have also used SMCE in a variety of contexts, including water conflicts (Kallis et al., 2006), energy-related public decisions (Munda & Russi, 2008; Gamboa & Munda, 2007) and coastal resources planning (Garmendia & Gamboa, 2012; Garmendia et al., 2010).

Scholars of the School working on knowledge systems alternative to science, usually known as traditional ecological knowledge, have moved from documenting their erosion to highlighting the factors (e.g., Reyes-García et al., 2014) and processes (e.g., Gómez-Baggethun et al., 2010) that both undermine and reinvigorate them. They have not only undertaken research in the Global South (e.g., Beyei et al., 2020) but also in the Global North, and expanded the scope from an interest from customary forms of knowledge into an interest into digital knowledge and citizen science (e.g., Calvet-Mir et al., 2018).

3.2.3 *The Diversity of Languages of Valuation*

Differences in perceptions and values are often grounded and expressed as multiple languages of valuation (Avci et al., 2010). Ecological distribution conflicts can indeed be also viewed as conflicts arising out of different attitudes and meanings given to nature by different cultures (Escobar, 2006). Multiple languages of valuation derive from the incommensurability of environment values, which presupposes value pluralism (Martinez-Alier et al., 1998). While this renders the values weakly comparable, they are, however, amenable to multi-criteria valuation (Munda, 2008). languages such as "sacredness," "community life and livelihood" and "ethnic identity" arise commonly in socio-environmental conflicts in response to a predominantly monetary language that justifies the extraction of the resource at stake (Martinez-Alier, 2002; Avci et al., 2010). Attributing sacred values to nature is a common practice, especially in the Indian subcontinent. Examples are the sacred groves in Khasi hills, the Western Ghats and the Aravalli hills in the state of Rajasthan, where nature is deemed to remain pristine, with even the collection of dead firewood prohibited (Gadgil & Vartak, 1975). "Sacredness" as a language of valuation, for example, has appeared in several distributional conflicts involving access and rights on ecosystems (Gerber, 2011; Avci et al., 2010; Rival, 2010; Temper & Martinez- Alier, 2013). Temper and Martinez-Alier (2013) highlight the case of Niyamgiri Hill in Odisha state of India, which the Dongria Kondh tribe consider as sacred; the hill is considered to be the abode of Niyam Raja. This was the site of a conflict with a bauxite mining company that justified the activity by the monetary valuation of the benefits versus costs of extracting the large quantities of bauxite ore lying under the mountain.

When socio-environmental conflicts occur, and in general when opposing worldviews are at stake, making use of an extended peer community could be a way to gain legitimacy in public decisions (Turnpenny et al., 2011). Political ecology scholars follow this premise and have advocated the acceptance of different perceptions and values around conflicts and the need to take them into account through genuine participatory processes (Adger et al., 2001; Martinez-Alier, 2009). Within the Barcelona School, this has translated into a fair number of applications of social multi-criteria evaluation in situations of conflict (Gerber et al., 2012; Zografos & Rodríguez-Labajos, 2014; Corral & Acosta, 2017; Walter et al., 2016; Corzo & Gamboa, 2018). In a recent application, Corzo and Gamboa (2018) examine the environmental effects of mining liabilities and small-scale mining on peasant communities. After a measurement of critical water quality parameters, SCME was used to reveal key social actors and their perceptions regarding tailing problems.

3.2.4 Activism Mobilizing Science

The interpretation of reality, including the scope and causes of socio-environmental problems and conflicts, is indeed very much dependent on the prevailing worldview in a given social group. There is, therefore, an intertwined relationship between cognition (overarching values framing the way we interpret the world), knowledge and decisions (action). Indeed, ideological frameworks are unavoidable and very relevant across any academic field, but especially among social sciences. It should be an ethical requirement in science to make those ideological positions explicit. An important number of authors of the Barcelona School have not only done so but also engaged in what has been coined as "activism-mobilizing science" (Conde, 2014). This refers to a self-reflective epistemological stand of scholars committed with the co-production of knowledge that can be mobilized in socio-environmental causes and transformations, usually embedded in environmental justice challenges. Such an engagement could be controversial in academic circles, since some authors could argue that it could compromise the commitment with the notion of truth. Nonetheless, this approach has the undeniable merit of making clear and explicit the underlying ideological and ethical basis of the academic work. Furthermore, through activism-mobilizing science, it is also possible to build strategic partnerships with key social actors, gaining insights and points of view that academics usually do not have access to (Tallapragada, 2018). This would be particularly the case in the study of socio-environmental conflicts through the knowledge-based support of social (e.g., environmental justice) movements (McCormick, 2007).

3.3 Building Bridges: Cross-Fertilization Between Ecological Economics and Political Economy

This section provides some examples on how scholars at the Barcelona School have advanced on the development of fruitful interactions between the concepts, notions and approaches outlined above.

3.3.1 Ecological Asymmetries, Distributional Conflicts and the Environmental Justice Atlas

Addressing the interface between social metabolism and environmental justice has been a core concern of the Barcelona School since its origin (Martinez-Alier et al., 2016; see also M'Gonigle, 1999 for a pioneering effort). Ecological distribution conflicts arise from the unequal distribution of benefits and burdens of economic activities derived from changes in the metabolism of societies (Martinez-Alier, 1993; Martinez-Alier & O'Connor, 1996; Martinez-Alier et al., 2016). Social metabolism has been instrumental to illustrate economic and environmental asymmetries and conflicts (see Oppon et al., 2018; Oulu, 2015; Infante-Amate & Krausmann, 2019 for recent examples) and adding precision to the claims of environmental justice organizations and the quantification of injustices (Hornborg & Martinez-Alier, 2016). This approach (combining social metabolism and conflict analyses) has enabled a nuanced understanding of ecological distribution conflicts involving mining, biomass or waste disposal conflicts in Latin America (see the Chap. 11 by M. Perez-Rincón) and other Global South regions (Demaria, 2010; Gerber et al., 2009; Kronenberg, 2013).

The interest in plural values has given visibility to the needs and visions of communities in non-Western cultures and justified the use of multi-criteria valuation techniques in the context of socio-environmental conflicts. The School has also been keen on combining social metabolic analyses with the analysis of the political economy and institutions that govern modes of appropriation, distribution and disposal of materials and energy within societies. The idea of socio-metabolic configurations and their current linkages with capitalism and resource distribution among different social groups captures such spirit of integration of metabolic and political ecology concerns (Muradian et al., 2012; Scheidel et al., 2018). The current ecological crisis has to do with the metabolic configuration of globalized, industrial and capitalist societies, whose dynamics are not only behind the acceleration of resource degradation but also driving the dispossession of large numbers of people from basic living conditions (see Chap. 16 by A. Scheidel in this book).

Scholars of the Barcelona School have developed a fair amount of applications of social metabolic analyses, mostly in the Global South and around the ideas of ecological unequal exchange (EUE) and ecological debt. EUE states that resource exchange between high- and low-income and middle-income nations rich in natural resources is asymmetrical; it increases the economic growth of the former while producing environmental degradation in the latter, in the form of, e.g., biodiversity loss and water pollution (Givens et al., 2019). When measured in calories, for example, the EUE is reflected in a loss of self-sufficiency in food and the quality of diets (higher-rated calories in nutritional terms – such as fruit – are exported and poorly rated calories – such as oils and fats – are imported) (Falconi et al., 2017). Findings from this research program have led some scholars to argue that resource-rich Global South countries are indeed creditors of an "ecological debt" (see Givens, 2018 for a review). This concept was developed initially by Latin American environmental justice organizations (EJOs), already in the 1990s, but it was uptaken by scholars later on (Martinez-Alier et al., 2014; Hornborg & Martinez-Alier, 2016). Studies in the Latin-American context have indeed profiled strongly within the School (see Chap. 11 by M. Perez-Rincón for a summary). Recent insights from this literature, for example, point to the relationship between the terms of trade between world regions and environmental degradation (Infante-Amate et al., 2020; Samaniego et al., 2017).

Most recent developments within the School around all the above include claims around the existence of a global environmental justice movement (Martinez-Alier et al., 2016). EUE is indeed considered as the underlying source of most of the environmental distribution conflicts in our time (Hornborg & Martinez-Alier, 2016), to the point of equating the theory of EUE to a theory of global environmental injustice that links justice research with global structural dynamics (Givens et al., 2019, see also Chap. 33 by Falconi et al. in this book).

The Environmental Justice Atlas, which has been an integral part of the School since 2012, shows the current interest in ecological distribution conflicts and environmental justice movements around the world. The Atlas, which in essence is a compilation and categorization of socio-environmental conflicts and movements, was designed as a tool to co-produce knowledge between scholars and activists and help denounce cases of environmental injustice, encourage learning and exchange of experiences, sensitize the media, opinion-makers and public opinion, and to put pressure on politicians and policy-makers, among other motivations (Temper et al., 2015). Additionally, the Atlas can be understood as an effort to advance a comparative environmental justice research program that unveils commonalities among conflicts and their connections to the larger systemic dynamics that the EUE theory captures.

Key categories of the Atlas cover the material/metabolic basics of conflicts, ranging from nuclear, fossil fuels, mineral ores and building material extraction to waste management, biomass and land, water issues, infrastructure and built environment, tourism and recreation, or biodiversity conservation; and the types of actors (e.g., extractive companies, governments, local communities) involved in the conflicts and/or movements (Temper et al., 2015). The large number of cases included in the Atlas has permitted a new series of comparative statistics analyses synthesizing patterns of conflict and resistance. Examples include dam-building projects (Del Bene et al., 2018), land grabbing (Dell'Angelo et al., 2021), the role of women or working-class communities (Le Tran et al., 2020; Navas et al., 2022), fossil fuel and low-carbon energy projects (Temper et al., 2020) and a variety of other conflicts (Scheidel et al., 2020).

The large comparative effort carried out under the umbrella of the Atlas can be framed within a broader interest in developing a comparative political ecology of themes traditionally connected with the School, such as traditional ecological knowledge (Gómez-Baggethun et al., 2013; Reyes-García et al., 2019), as well as more recent themes around alternatives to mainstream development (Temper et al., 2018; Villamayor-Tomas & Garcia-López, 2018). It is worth mentioning the work undertaken on local indicators of climate change impacts (LICCI), which has translated into a series of field data collection protocols with worldwide applicability potential (Reyes-García et al., 2020); and the workaround green locally unwanted land uses (Green LULUs), which has resulted in new methodologies to quantitatively assess and compare green gentrification effects across cities (Connolly, 2019), among other studies.

3.3.2 The Transformative Power of Environmental Justice Movements

Some scholars of the Barcelona School argue that environmental justice movements have the potential to create transformative change beyond specific struggles (Temper et al., 2018; Scheidel et al., 2018; Demaria & Kothari, 2017; Kothari et al., 2019; Villamayor-Tomas & Garcia-Lopez, 2018). In one of the last and most promising contributions in this direction, Scheidel et al. (2018) propose a framework that connects social metabolism configurations with ecological distribution conflicts, the agency of social movements to push for alternatives, and sustainability transitions. As they point out, ecological distribution conflicts of the kind resulting from ecological unequal exchange bring to light conflicting values over the environment as well as unsustainable resource uses affecting people and the planet. Environmental justice organizations are key actors in politicizing such unsustainable resource uses and prefiguring more sustainable alternatives that can ultimately be scaled up and out.

Some works within the School have highlighted the intricate connections between environmental justice movements, community-based natural resource management and commoning processes in the consolidation of sustainability alternatives to mainstream development (see García-López et al., 2017; Villamayor-Tomas & García-López, 2021; Villamayor-Tomas et al., 2022 and the Chap. 19 by S. Villamayor-Tomas, G. García-López and G. D'Alisa in this book). As pointed out by these authors, "commons movements" can help create and strengthen institutions and discourses favoring collective action, up-scaling it horizontally and vertically; while commons institutions and commoning can serve as the basis of social mobilization and a key frame for social movements.

Contributions around alternative ways of organizing the social life have benefited from discussions around the decommodification of nature and languages of valuation. As put by Kallis (2013), mainstream valuation processes, usually encoded in monetary terms, are part of a broader process of commodification, and in turn of the broader process of capitalist expansion into new social and environmental domains. This, however, does not mean that monetary valuation should be totally dismissed, as some forms of monetary evaluations can enhance the weight of environmental values in social decisions, reduce inequalities, and respect other languages of valuation and non-commodified environmental amenities and resources (Kallis, 2013).

Finally, the research on alternative livelihoods is intricately related to previous and current works around knowledge plurality. Much of the School's thinking around alternative social organization finds inspiration from and embodies values encoded in the knowledge and practices of traditional communities (Demaria & Kothari, 2017). Furthermore, as pointed by some scholars from the School, traditional and citizen's knowledge is itself a common good and should be studied and practiced as such (Calvet-Mir et al., 2018; Benyei et al., 2020).

3.3.3 The Challenge of Degrowth

Some scholars from the Barcelona School have focused on the possibility of slowing consumption or economic growth as a way to ameliorate self-destructive social metabolic patterns and reduce conflicts and injustices. Proposals around degrowth have been a part of the School since the early 2010s and proponents call both for an equitable and democratic transition to smaller economies (at least in the Global North) and moving away from excessive consumption and extraction (i.e., in the Global South) (Sekulova et al., 2013).

Degrowth and social metabolism thinking are intrinsically connected (Kallis et al., 2014). Sustainable degrowth has been defined "a socially sustainable and equitable reduction of society's throughput (or metabolism)" (Kallis, 2011, p. 874). Social metabolic analyses have also been instrumental in the call for degrowth by raising flags about the limits of growth and the consequences of reaching those limits in the form of resource shortages, price fluctuations, inequality and inefficiencies (Scheidel & Schaffartzik, 2019). In their global analysis of material flows, for example, Schaffartzik and Pichler (2017) show that growth-led capitalist expansion has relied on extractivism, dispossession and the loss of livelihoods in the places of resource extraction. Infante-Amate and Gonzáled de Molina (2013), in turn, illustrate the disproportionate use of energy by the agri-food system (production, preservation, packaging, and transportation of food) in Spain as compared to the energy that is finally consumed by residents; and plea for a degrowth strategy based on reducing that difference through organic production and re-territorialization of value chains.

Analyses have also addressed the implications, feasibility and desirability of possible trajectories of downscaling growth. D'Alisa and Cattaneo (2013), for example, combine a time and energy analysis of paid and unpaid work in Catalonia and suggest a degrowth strategy based on re-allocating some services and goods from the market to the household and the promotion of work sharing at the household and neighborhood levels. However, Sorman and Giampietro (2013) analyzed the metabolic pattern of a sample of developed countries, and conclude that some assumptions and recipes of the degrowth movement are problematic, including the possibility to reduce working hours and individual energy consumption (see Kallis, 2013 for a response). A common understanding across all applications is that each socio-metabolic context may require specific degrowth aims and strategies (Scheidel & Schaffartzik, 2019).

Debates within the School around the relevance of degrowth (as compared to "agrowth"; van den Bergh & Kallis, 2012) have also given visibility to a shared interest in the need to move beyond outdated welfare measurements (like GDP) and the importance to advance politically feasible solutions to stall the current consumerism trend. Some of these solutions could be applied to foster a new organization of labor and work time that include shorter average working weeks, more stringent regulations of commercial advertisement or radical efforts (i.e., at schools and media) in consumer information and communication that promote taking advantage of low-cost pro-environmental behavior (van den Bergh, 2011).

Environmental justice and degrowth share the overall concern for justice and sustainability and face the same obstacle posed by growth-led development, but operate in different contexts (Scheidel & Schaffartzik, 2019; Akbulut et al., 2019; Kallis et al., 2018). Subsistence-oriented local communities that struggle in environmental justice conflicts might not think about their struggle as one for degrowth but as one aiming to defend their (sometimes precarious) customary livelihoods (Rodríguez-Labajos et al., 2019). Alliances are however possible. Environmental justice scholars can facilitate information to degrowth groups that document the adverse impacts of growth-led development; and they can provide early warnings of resource shortages, price fluctuations, or shifts in demand that may induce the expansion of the extractive frontier and give rise to new environmental injustices (Scheidel & Schaffartzik, 2019; Akbulut et al., 2019).

3.4 Final Remarks

This overarching chapter aims at giving some logical structure to the present book. It shows the intricate ways that ideas like social metabolism, environmentalism of the poor, ecological distribution conflicts, traditional ecological knowledge, the commons, degrowth or activism mobilizing science are connected with each other and how they conform to a relatively cohesive way of understanding human-environment interactions. It also shows how such interaction can contribute to the social construction of fairer and more sustainable social and ecological futures.

It is worth mentioning that this chapter (and the whole book) does not do justice to all the work that has been carried out by the scholars of the Barcelona School. It does not cover, for example, nascent themes within the School like urban ecosystem services (e.g., Baró et al., 2015), food transitions (e.g., Calvet-Mir et al., 2018), or attitudes and behavior in the context of incentive-based policy instruments (e.g., Drews & van den Bergh, 2016; Moros et al., 2019). We hope that, despite its limitations, this chapter and the whole book will help to connect scholars, both within and outside the School. In that sense, we want to emphasize our commitment with the intellectual openness and heterodoxy of both ecological economics and political ecology, which allows the type of cross-fertilization between disciplines, approaches and methods here described. The Barcelona School is by no means a "closed space" epistemological or methodologically speaking. On the contrary, as shown here, it is in continuous evolution and committed to building bridges across moving disciplinary shores.

References

Adger, N., Benjaminsen, T., Brown, K., & Svarstad, H. (2001). Advancing a political ecology of global environmental discourses. *Development and Change, 32*(4), 681–715.

Akbulut, B., Demaria, F., Gerber, J. F., & Martínez-Alier, J. (2019). Who promotes sustainability? Five theses on the relationships between the degrowth and the environmental justice movements. *Ecological Economics, 165*, 106418.

Aligica, P. G., & Boettke, P. (2009). *Challenging institutional analysis and development: The Bloomington school*. Routledge. 176 pp.

Avci, D., Adaman, F., & Özkaynak, B. (2010). Valuation languages in environmental conflicts: How stakeholders oppose or support gold mining at Mount Ida, Turkey. *Ecological Economics, 70*, 228–238.

Baró, F., Haase, D., Gómez-Baggethun, E., & Frantzeskaki, N. (2015). Mismatches between ecosystem services supply and demand in urban areas: A quantitative assessment in five European cities. *Ecological Indicators, 55*, 146–158.

Benyei, P., Arreola, G., & Reyes-García, V. (2020). Storing and sharing: A review of indigenous and local knowledge conservation initiatives. *Ambio, 49*(1), 218–230.

Benyei, P., Calvet-Mir, L., Reyes-García, V., & Rivera-Ferre, M. (2020). Resistance to traditional agroecological knowledge erosion in industrialized contexts: A study in La Plana de Vic (Catalonia). *Agroecology and Sustainable Food Systems, 44*(10), 1309–1337.

Calvet-Mir, L., Benyei, P., Aceituno-Mata, L., Pardo-de-Santayana, M., López-García, D., Carrascosa-García, M., et al. (2018). The contribution of traditional agroecological knowledge as a digital commons to agroecological transitions: The case of the CONECT-e platform. *Sustainability, 10*(9), 3214.

Cattaneo, C., Marull, J., & Tello, E. (2018). Landscape agroecology. The dysfunctionalities of industrial agriculture and the loss of the circular bioeconomy in the Barcelona region, 1956–2009. *Sustainability, 10*(12), 4722.

Conde, M. (2014). Activism mobilising science. *Ecological Economics, 105*, 67–77.

Connolly, J. J. (2019). From Jacobs to the Just City: A foundation for challenging the green planning orthodoxy. *Cities, 91*, 64–70.

Corral, S., & Acosta, M. (2017). Social sensitivity analysis in conflictive environmental governance: A case of forest planning. *Environmental Impact Assessment Review, 65*, 54–62.

Corzo, A., & Gamboa, N. (2018). Environmental impact of mining liabilities in water resources of Parac micro-watershed, San Mateo Huanchor district, Peru. *Environment, Development and Sustainability, 20*(2), 939–961.

D'Alisa, G., & Cattaneo, C. (2013). Household work and energy consumption: A degrowth perspective. Catalonia's case study. *Journal of Cleaner Production, 38*, 71–79.

Del Bene, D., Scheidel, A., & Temper, L. (2018). More dams, more violence? A global analysis on resistances and repression around conflictive dams through co-produced knowledge. *Sustainability Science, 13*(3), 617–633.

Dell'Angelo, J., Navas, G., Witteman, M., D'Alisa, G., Scheidel, A., & Temper, L. (2021). Commons grabbing and agribusiness: Violence, resistance and social mobilization. *Ecological Economics, 184*, 107004.

Demaria, F. (2010). Shipbreaking at Alang-Sosiya (India): An ecological distribution distribution conflict. *Ecological Economics, 70*(2), 250–260.

Demaria, F., & Kothari, A. (2017). The post-development dictionary agenda: Paths to the pluriverse. *Third World Quarterly, 38*(12), 2588–2599.

Drews, S., & van den Bergh, J. C. J. M. (2016). What explains public support for climate policies? A review of empirical and experimental studies. *Climate Policy, 16*(7), 855–876.

Escobar, A. (2006). Difference and conflict in the struggle over natural resources: A political ecology framework. *Development, 49*(3), 6–13.

Falconí, F., Ramos-Martin, J., & Cango, P. (2017). Caloric unequal exchange in Latin America and the Caribbean. *Ecological Economics, 134*, 140–149. https://doi.org/10.1016/j.ecolecon.2017.01.009

Farrell, K. N. (2018). The Vienna school of ecological economics. *Real-World Economics Review, 163*, 163–171.

Fischer-Kowalski, M., & Haberl, H. (Eds.). (2007). *Socioecological transitions and land-use change socioecological transitions and global change: Trajectories of social metabolism and land use*. Edward Elgar. 263 pp.

Fischer-Kowalski, M., & Weisz, H. (2016). The archipelago of social ecology and the island of the Vienna school. *Social Ecology, 5*, 3–28.

Funtowicz, S., & Ravetz, J. (1993). Science for the post-normal age. *Futures, 25*(7), 739–755.

Gadgil, M., & Vartak, V. D. (1975). Sacred groves of India- a plea for continued conservation. *Journal of the Bombay Natural History Society, 72*(2), 314–320.

Gamboa, G., & Munda, G. (2007). The problem of windfarm location: A social multi-criteria evaluation framework. *Energy Policy, 35*(3), 1564–1583.

García López, G. A., Velicu, I., & D'Alisa, G. (2017). Performing counter-hegemonic common (s) senses: Rearticulating democracy, community and forests in Puerto Rico. *Capitalism Nature Socialism, 28*(3), 88–107.

Garmendia, E., & Gamboa, G. (2012). Weighting social preferences in participatory multi-criteria evaluations: A case study on sustainable natural resource management. *Ecological Economics, 84*, 110–120.

Garmendia, E., Gamboa, G., Franco, J., Garmendia, J. M., Liria, P., & Olazabal, M. (2010). Social multi-criteria evaluation as a decision support tool for integrated coastal zone management. *Ocean and Coastal Management, 53*(7), 385–403.

Gerber, J.-F. (2011). Conflicts over industrial tree plantations in the south: Who, how and why? *Global Environmental Change, 21*(1), 165–176.

Gerber, J.-F., & Scheidel, A. (2018). In search of substantive economics: Comparing today's two major socio-metabolic approaches to the economy– MEFA and MuSIASEM. *Ecological Economics, 144*, 186–194.

Gerber, J.-F., Veuthey, S., & Martinez-Alier, J. (2009). Linking political ecology with ecological economics in tree plantation conflicts in Cameroon and Ecuador. *Ecological Economics, 68*(12), 2885–2889.

Gerber, J.-F., Rodríguez-Labajos, B., Yánez, I., Branco, V., Roman, P., Rosales, L., & Johnson, P. (2012). *Guide to multicriteria evaluation for environmental justice organisations* (EJOLT Report No. 8). 45 p.

Givens, J. E. (2018). Ecologically unequal exchange and the carbon intensity of well-being, 1990–2011. *Environmental Sociology, 4*(3), 311-324.

Givens, J. E., Huang, X., & Jorgenson, A. K. (2019). Ecologically unequal exchange: A theory of global environmental injustice. *Sociology Compass, 13*(5), e12693.

Gómez-Baggethun, E. R. I. K., Mingorria, S., Reyes-García, V., Calvet, L., & Montes, C. (2010). Traditional ecological knowledge trends in the transition to a market economy: Empirical study in the Doñana natural areas. *Conservation Biology, 24*(3), 721–729.

Gómez-Baggethun, E., Corbera, E., & Reyes-García, V. (2013). Traditional ecological knowledge and global environmental change: Research findings and policy implications. *Ecology and Society: A Journal of Integrative Science for Resilience and Sustainability, 18*(4), 72.

González de Molina, M., & Toledo, M. (2014). *The social metabolism: A socio-ecological theory of historical change*. Springer. 355 pp.

Guha, R., & Alier, J. M. (2013). *Varieties of environmentalism: Essays North and South*. Routledge.

Hornborg, A., & Martinez-Alier, J. (2016). Ecologically unequal exchange and ecological debt. *Journal of Political Ecology, 23*(1), 328–333.

Infante-Amate, J., & De Molina, M. (2013). Sustainable de-growth in agriculture and food: An agro-ecological perspective on Spain's agri-food system (year 2000). *Journal of Cleaner Production, 38*, 27–35.

Infante-Amate, J., & Krausmann, F. (2019). Trade, ecologically unequal exchange and colonial legacy: The case of France and its former colonies (1962–2015). *Ecological Economics, 156*, 98–109.

Infante-Amate, J., Urrego-Mesa, A., & Tello-Aragay, E. (2020). Las venas abiertas de América Latina en la era del antropoceno: un estudio biofísico del comercio exterior (1900-2016). *Diálogos Revista Electrónica de Historia, 21*(2), 177–214.

Kallis, G. (2011). In defence of degrowth. *Ecological Economics, 70*(5), 873–880.

Kallis, G. (2013). Societal metabolism, working hours and degrowth: A comment on Sorman and Giampietro. *Journal of Cleaner Production, 38*, 94–98.

Kallis, G., Videira, N., Antunes, P., Guimarães Pereira, A., Spash, C., Coccossis, H. C., Quintana, S., del Moral, L., Hatzilacou, D., Lobo, G., Mexa, A., Paneque, P., Pedregal, B., & Santos, R. (2006). Participatory methods for water resources planning and governance. *Environment and Planning C: Government and Policy, 24*, 215–234.

Kallis, G., Demaria, F., & D'Alisa, G. (2014). Introduction: Degrowth. In *Degrowth* (pp. 29–46). Routledge.

Kallis, G., Kostakis, V., Lange, S., Muraca, B., Paulson, S., & Schmelzer, M. (2018). Research on degrowth. *Annual Review of Environment and Resources, 43*, 291–316.

Kothari, A., Salleh, A., Escobar, A., Demaria, F., & Acosta, A. (2019). *Pluriverse. A post-development dictionary*. Tulika Books.

Kronenberg, J. (2013). Linking ecological economics and political ecology to study mining, glaciers and global warming. *Environmental Policy and Governance, 23*(2), 75–90.

Le Tran, D., Martinez-Alier, J., Navas, G., & Mingorria, S. (2020). Gendered geographies of violence: A multiple case study analysis of murdered women environmental defenders. *Journal of Political Ecology, 27*(1), 1189–1212.

M'Gonigle, M. (1999). Ecological economics and political ecology: Towards a necessary synthesis. *Ecological Economics, 28*(1), 11–26.

Martines-Alier, J. (1995). Political ecology, distributional conflicts, and economic inconmensurability. *New Left Review, 0*(211), 70–88.

Martinez-Alier, J. (1993). *Ecological economics: Energy, environment and society*. Blackwell Publishers. 287 pp.

Martinez-Alier, J. (2002). *The environmentalism of the poor: A study of ecological conflicts and valuation*. Edward Elgar. 328 pp.

Martinez-Alier, J. (2009). Social metabolism, ecological distribution conflicts, and languages of valuation. *Capitalism Nature Socialism, 20*(1), 58–87.

Martinez-Alier, J., & O'Connor, M. (1996). Ecological and economic distribution conflicts. In R. Costanza, J. Martinez-Alier, & O. Segura (Eds.), *Getting down to earth: Practical applications of ecological economics*. Island Press/ISEE.

Martinez-Alier, J., Munda, G., & O'Neill, J. (1998). Weak comparability of values as a foundation for ecological economics. *Ecological Economics, 26*(3), 277–286.

Martinez-Alier, J., Anguelovski, I., Bond, P., Del Bene, D., Demaria, F., Gerber, J.-F., Greyl, L, Haas, W, Healy, H., Marín-Burgos, V., Ojo, G., Porto, M., Rijnhout, L., Rodríguez-Labajos, B., Spangenberg, J., Temper, L., Warlenius, R., & Yánez, I. (2014). Between activism and science: Grassroots concepts for sustainability coined by Environmental Justice Organizations. *Journal of Political Ecology, 21*, 19–60.

Martinez-Alier, J., Temper, L., Del Bene, D., & Scheidel, A. (2016). Is there a global environmental justice movement? *The Journal of Peasant Studies, 43*(3), 731–755.

Marull, J., Tello, E., Bagaria, G., Font, X., Cattaneo, C., & Pino, J. (2018). Exploring the links between social metabolism and biodiversity distribution across landscape gradients: A regional-scale contribution to the land-sharing versus land-sparing debate. *Science of the Total Environment, 619*, 1272–1285.

McCormick, S. (2007). Democratizing science movements: A new framework for mobilization and contestation. *Social Studies of Science, 37*(4), 609–623.

Moros, L., Vélez, M. A., & Corbera, E. (2019). Payments for ecosystem services and motivational crowding in Colombia's Amazon Piedmont. *Ecological Economics, 156*, 468–488.

Munda, G. (2008). *Social multi-criteria evaluation for a sustainable economy*. Springer.

Munda, G., & Russi, D. (2008). Social multicriteria evaluation of conflict over rural electrification and solar energy in Spain. *Environment and Planning C: Government and Policy, 26*, 712–727.

Muradian, R., & Martinez-Alier, J. (2001). Trade and the environment: From a 'Southern' perspective. *Ecological Economics, 36*(2), 281–297.

Muradian, R., Walter, M., & Martinez-Alier, J. (2012). Hegemonic transitions and global shifts in social metabolism: Implications for resource-rich countries. Introduction to the special section. *Global Environmental Change, 22*(3), 559–567.

Navas, G., D'Alisa, G. J., & Martínez-Alier. (2022). The role of working-class communities and the slow violence of toxic pollution in environmental health conflicts: A global perspective. *Global Environmental Change, 73*, 102474.

Oppon, E., Acquaye, A., Ibn-Mohammed, T., & Koh, L. (2018). Modelling multi-regional ecological exchanges: The case of UK and Africa. *Ecological Economics, 147*, 422–435.

Oulu, M. (2015). The unequal exchange of Dutch cheese and Kenyan roses: Introducing and testing an LCA-based methodology for estimating ecologically unequal exchange. *Ecological Economics, 119*, 372–383.

Paavola, J., & Adger, W. N. (2005). Institutional ecological economics. *Ecological Economics, 53*(3), 353–368.

Reyes-García, V., Aceituno-Mata, L., Calvet-Mir, L., Garnatje, T., Gómez-Baggethun, E., Lastra, J. J., et al. (2014). Resilience of traditional knowledge systems: The case of agricultural knowledge in home gardens of the Iberian Peninsula. *Global Environmental Change, 24*, 223–231.

Reyes-García, V., García-del-Amo, D., Benyei, P., Fernández-Llamazares, Á., Gravani, K., Junqueira, A. B., et al. (2019). A collaborative approach to bring insights from local observations of climate change impacts into global climate change research. *Current Opinion in Environmental Sustainability, 39*, 1–8.

Reyes-García, V., García del Amo, D., Benyei, P., Junqueira, A. B., Labeyrie, V., Li, X., et al. (2020). *Local indicators of climate change impacts. Data collection protocol*. figshare. Book. https://doi.org/10.6084/m9.figshare.11513511.v3

Rival, L. (2010). Ecuador's Yasuní-ITT initiative: The old and new values of petroleum. *Ecological Economics, 70*(2), 358–365.

Rodríguez-Labajos, B., Yánez, I., Bond, P., Greyl, L., Munguti, S., Ojo, G. U., & Overbeek, W. (2019). Not so natural an alliance? Degrowth and environmental justice movements in the global south. *Ecological Economics, 157*, 175–184. https://doi.org/10.1016/j.ecolecon.2018.11.007

Samaniego, P., Vallejo, M. C., & Martínez-Alier, J. (2017). Commercial and biophysical deficits in South America, 1990–2013. *Ecological Economics, 133*, 62–73.

Schaffartzik, A., & Pichler, M. (2017). Extractive economies in material and political terms: Broadening the analytical scope. *Sustainability, 9*(7), 1047.

Scheidel, A., & Schaffartzik, A. (2019). A socio-metabolic perspective on environmental justice and degrowth movements. *Ecological Economics, 161*, 330–333.

Scheidel, A., Temper, L., Demaria, F., & Martínez-Alier, J. (2018). Ecological distribution conflicts as forces for sustainability: An overview and conceptual framework. *Sustainability Science, 13*, 585–598.

Scheidel, A., Del Bene, D., Liu, J., Navas, G., Mingorría, S., Demaria, F., et al. (2020). Environmental conflicts and defenders: A global overview. *Global Environmental Change, 63*, 102104.

Sekulova, F., Kallis, G., Rodríguez-Labajosa, B. And F. Schneider. (2013). Degrowth: From theory to practice. Journal of Cleaner Production 38: 1–6.

Sorman, A. H., & Giampietro, M. (2013). The energetic metabolism of societies and the degrowth paradigm: Analyzing biophysical constraints and realities. *Journal of Cleaner Production, 38*, 80–93.

Spash, C. L. (2011). Social ecological economics: Understanding the past to see the future. *American Journal of Economics and Sociology, 70*(2), 340–375.

Tallapragada, M. (2018). A new research agenda: Instructional practices of activists mobilizing for science. *Communication Education, 67*(4), 467–472.

Temper, L., & Martinez-Alier, J. (2013). The god of the mountain and Godavarman: Net Present Value, indigenous territorial rights and sacredness in a bauxite mining conflict in India. *Ecological Economics, 96*, 79–87.

Temper, L., Del Bene, D., & Martinez-Alier, J. (2015). Mapping the frontiers and front lines of global environmental justice: The EJAtlas. *Journal of Political Ecology, 22*(1), 255–278.

Temper, L., Walter, M., Rodriguez, I., Kothari, A., & Turhan, E. (2018). A perspective on radical transformations to sustainability: Resistances, movements and alternatives. *Sustainability Science, 13*(3), 747–764.

Temper, L., Avila, S., Del Bene, D., Gobby, J., Kosoy, N., Le Billon, P., et al. (2020). Movements shaping climate futures: A systematic mapping of protests against fossil fuel and low-carbon energy projects. *Environmental Research Letters, 15*(12), 123004.

Turnpenny, J., Jones, M., & Lorenzoni, I. (2011). Where now for post-normal science? A critical review of its development, definitions, and uses. *Science, Technology, & Human Values, 36*(3), 287–306.

Van den Bergh, J. C. (2011). Environment versus growth—A criticism of "degrowth" and a plea for "a-growth". *Ecological Economics, 70*(5), 881–890.

Van den Bergh, J. C., & Kallis, G. (2012). Growth, a-growth or degrowth to stay within planetary boundaries? *Journal of Economic Issues, 46*(4), 909–920.

Villamayor-Tomas, S., & García-López, G. (2018). Social movements as key actors in governing the commons: Evidence from community-based resource management cases across the world. *Global Environmental Change, 53*, 114–126.

Villamayor-Tomas, S., & García-López, G. A. (2021). Commons movements: Old and new trends in rural and urban contexts. *Annual Review of Environment and Resources, 46*, 511–543.

Villamayor-Tomas, S., García-López, G., & D'Alisa, G. (2022). Social movements and commons: In theory and in practice. *Ecological Economics, 194*, 107328.

Walter, M., Tomás, S. L., Munda, G., & Larrea, C. (2016). A social multi-criteria evaluation approach to assess extractive and non-extractive scenarios in Ecuador: Intag case study. *Land Use Policy, 57*, 444–458.

Zimmerer, K. S. (2015). Methods and environmental science in political ecology. In *The Routledge handbook of political ecology* (pp. 172–190). Routledge.

Zografos, C., Rodriguez-Labajos, B. A. O., Aydin, C. A., Cardoso, A., Matiku, P., Munguti, S., O'Connor, M., Ojo, G., Özkaynak, B., Slavov, T., Stoyanova, D., & Živčič, L. (2014). *Economic tools for evaluating liabilities in environmental justice struggles*, the EJOLT experience. In EJOLT Report No. 16.

Open Access This chapter is licensed under the terms of the Creative Commons Attribution 4.0 International License (http://creativecommons.org/licenses/by/4.0/), which permits use, sharing, adaptation, distribution and reproduction in any medium or format, as long as you give appropriate credit to the original author(s) and the source, provide a link to the Creative Commons license and indicate if changes were made.

The images or other third party material in this chapter are included in the chapter's Creative Commons license, unless indicated otherwise in a credit line to the material. If material is not included in the chapter's Creative Commons license and your intended use is not permitted by statutory regulation or exceeds the permitted use, you will need to obtain permission directly from the copyright holder.

Part II
Epistemological Foundations

Chapter 4
Metaphysical Midwifery and the Living Legacy of Nicholas Georgescu-Roegen

Katharine N. Farrell

4.1 Introduction

Regard for the dignity of life is a central and persistent theme throughout Joan Martínez-Alier's work and a distinguishing feature of The Barcelona School of ecological economics and political ecology. It will serve here as a narrative thread for this brief essay in Joan's honour, elaborated from an ecological economics perspective.

I once heard Joan refer to himself as the mid-wife of ecological economics: a remark that reveals a great deal about him and about the works he has created, recovered, recorded, represented and inspired over the years. A mid-wife's work is humble work: supporting another in the sacred act of bringing life into the world. It is also woman's work – and like so much of the work done mostly by women or by most women, it is often taken for granted and overlooked, precisely because it is of such fundamental worth (Mellor, 1997), making, as it does, the very act of being possible.

As an historian, communicator, activist and mentor, Joan Martínez-Alier has, in many ways, made the very act of being possible for ecological economics: helping to create, through his writings and activism and through establishing intellectual and physical spaces, including the International Society for Ecological Economics (ISEE) and L'Institut de Ciència i Tecnologia Ambientals (ICTA), where a host of counter-hegemonic narratives about environmental values have been able to gestate, be born and flourish. He has also been, personally, a pioneer of third-wave feminism (Plumwood, 1992), highlighting and integrating throughout his work ecofeminist

---

K. N. Farrell (✉)
Universidad del Rosario, Bogotá, Colombia
e-mail: katharine.farrell@qub.ac.uk

© The Author(s) 2023
S. Villamayor-Tomas, R. Muradian (eds.), *The Barcelona School of Ecological Economics and Political Ecology*, Studies in Ecological Economics 8,
https://doi.org/10.1007/978-3-031-22566-6_4

insights related to the embodied embeddedness and complex value of life, serving as a role model for hegemonically advantaged European male scholars, showing the way, with collaborations across experiential perspectives (be they gendered, colonial or of other forms) that are neither paternalistic nor dismissive.

His success in this regard is clear to see in the composition of authors in this collection, with a whopping, albeit still abysmal, 30+ percent of the lead authors being wymyn and a better than average proportion of contributors, almost 10%, based in the majority world. As both he and Val Plumwood remind us, the work of paying due regard to the dignity of life is a team endeavour and 'tis in that spirit that the following reflections are presented: a positioned perspective, on positioned perspectives.

4.2 Context

Ecological economics was originally conceived of, in the late 1980s, as a *transdiscipline* (Costanza, 1989, 1991), situated at the frontier between, but by no means limited to, the modern eurodescendent academic discourses on economics and ecology (Costanza, 1989; Giampietro & Mayumi, 2001; Norgaard, 1989), going "beyond our normal conceptions of scientific disciplines… to integrate and synthesize many different disciplinary perspectives" (Costanza, 1991: 3). With one of its central tasks being "to investigate how sustainable development is possible" (Faber et al., 2002: 324), ecological economics is practiced under post-normal science conditions (Farrell, 2011b; Funtowicz & Ravetz, 1991, 1993, 1994) where "[e]xperts from government and firms are challenged by untitled 'experts' from environmental groups, or indigenous groups, or local groups of neighbours…" (Martinez-Alier, 1999: 137).

Concerned with the production of quality knowledge (Funtowicz & O'Connor, 1999; Funtowicz & Ravetz, 1990, 1997) fit for the purpose of doing ecological economics, the Barcelona School incorporates, as Giampietro and Mayumi (2001: 2) put it: "… ability to understand and scientifically represent ecological processes… [and] socioeconomic processes… ability to integrate the two systems of scientific representations in a way that makes possible to improve both understanding and representing the predicament of sustainability in a holistic way…" and, because humans are endowed with awareness and intent, the ability to describe, understand and engage with the processes through which humans translate understanding of that predicament into collective action.

At its best, ecological economics employs a range of expertise in collaborations among and between academics and other social actors, in order to address a common matter of concern:

> how to halt, reverse and replace with something better, the current rampant destruction of the biological substrate of life on earth that is being caused by modern industrialized human activity?

We might, today, call this the "'how to' question of the Anthropocene" (Farrell, 2020). But, as Joan points out (see Martinez-Alier, 2002; Martinez-Alier & with Schlüpmann, 1987), it arose almost in parallel with modern industrialization and was clearly and poignantly articulated at that time (Goethe, 1996). Still, the fruits of industrialization proved a powerful tranquilizer (Marcuse, 1969, 1978, 1991 [1964]), and the discourse waned, coming back to the fore with the historic publication of Carson's (1963) *Silent Spring* and giving rise, in due course, to the environmental studies discourses, and in that to ecological economics.

While this discourse remained marginal for most of the twentieth century, today, being green is sexy. As climate calamities and ecological collapses have come to characterize the early twenty-first century, now the Pollyanna proposition that everything is going to be fine is becoming marginal. Questions of fairness and equity are also now gaining prominence in sustainability discourses (Farrell & Löw Beer, 2019; Kehoe et al., 2020), as the SARS-CoV-2 pandemic makes ever more apparent the inhumanity of leaving the majority of living beings on this planet (including huge numbers of humans) to bear the entropic burden of a wanton and conspicuous consumption practiced by an elite minority.

4.3 Courage

Among the proto-ecological economists who first attempted to address, in a comprehensive and rigorous way, this now fashionable 'how to' question of the Anthropocene was Nicholas Georgescu-Roegen: a Romanian mathematical genius, polyglot and erudite economist, shunned by his Harvard colleagues for having demonstrated, in forensic detail (Georgescu-Roegen, 1999 [1971]), the ontological paucity of their physical mechanics–based economic models and analytics. Both that censure, which lasted from the early 1970s until his death in the mid-1990s, and a recent renaissance of interest in Georgescu-Roegen's work, help to illustrate something exceptional about The Barcelona School; openly critiquing the systematic disregard for life (human and non-human) which has accompanied the fetishization of technology in late-industrial global society is an act of both moral and professional integrity.

It is an act of moral integrity because ignoring this disregard means ignoring the lives and livelihoods it destroys, and it is an act of professional integrity because the ecological and social deterioration of the Anthropocene is not accidental. Rather, it is a direct and logical consequence of the same political economy of accumulation that structures and finances the universities and institutes within which we work. Practicing this critique at all requires courage, because contemporary academic institutions and culture do not reward stepping out of the disciplinary matrix. Practicing it responsibly requires, in addition, (i) systematic and structured attention to the historicity of contemporary socio-ecological situations, because they are shaped by their pasts and (ii) formal attention to constructing what Michel Foucault called, in response to Immanuel Kant (1990), "a critical ontology of ourselves"

(Foucault, 1984: 47), because the eurodescendent culture of scientific knowledge production is intimately bound up with that historicity (Farrell, 2008, 2011a, 2020; Marcuse, 1991 [1964]; Ravetz, 1971).

All three of these, courage, historicity and a critical ontology, can be found in Georgescu-Roegen's (1999 [1971]) work, of which Joan has long been a champion, and in related ecological economics discourses that Joan himself has developed and that he has encouraged in the work of others.

4.4 Historicity

Georgescu-Roegen, who worked closely with both Schumpeter and Samuelson, was highly attentive to historicity in general and, in particular, to the role that institutions, an inherently historical, anthropogenic phenomenon, play in structuring both the social *and* material dimensions of economic processes (Farrell, 2018). In this respect, he was one of the first institutional ecological economists (Farrell, 2018; Paavola & Adger, 2005), calling to our attention the inescapable historicity of human acts intended to bring about qualitative change in material manifestations of the physical world: i.e. in processes of production. However, his work tends to be employed today *either* in quantitative *or* qualitative contexts (Farrell & Mayumi, 2009), leaving out his attention to the relationship between the two, which he discusses in terms of institutions, obscuring, in the process, this aspect.

Notable exceptions, which simultaneously employ quantitative *and* qualitative aspects of Georgescu-Roegen's work, tend to come from the Barcelona School (Farrell, 2018; Farrell & Löw Beer, 2019; Farrell & Mayumi, 2009; Farrell & Silva-Macher, 2017; Giampietro et al., 2006; Giampietro & Mayumi, 2000a, b; Mayumi & Giampietro, 2006; Mayumi et al., 1999; Moreau et al., 2017; Ramos-Martin et al., 2007; Scheidel & Farrell, 2015; Scheidel et al., 2014; Silva-Macher & Farrell, 2014). The place of historicity within his theoretical work gives some clues as to why this should be the case (Mayumi, 1995). For Georgescu-Roegen, historicity is a defining feature of living systems, which, in persisting as organized systems (or organisms), are constantly appropriating energy and materials and generating residuals (Schrödinger, 1948 [1944]). Being alive is a complex, thermodynamically open process, positioned along the irreversible passage of experienced time (Georgescu-Roegen, 1999 [1971]: C.5; Prigogine, 1997). Products of the combination of their own constitution, circumstances and choices, theirs and those of others, living systems are characterized by novelty (Prigogine, 1997). This is what makes it possible for a living organism to persist whilst their environment changes. However, this feature complicates the work of tracing causalities, since novelty cannot be predicted based on purely quantitative referents.

Georgescu-Roegen's strategy is to consider the historicity of organisms and organizations, which includes both qualitative and quantitative data concerning from whence they came and toward whence they are likely going. Arguing that the biologically contextualised character of early twentieth century eastern European rural village institutions had mitigated against a successful Bolshevik transition from village based to industrialised agricultural production, Georgescu-Roegen proposed that the productive flexibility and reliability afforded by the complex social organisation of village-based peasant agriculture be understood as a *biological* characteristic of that economic process (Georgescu-Roegen, 1965a).

He identified this social complexity as a defining characteristic of the successful ecologically dependent agricultural production processes of those villages, which he described as organisms, defining them as "social forms in the superorganic domain" (Georgescu-Roegen, 1999 [1971]: 13): i.e. the domain of living systems, where organic materials are combined together into living organisms, which are able to maintain low entropy at the organism level, in spite of being constantly active, by processing low entropy inputs and disposing of high entropy residuals (Schrödinger, 1948 [1944]).

The organism to which he is referring, in this context, is an ecologically embedded combination of (i) environment-attentive actors constructing, performing and being guided in their practices by institutions (Bromley, 2006; Vatn, 2005) and (ii) locally designed and operated organizations that have evolved over time, within an historical, biological context (Beer, 1994 [1966]; Maturana & Varela, 1994 [1973]; Varela et al., 1974): understood to be simultaneously both a product of and producing its own economic history.

By focusing on the historicity of environmental problems and solutions, Georgescu-Roegen is able to shift the focus from symptoms to causes, calling upon us to think ahead in evolutionary time, to be conscious not only of the immediate but also of the evolutionary implications of our choices, as we participate in configuring the physical and cultural fitness landscapes to which we and our successors will be adapting in the future (Bahro, 1977; Farrell, 2009 [2005]; Laland et al., 2000; Lewontin, 2000).

4.5 Ontology

While his flow-fund theory introduces a number of new economic analysis categories, not least, of course, those of flow and fund elements, it is with the concept of "economic *Anschuaung*" or "cultural propensities" that Georgescu-Roegen (1999 [1971]: 362–363) introduces the role of the purposive economic actor – Kant's enlightened rational being, Foucault's aspiring adult, today's co-author of the Anthropocene, as it were – into his bioeconomics ontology. This adjustment makes it possible for him to formalise analyses that explicitly link together the institutionalisation of economic decision-making processes (and with this, questions of responsibility and environmental justice so central to Joan's work) and the material manifestations of the associated decisions taken, which are of such pressing importance for halting the course of currently looming irreversible cascades of ecological collapse (Steffen et al., 2018).

While the role of economic *Anschauung* in the ontology of *The Entropy Law and the Economic Process* is easy to miss, as the concept is introduced quite late in the text, Georgescu-Roegen (1999 [1971]: 363) is very explicit as regards its importance: "...the complete data in any economic problem must include the cultural propensities [economic *Anschauung*] as well." Reaching back to his earlier work helps illustrate the role of this concept within his flow-fund theory, its contribution towards both his moral and professional postures and its legacy in the Barcelona School of ecological economics and political ecology.

In a number of early texts, where he wrote extensively about institutions, Georgescu-Roegen (Georgescu-Roegen, 1960, 1965a, b, 1969, 1988) was concerned with the question of how to represent the influence that self-understanding of social-ecological place (Farrell & Thiel, 2013) has on the organisation and outcomes of agricultural production practices. There he also explored what this implies for the study of interplay between institutional and material aspects of production. His conclusion, that the material configuration and associated economic and ecological products of these practices are mediated by the cultural propensities of their practitioners, underpins his position regarding the need to specify the economic process, ontologically, as an activity of purposive human society understood as a life-bearing organism.

Under flow-fund theory, the social-ecological composition and performance of a village or society undertaking an economic process is understood to be attenuated by the more or less autonomously selected teleology (final cause) assigned to that process, manifest in the form of an economic *Anschauung*.

> No doubt, the only reason why thermodynamics initially differentiated between the heat contained in the ocean waters and that inside a ship's furnace is that we can use the latter but not the former... however, ...while in the material environment there is only shuffling, in the economic process there is also sorting... [which] must feed on low entropy. Hence, the economic process actually is more efficient than automatic shuffling in producing higher entropy, i.e., waste. What could then be the *raison d'etre* of such a process? The answer is that the true 'output' of the economic process is not a physical outflow of waste, but *the enjoyment of life*... (Georgescu-Roegen, 1999 [1971]: 282 emphasis original).

The role of purpose in delimiting analytical boundaries pertains, in Georgescu-Roegen's work, both to the objects of study and to the work of designing tools for scholarly analysis. This brings us then, full circle, back to the moral and professional commitment associated with being a member of the Barcelona School. On the one hand, in order to ensure that one is producing quality, fit for purpose, ecological economic and political ecology analysis, one must never lose sight of the particular economic *Anschuaungen* of one's study objects, which shape the purposive intents of individuals, communities and organizations. On the other hand, and of equal importance, one must never lose sight of one's own economic *Anschuaung*, that it is present, that it plays a role in shaping how one observes the study space, and that it shapes the purposive intent of one's own work as a scholar and/or activist. In as much, on empirical as well as moral and ethical grounds, we are obliged to always ask: What is the end being served by the way in which my research is designed? What has been left out? What included? And why?

In other words, to fail to explicitly ask 'what is the good life?' "El buen vivir?" (Acosta, 2013), for ourselves and for the worlds we study, is, for Georgescu-Roegen, and for the Barcelona School, to generate ecological economic analysis and political ecology that is less than fit for purpose.

4.6 Conclusions

Making explicit the underdetermined character of social-ecological complexity, which is always shaped by cultural propensities and cosmologies, and the morally entailed ontological choices that this implies does not always make for easily digestible soundbites and it can get one into trouble with the hegemon. Remaining committed to stay with this trouble, to paraphrase Haraway (2016), means placing the quality of the scholarly work above and before personal gain and professional acceptance. With this commitment, to give due regard to the dignity of life, in all its irreducible, splendorous complexity, the Barcelona School of ecological economics and political ecology is helping to shed light upon viable, if at times awkward, paths into the future: paths that hold out the promise of preserving not only the biological diversity and viable ecological systems of our planet but also, and perhaps more importantly, our humanity.

Practicing regard for the dignity of life is a choice – and what's more, while it depends upon the commitment of individuals, it is a collective choice. In the Barcelona School of ecological economics and political ecology, this regard is treated as a matter of both empirical and moral concern, manifest in ontological choices.

Paying explicit attention to the ways in which the past has configured the present and to what that implies for the decisions taken today and their implications for the qualities of tomorrows is not only a technical but also a moral task. When we are conscious of the post-normal science conditions within which the work of ecological economics and political ecology is inevitably practiced, we can take responsibility for our actions, as we engage in the historically attenuated work of negotiating "who has the power to simplify complexity" (Martinez-Alier, 2002: 149).

References

Acosta, A. (2013). *El Buen Vivir: Sumak Kawsay, una oportunidad para imaginar otros mundos*. Icaria.
Bahro, R. (1977). *Die Alternative: zur Kritik der real existierenden Sozialismus*. Europäische Verlaganstalt.
Beer, S. (1994 [1966]). *Decision and control: The meaning of operational research and management cybernetics*. Wiley.
Bromley, D. W. (2006). *Sufficient reason: Volitional pragmatism and the meaning of economic institutions*. Princeton University Press.

Carson, R. (1963). *Silent spring*. Hamish Hamilton.
Costanza, R. (1989). What is ecological economics? *Ecological Economics, 1*, 1–7.
Costanza, R. (1991). *Ecological economics: The science and management of sustainability*. Columbia University Press.
Faber, M., Petersen, T., & Schiller, J. (2002). Homo oeconomicus and homo politicus in ecological economics. *Ecological Economics, 40*, 323–333.
Farrell, K. N. (2008). The politics of science and sustainable development: Marcuse's new science in the 21st century. *Capitalism Nature Socialism, 19*, 68–83.
Farrell, K. N. (2009 [2005]). *Making good decisions well: A theory of collective ecological management*. Shaker Verlag GmbH.
Farrell, K. N. (2011a). The politics of science: Has Marcuse's new science finally come of age? In A. Biro (Ed.), *Critical ecologies: The Frankfurt school and contemporary environmental crises* (pp. 73–107). University of Toronto Press.
Farrell, K. N. (2011b). Snow White and the Wicked Problems of the West: A look at the lines between empirical description and normative prescription. *Science, Technology & Human Values, 36*, 334–361.
Farrell, K. N. (2018). The analytical economics of living well: concepts and applications for the combined study of value, production and the ecological contribution of economic processes. In: L. F. d. H.-U. z. Berlin (Ed.), *Habilitationsschrift zur Erlangung der Lehrbefähigung für das Fach: Agrarökonomie ed*. Humboldt-Universität zu Berlin.
Farrell, K. N. (2020). Untrol: Post-truth and the new normal of post-normal science. *Social Epistemology, 34*, 330–345.
Farrell, K. N., & Löw Beer, D. (2019). Producing the ecological economy. *Ecological Economics, 165*, 106391.
Farrell, K. N., & Mayumi, K. (2009). Time horizons and electricity futures: An application of Nicholas Georgescu-Roegen's general theory of economic production. *Energy, 34*, 301–307.
Farrell, K. N., & Silva-Macher, J. C. (2017). Exploring futures for Amazonia's Sierra del Divisor: An environmental valuation triadics approach to analyzing ecological economic decision choices in the context of major shifts in boundary conditions. *Ecological Economics, 141*, 166–179.
Farrell, K. N., & Thiel, A. (2013). Nudging evolution? *Ecology and Society, 18*.
Foucault, M. (1984). What is enlightment? In P. Rabinow (Ed.), *The Foucault reader: An introduction to Foucault's thought* (pp. 32–50). Penguin.
Funtowicz, S. O., & O'Connor, M. (1999). The passage from entropy to thermodynamic indeterminacy: A social and science epistemology for sustainabiliy. In K. Mayumi & J. M. Gowdy (Eds.), *Bioeconomics and sustainability: Essays in Honor of Nicholas Georgescu-Roegen* (pp. 257–286). Edward Elgar.
Funtowicz, S. O., & Ravetz, J. R. (1990). *Uncertainty and quality in science for policy*. Kluwer Academic Publishers.
Funtowicz, S. O., & Ravetz, J. R. (1991). A new scientific methodology for global environmental issues. In R. Costanza (Ed.), *Ecological economics: The science and management of sustainability* (pp. 137–152). Columbia University Press.
Funtowicz, S. O., & Ravetz, J. R. (1993). Science for the post-normal age. *Futures, 25*, 739–755.
Funtowicz, S. O., & Ravetz, J. R. (1994). The worth of a songbird: Ecological economics as a post-normal science. *Ecological Economics, 10*, 197–207.
Funtowicz, S. O., & Ravetz, J. R. (1997). The poetry of thermodynamics – Energy, entropy/exergy and quality. *Futures, 29*, 791–810.
Georgescu-Roegen, N. (1960). Economic theory and Agrarian economics (with Postscripts from 1966 and 1975). In N. Georgescu-Roegen (Ed.), *Energy and economic myths: Institutional and analytical economic essays* (pp. 103–145). Pergamon Press.
Georgescu-Roegen, N. (1965a). The institutional aspects of peasant communities. In N. Georgescu-Roegen (Ed.), *Energy and economic myths: Institutional and analytical economic essays* (pp. 199–231). Pergamon Press.

Georgescu-Roegen, N. (1965b). Process in farming versus process in manufacturing: A problem of balanced development. In N. Georgescu-Roegen (Ed.), *Energy and economic myths: Institutional and analytical economic essays* (pp. 71–102). Pergamon Press.

Georgescu-Roegen, N. (1969). The economics of production. In N. Georgescu-Roegen (Ed.), *Energy and economic myths: Institutional and analytical economic essays* (pp. 61–69). Pergamon Press.

Georgescu-Roegen, N. (1988). The interplay between institutional and material factors: The problem and its status. In J. A. Kregel, E. Matzner, & A. Roncaglia (Eds.), *Barriers to full employment* (pp. 297–326). Macmillan Press.

Georgescu-Roegen, N. (1999 [1971]). *The entropy law and the economic process*. Harvard University Press, Distributed by Oxford University Press.

Giampietro, M., & Mayumi, K. (2000a). Multiple-scale integrated assessment of societal metabolism: Integrating biophysical and economic representations across scales. *Population and Environment, 22*, 155–210.

Giampietro, M., & Mayumi, K. (2000b). Multiple-scale integrated assessment of societal metabolism: Introducing the approach. *Population and Environment, 22*, 109–153.

Giampietro, M., & Mayumi, K. (2001). *Integrated assessment of sustainability trade-offs: Methodological challenges for ecological economics*. Frontiers in Ecological Economics.

Giampietro, M., Allen, T. F. H., & Mayumi, K. (2006). The epistemological predicament associated with purposive quantitative analysis. *Ecological Complexity, 3*, 307–327.

Goethe, J. W. V. (1996). *Faust: Der Tragödie erster und zweiter Teil, Urfaust*. Beck.

Haraway, D. (2016). *Staying with the trouble: Making Kin in the Chthulucene*. Duke University Press.

Kant, I. (1990). An answer to the question. What is enlightment? In H. Reiss (Ed.), *Kant: Political writings* (pp. 54–60). Cambridge University Press.

Kehoe, L., Reis, T. N. P. D., Meyfroidt, P., Bager, S., Seppelt, R., Kuemmerle, T., Berenguer, E., Clark, M., Davis, K. F., Ermgassen, E. K. H. J. Z., Farrell, K. N., Friis, C., Haberl, H., Kastner, T., Murtough, K. L., Persson, U. M., Romero-Muñoz, A., O'Connell, C., Schäfer, V. V., Virah-Sawmy, M., Waroux, Y. I. P. D., & Kiesecker, J. (2020). Inclusion, transparency, and enforcement: How the EU-Mercosur trade agreement fails the sustainability test. *One Earth, 3*, 268–272.

Laland, K. N., Odling-Smee, J., & Feldman, M. W. (2000). Niche construction, biological evolution, and cultural change. *Behavioral and Brain Sciences, 23*, 131–175.

Lewontin, R. (2000). *The Triple Helix: Gene, organism and environment*. Harvard University Press.

Marcuse, H. (1969). *An essay on liberation*. Allen Lane The Penguin Press.

Marcuse, H. (1978). Protosozialismus und Spätkapitalismus – Versuch einere revolutionstheoretischen Synthese von Bahros Anzatz. In D. Claussen (Ed.), *Spuren der Befreiung – Herbert Marcuse* (pp. 89–116). Luchterhand [reprinted from the original in Kritik, Zeitschift für sozialistische Diskussion, 19/later published as (1980) Protosocialism and late capitalism: Toward a theoretical synthesis based on Bahro's analysis. *International Journal of Politics, 10*(2/3) Rudolf Bahro: Critical Responses:25–48], Darmstadt und Neuwind.

Marcuse, H. (1991 [1964]). *One-dimensional man: Studies in the ideology of advanced industrial society*. Routledge.

Martinez-Alier, J. (1999). The socio-ecological embeddedness of economic activity: The emergence of a transdisciplinary field. In E. Becker & T. Jahn (Eds.), *Sustainability and the Social Sciences: A cross-disciplinary approach to integrating environmental considerations into theoretical reorientation* (pp. 112–139). Zed Books.

Martinez-Alier, J. (2002). *The environmentalism of the poor*. Edward Elgar.

Martinez-Alier, J., & with Schlüpmann, K. (1987). *Ecological economics: Energy, environment and society*. Blackwell.

Maturana, H., & Varela, F. (1994 [1973]). De Máquinas y Seres Vivos Autopoeisis: la organisación de lo vivo. Grupo Editorial Lumen [Coedición de Editorial Universitaria, Santiago de Chile con Editorial Lumen, 2004].

Mayumi, K. (1995). Nicholas Georgescu-Roegen (1906–1994): An admirable epistemologist. *Structural Change and Economic Dynamics, 6*, 261–265.

Mayumi, K., & Giampietro, M. (2006). The epistemological challenge of self-modifying systems: Governance and sustainability in the post-normal science era. *Ecological Economics, 57*, 382–399.

Mayumi, K., Giampietro, M., & Gowdy, J. (1999). Georgescu-Roegen/Daly versus Solow/Stiglitz revisited. *Ecological Economics, 27*, 115–117.

Mellor, M. (1997). Women, nature and the social construction of 'economic man'. *Ecological Economics, 20*, 129–140.

Moreau, V., Sahakian, M., van Griethuysen, P., & Vuille, F. (2017). Coming full circle: Why social and institutional dimensions matter for the circular economy. *Journal of Industrial Ecology, 21*, 497–506.

Norgaard, R. B. (1989). The case for methodological pluralism. *Ecological Economics, 1*, 37–57.

Paavola, J., & Adger, N. W. (2005). Institutional ecological economics. *Ecological Economics, 53*, 353–368.

Plumwood, V. (1992). Feminism and ecofeminism: Beyond the dualistic assumptions of women, men and nature. *The Ecologist, 22*, 8–13.

Prigogine, I. (1997). *The end of the certainty – Time, chaos, and the new laws of the nature*. Free Press.

Ramos-Martin, J., Giampietro, M., & Mayumi, K. (2007). On China's exosomatic energy metabolism: An application of multi-scale integrated analysis of societal metabolism (MSIASM). *Ecological Economics, 63*, 174–191.

Ravetz, J. R. (1971). *Scientific knowledge and its social problems*. Clarendon Press.

Scheidel, A., & Farrell, K. N. (2015). Small-scale cooperative banking and the production of capital: Reflecting on the role of institutional agreements in supporting rural livelihood in Kampot, Cambodia. *Ecological Economics, 119*, 230–240.

Scheidel, A., Farrell, K. N., Ramos-Martin, J., Giampietro, M., & Mayumi, K. (2014). Land poverty and emerging ruralities in Cambodia: Insights from Kampot province. *Environment, Development and Sustainability, 16*, 823–840.

Schrödinger, E. (1948 [1944]). *What is life?* Cambridge University Press.

Silva-Macher, J. C., & Farrell, K. N. (2014). The flow/fund model of Conga: Exploring the anatomy of environmental conflicts at the Andes-Amazon commodity frontier. *Environment, Development and Sustainability*, 1–22.

Steffen, W., Rockström, J., Richardson, K., Lenton, T. M., Folke, C., Liverman, D., Summerhayes, C. P., Barnosky, A. D., Cornell, S. E., Crucifix, M., Donges, J. F., Fetzer, I., Lade, S. J., Scheffer, M., Winkelmann, R., & Schellnhuber, H. J. (2018). Trajectories of the Earth system in the anthropocene. *Proceedings of the National Academy of Sciences, 115*, 8252.

Varela, F. G., Maturana, H. R., & Uribe, R. (1974). Autopoiesis: The organization of living systems, its characterization and a model. *Biosystems, 5*, 187–196.

Vatn, A. (2005). *Institutions and the environment*. Edward Elgar.

Open Access This chapter is licensed under the terms of the Creative Commons Attribution 4.0 International License (http://creativecommons.org/licenses/by/4.0/), which permits use, sharing, adaptation, distribution and reproduction in any medium or format, as long as you give appropriate credit to the original author(s) and the source, provide a link to the Creative Commons license and indicate if changes were made.

The images or other third party material in this chapter are included in the chapter's Creative Commons license, unless indicated otherwise in a credit line to the material. If material is not included in the chapter's Creative Commons license and your intended use is not permitted by statutory regulation or exceeds the permitted use, you will need to obtain permission directly from the copyright holder.

Chapter 5
Languages of Valuation

Christos Zografos

5.1 Introduction

In this chapter, I look at the analytical concept of languages of valuation and specifically at the work of the Barcelona School of environmental social science, which uses it to study environmental conflicts and governance. The genealogy of the concept goes back to the claim advanced by Joan Martínez-Alier that many environmental conflicts are conflicts over different languages used to place a value on the environment, which are regularly expressed in the context of unequal distributions of material costs and benefits generated by environmental transformations. Beyond being a 'real life' issue, I understand the languages of valuation concept as an analytical device for examining environmental conflicts, one distinctively advanced by Martínez-Alier and subsequently by researchers and scholars connected to the Barcelona School.

I follow the concept as it passes through the Barcelona School in the roughly 30-year period to 2020. I trace this trajectory in a selected number of doctoral and postdoctoral work of researchers connected with the Barcelona School and some of their collaborations with scholars outside it. The starting point for that work is the ecological economics criticism of monetary valuation of the environment for its reductionism and exclusion of certain sets of environmental values and the ecological economics espousal of value diversity, incommensurability and plurality in environmental decision-making.

I classify into themes and present contributions from the Barcelona School that are informed by this framework of analysis and which have discussed and employed the concept of languages of valuation to advance understanding of environmental conflicts, justice, movements and decision-making. I conclude by drawing some lessons from that literatureand present my reflections on promising research

C. Zografos (✉)
Universitat Pompeu Fabra, Barcelona, Spain

avenues. I believe that future research should seek to expand links between languages of valuation and the pluriverse project in an effort to both advance knowledge about decoloniality and contribute to much-needed radical socio-ecological transformations, particularly in the face of the climate crisis.

5.2 Languages of Valuation

In 1995, economists Clive Spash and Nick Hanley published an article in *Ecological Economics* that reported the results of a Willingness-to-Pay (WTP) survey for biodiversity preservation. The study found that almost a quarter of the general public sample refused giving a WTP amount, on the grounds "that animals/ecosystems/plants should be protected irrespective of the costs" (Spash & Hanley, 1995, p. 203). The authors explained refusal to trade off nature for money as an expression of what neo-classical economic theory terms as lexicographic preferences. As with dictionary (lexicon) entries where a word that starts with a letter earlier in the alphabet (e.g. 'a') is *always* given priority (comes first in the dictionary) over one that starts with a letter that follows (e.g. 'c'), an agent holding lexicographic preferences will always prioritise one good over another and reject making trade-offs between two different goods. When one of those goods is money, as in the case of monetary valuation of the environment, without such trade-offs one cannot produce a utility curve and hence meaningfully arrive at a money value for that good (e.g. biodiversity). Spash and Hanley concluded that the existence of lexicographic preferences towards biodiversity in a considerable part of the general population raised significant concerns about the acceptability of using contingent valuation to value biodiversity protection and decision-making. One of those concerns is that the use of monetary valuations as input to decision-making could leave out some people's values from that process, implying that monetary valuations can become instruments of exclusion (Zografos, 2015a, b).

The lexicographic preferences argument came to add to a battery of arguments advanced around that time in ecological economics in the context of an ardent criticism of monetary valuation. This is not the space to explain those arguments in detail, but it's worth briefly mentioning some of them: value incommensurability, emphasising that environmental values are not always commensurable and that they cannot be measured in the same unit (Martinez-Alier et al., 1997); value pluralism, claiming that there is a plurality of beliefs about what is of value (O'Neill & Spash, 2000), which in concert with incommensurability calls for multiple means of valuation to be brought into the resolution of environmental conflicts and decision-making (Martinez-Alier et al., 1997); value articulating institutions, referring to frames (such as, but not limited to, cost-benefit analysis) invoked in the process of expressing values and which regulate and influence which values come forward and which are excluded (Vatn, 2005); multiple rationalities, that beyond homo economicus, i.e. the model of rationality upon which monetary valuation of the environment is premised, human rationality can also be consequentialist, deontological and procedural in its outlook (Zografos & Paavola, 2008).

All that work and, overall, the field of ecological economics were establishing a case for diversity, inclusion and plurality, which claimed that multiple rationalities, values and ethics are relevant when valuing the environment and 'resolving' or understanding environmental conflicts. What is more, some scholars in ecological economics were working towards finding ways to operationalise those principles, such as Munda's development of a model of social multi-criteria analysis that permits operationalising 'weak comparability' of environmental values expressed in different units through his NAIADE model (Munda, 1995); or adopting non-positivist, mixed-methods and interpretive approaches such as Q methodology for analysing environmental policy (Barry & Proops, 1999) and values (Zografos, 2007). Additionally, post-normal science (Funtowicz & Ravetz, 1994) with its emphasis on the importance of extended peer review communities for democratising both expertise and public decision-making exerted influence over ecological economists who sought to apply such tools for improving environmental decisions. Those trends also combined with increased calls in the field for considering the relevance of democratic deliberation in environmental policy-making (O'Neill & Spash, 2000).

The concept of languages of valuation appears in that climate of intellectual ebullition in ecological economics. For Martinez-Alier, environmental conflicts are ecological distribution conflicts, that is conflicts concerning the unequal distribution of 'goods' and 'bads' from environmental change (Martinez-Aier, 2002). Ecological distribution conflicts involve unequal cost-shifts (Kapp, 1975) (Aguilera-Klink & Alcántara, 1994) of the harmful impacts that result from expanding the social metabolism of materially abundant societies and economies. In economies that seek to grow, this expansion is inevitable because the second law of thermodynamics, which establishes that energy is dissipated and cannot be recovered, implies that circular notions of the economy are in practice unrealistic (Martínez Alier, 2020). Such unequal distributions often occur in the context of an expansion of commodity frontiers, that is the arrival in certain locations, communities and ecosystems of contaminating activities that result from the quest to reduce production costs (e.g. by developing mining activity in places where it is poorly regulated) or generate new opportunities for profits (e.g. via the mining of materials necessary for the production of new, profitable commodities, such as lithium for the green economy). In those situations, local communities or environmental justice organisations seek to confront inequality by recurring to ways of valuing nature and their relation to nature that cannot always be captured or directly compared to the language of monetary value. Some examples of those languages are the sacredness of nature, the rights of nature, national or local sovereignty, territorial rights, environmental and social justice and livelihoods – languages that cannot be readily translated into a price tag. In that context, it becomes impossible to internalise externalities and so offer money compensation for the loss of certain values to either prevent conflict from happening (Temper et al., 2018) or arrive at a fair conflict settlement. What is more, imposing either monetary valuation as the single procedure or monetary value as the single language of valuation amounts to a questionable exercise of 'procedural power', i.e. the power to determine the bottom line in deciding over ecological distribution conflicts in the face of complexity (Martinez-Aier, 2002).

This 'clash' between the expression of environmental value in monetary terms versus its expression in non-monetary terms came to define the approach taken by the Barcelona School in its research and scholarship on environmental conflict, environmental justice and inequality, environmental policy and decision-making.

5.3 Languages of Valuation and the Barcelona School

5.3.1 The Clash

A key outlet of Martinez-Alier's work since 2012 has been the Environmental Justice Atlas. Together with a dedicated core team of younger researchers at ICTA-UAB and the help of several ICTA-UAB postgraduate students and numerous volunteers and environmental NGOs around the world, they have created an online map of more than 3000 ecological distribution conflicts across the globe. Among other data, the EJAtlas records hundreds of cases of different languages of valuation used in those conflicts, trying to capture how local communities and protest groups frame their claims and languages of valuation (EJOLT, n.d.).

The Atlas is a project of comparative environmentalism that records commonalities and differences of environmentalism across locations and the characteristics of an incipient global movement for environmental justice (Temper et al., 2018). This includes ways in which different languages of valuation, such as livelihood, sacredness, ecological values, territorial rights or economic compensation, are deployed in ecological distribution conflicts (Temper et al., 2018). A 2018 special issue in the journal *Sustainability Science* analysed several instances of value system contests in ecological distribution conflicts, where the assumption that externalities can have a price tag is questioned (Temper et al., 2018).

The idea that non-monetary valuation languages stand *in opposition* to monetary valuation, is probably the most common focus of published studies whose analysis inter alia looks at languages of valuation. Numerous examples of that opposition have been presented by the Barcelona School. Those include: the case of the conservation movement that has favoured monetary valuation of ecosystems in contrast to the environmentalism of the poor which appeals more to non-economic values (Rodríguez-Labajos & Martínez-Alier, 2013); cases of urban community gardens advancing languages of valuation that combine historic and cultural preservation, the repair of fragmented communities, community cohesion and defence of traditional land and territory, in contrast to languages of green consumption or compact cities (Anguelovski & Martínez Alier, 2014); the case of commercial logging in Cameroon, where defence of livelihood, customary institutions and sacredness are mobilised against economic growth and the language of monetary valuation

(Veuthy & Gerber, 2011); 'energy sovereignty' used by anti-dam resistance movements in India, which deploy the vocabulary of environmental justice to reclaim popular control over territories and energy models (Del Bene, 2018); the idioms of 'ecological balance' and 'environmental quality' used by communities in Turkey to oppose gold-mining projects, where monetary and technical compensatory schemes fail resolving disagreements (Avcı et al., 2010); and the use of civil and human rights language in mining (Urkidi & Walter, 2011) or oil palm and sugarcane plantation conflicts (Mingorria Martinez, 2017) in Latin America.

In some cases, the mobilisation of non-monetary valuation languages can be relatively *successful*. In Mexico, indigenous groups' use of languages of valuation that diverged from those employed by state and corporations in Mexico opened up spaces of political organisation that enabled the creation of resistance networks (Avila-Calero, 2017). Still, others have questioned the effectiveness of plural and multiple valuation languages, precisely on the grounds that their diversity makes it difficult to establish paths for alliances among social actors (Cardoso, 2018).

In all cases, and although the economic language of valuation does not always carry the day (Martinez-Alier et al., 2010), most published studies present situations where monetary valuation is *imposed* and non-monetary valuation languages are excluded through either legal or illegal exercise of power (Martinez-Alier et al., 2010), including the murdering of environmental activists. States, municipalities and companies regularly try to impose a single valuation language (money) and emphasise the benefits of economic growth that will eventually 'trickle down' and compensate for any losses (Anguelovski & Martínez Alier, 2014). The Indian Supreme Court 2006 controversy over the dismantling of the ocean liner 'Blue Lady' is a typical case in which sustainability expressed as monetary benefit at the national scale prevailed over non-monetary languages of valuation expressed by contending social groups, allowing to shift the costs of development to poorer, disenfranchised communities and accumulation by contamination (Demaria, 2010).

But such clashes between languages of valuation do not always have to happen. The Barcelona School acknowledges that non-monetary valuation languages can – and are indeed – often used *in combination* with monetary valuation. Although they prefer remaining within other valuation standards, especially those concerning the environmental conditions of their productive activities, human and customary rights and infrastructure needs, rural communities in the global South may use monetary reparation as a language (Gerber et al., 2009). Grassroots organisations, indigenous communities, citizen groups and women activists may request monetary compensation for damages and simultaneously demand respect for human rights (e.g. to health), indigenous territorial rights and sacredness (Martinez Alier et al., 2010; Anguelovski & Martínez Alier, 2014). And the climate justice movement has supported the monetary calculation of the so-called ecological debt (Rodríguez-Labajos & Martínez-Alier, 2013). It looks likely that social movements employ the technical language of Western environmentalism for strategic reasons but also combine it with arguments about identity and culture (Temper et al., 2018). And in some cases, monetary-based policy tools such as PES-like schemes (e.g. the Yasuni ITT initiative) have managed to integrate diverse valuation languages (Kallis et al., 2013).

5.3.2 Resolving the Clash

One can probably identify two classes of attempts at resolving the clash between monetary and non-monetary languages of valuation in which Barcelona School researchers have been involved.

The first involves resolving the clash by entering into *the debate* about the relative merits of employing monetary and non-monetary valuation in ecological distribution conflicts.

A characteristic example of this is a debate between Barcelona School researchers that took place in the journal *Ecological Economics* on the occasion of the publication of an article titled 'To Value or Not to Value?' That article pondered that environmentalists regularly find themselves trapped in a dilemma when trying to defend the environment: Concede that money is a language understood by policy-makers and the general public – the language that speaks to dominant economic and political views (Brondízio et al., 2012) – and value monetarily nature in order to protect it; or maintain deeply held and incommensurable values but risk irrelevance in nature protection struggles?

With the 'To Value or Not to Value?' article, Kallis et al. (2013) attempted to go past this conundrum by reformulating the 'should we value' question into 'when and how to value with money?' and "under what conditions?". To do so, the authors mobilised an analytical approach that brought political ecology in conversation with ecological economics. The conclusion was that monetary valuation is acceptable if it forms part of processes that improve the environment while bringing more equality, including maintaining the relevance of plural valuation languages. In those cases where monetary valuation could suppress other languages and value-articulating institutions, it should be rejected. When monetary valuation expands its domain, colonises and displaces other values by becoming the dominant language through which values are expressed, value reductionism occurs and should be avoided.

A response to that paper was published (Gsottbauer et al., 2015), criticising Kallis et al. for approaching monetary valuation in a much more critical way than other languages. Gsottbauer et al. claimed that in real life, non-monetary considerations such as rights, safety and ethics overrule or preclude monetary assessments and advocated a more mixed approach, where monetary valuation helps strengthen the case of other valuation languages. Climate policy goals were presented as an example where economic values of climate damages can convince politicians, corporations and citizens that it is important to establish policies to halt climate change and be used in complementarity with non-monetary languages.

That interchange gave some new impetus to the old 'nature valuation debate' in ecological economics by establishing certain conditions or criteria for considering the use of monetary vs. non-monetary languages of valuation in ecological distribution conflicts and by nuancing arguments about the potential for complementarity between different languages.

Deliberative ecological economics represents a different approach to resolving 'the clash' by attempting to accommodate both monetary and non-monetary languages of valuation (value pluralism) through a formal procedure of deliberation integrated in environmental decision-making processes.

There are two main lines of work in deliberative ecological economics (Zografos & Howarth, 2008a, b). The first combines deliberation with either monetary (e.g. via choice experiments and group-based valuation) or non-monetary (e.g. multi-criteria analysis) decision-making tools in an effort to integrate multiple valuation languages and reach group decisions, either through monetising (deliberative monetary valuation) or by keeping with the incommensurability principle (deliberative multi-criteria analysis). This work shows that in order to achieve consensus, value plurality does not need to diminish (Lo, 2013), and that through deliberation, preferences can also converge in making ecosystem services obtain incommensurable values. Social learning through deliberation may even induce decision-makers to consider ecosystems as priceless and become unwilling to trade off ecosystem services for money (Kenter et al., 2011).

A second, more critical line of work investigates obstacles to the expression of plural perspectives and multiple valuation languages in environmental decision-making with a view to specifying conditions for inclusive sustainability politics (Zografos & Howarth, 2008a, b). It has shown how distributional inequalities may combine with informal elements of the decision-making process and technocratic planning tools to encourage instrumental rationalities and create procedural environmental injustice where multiple languages of valuation cannot be expressed and negotiated (Zografos & Martínez-Alier, 2009). Similarly, the idioms of 'common good' or 'public benefit' can silence certain voices in climate adaptation policy, which express their value claims in idioms that emphasise personal experience (e.g. land connections with ancestors) (Zografos, 2017).

Deliberation for integrating multiple valuation languages has been marred by criticisms. A main criticism is that deliberative monetary valuation pretends that two models with radically different ontological presuppositions such as deliberation (with its collectivist outlook) and monetary valuation (with its individualist outlook) can be combined or held in conjunction, which is not possible (Spash, 2008). A second criticism holds that the normative emphasis on deliberation ignores the practical context of power surrounding and pervading environmental decision-making. The deliberative emphasis on consensus reached via communicative reason can end up silencing the importance of conflict for democracy and privilege certain voices, in particular the voice of reason, as relevant for decision-making at the expense of emotional aspects of human experience (Zografos & Howarth, 2010). The under-representation of emotional aspects in deliberative processes has been linked to a Kantian view of enlightenment that stands at the origins of the deliberative approach and is very problematic as the appeal to emotions can significantly motivate public action (O'Neill, 2007) for radical socio-ecological change (Nelson, 2011).

Admittedly, deliberative ecological economics has not taken deep root with the Barcelona School. Nevertheless, partly in response to those criticisms, some work in this sub-field has been re-oriented towards examining the challenges that direct democracy (a process that involves deliberation between plural values via, e.g., assembly-based decision-making) faces as a vehicle for inclusive, radical socio-ecological change, such as degrowth (Zografos, 2015a, b) and post-development (Zografos, 2019).

5.4 Conclusions

At a fundamental level, the concept of languages of valuation is about diversity and exclusion. The intense ecological economics criticism raged against the methodological individualism and reductionism of monetary valuation which sits at the origin of the term, hinges upon the argument that there are multiple ways of not only valuing nature but also of expressing that value. And the normative implication that ecological economists conferred to that criticism, namely value pluralism in environmental decision-making, is an argument for inclusion and voice equality. It involves inclusion of various ways of understanding, expressing and valuing nature when it comes to deciding about human affairs that entangle nature, but also when it comes to studying why and how ecological distribution conflicts appear.

The Barcelona School has significantly advanced our knowledge concerning languages of valuation. It has taken what used to be a debate confined to disagreements between environmentally-minded economists about the capacity of money to 'capture' 'real' preferences and design environmental policy, to an ample and varied scholarship that connects discussions between environmental policy, social movements, sustainability governance, environmental philosophy and ethics, institutional economics, environmental history and political ecology-minded scholars. What is more, the wealth of cases around valuation language clashes documented with the EJ Atlas project and its related publications are a promising indication for future development in the field.

What the Barcelona School has considered less are connections between valuation languages' exclusions and ideology. For example, some scholars have pointed out that favouring instrumental environmental values while ignoring, relational, non-western languages perpetuates the historical, forced assimilation to settler narratives (Himes & Muraca, 2018). In a somehow flip side to that situation, researchers record cases where western notions and representations of 'harmonious' indigenous life within nature and corresponding environmental values may successfully advance the political causes of nature advocates but at the same time add to a long history of denying indigenous agency (Tanasescu, 2015). In both those cases, the language of valuation clashes have implications that go beyond the purely functional effect that marginalising certain languages produces in terms of resource dispossession. Indeed, environmental conflicts where different valuation languages clash can be conflicts about 'how one is allowed to feel, what one is allowed to enjoy (doing), how is one supposed to live (spend time)' (Velicu, 2015, p. 857). In the course of those struggles, subjects struggle to create visibilities for new things, objects and languages that have been downplayed by dominant political contexts, and in effect strive to advance democracy and equality by attempting what Ranciere calls a 'redistribution of the sensible' (Velicu, 2015).

Recently, Martinez-Alier has placed languages of valuation within the project of the pluriverse, a 'process of intellectual, emotional, ethical, and spiritual decolonization …[that seeks deconstructing]…the idea of "development as progress"…to open a way for cultural alternatives that nurture and respect life on Earth' (Kothari et al., 2019, p. xvii). He asserts that the coining and use of terms such as biopiracy, sacrifice zones, green deserts, etc., permits environmental movements to push for alternative social transformation by deploying new vocabularies in the course of environmental justice struggles (Martínez-Alier, 2019).

Decoloniality, which sits at the basis of the pluriverse project, is certainly a relevant and promising context for future discussions of languages of valuation and their political significance. Yet, not all languages of valuation hold the potential, or indeed have the aspiration to help 'learning to unlearn' 'what imperial/colonial designs have naturalized as the only way to know and the only way to be' (Tlostanova & Mignolo, 2012, p. 22) which is the hallmark of decolonial pedagogy. How do different languages of valuation contribute to learning to unlearn colonial ways of knowing and registering nature? What histories, actors, contingencies, politics and power configurations play out when it comes to such contributions to decolonial pedagogy? What alternative 'buried epistemologies' (Willems–Braun, 1997) do different languages of valuation bring to light and how might these contribute to radical socio-ecological transformations? Research asking such questions related to decoloniality of knowledge, while linking those inquiries to contemporary capacities for transformation could channel the wealth of the School's work on languages of valuation in a way that advances theory and our understanding of the politics of socio-ecological transformations.

References

Aguilera-Klink, F., & Alcántara, V. (1994). *De la economía ambiental a la economía ecológica*. Universidad Privada del Norte.

Anguelovski, I., & Martínez Alier, J. (2014). The 'Environmentalism of the poor' revisited: Territory and place in disconnected glocal struggles. *Ecological Economics, 102*, 167–176. https://doi.org/10.1016/j.ecolecon.2014.04.005

Avcı, D., Adaman, F., & Özkaynak, B. (2010). Valuation languages in environmental conflicts: How stakeholders oppose or support gold mining at Mount Ida, Turkey. *Ecological Economics, Special Section: Ecological Distribution Conflicts, 70*, 228–238. https://doi.org/10.1016/j.ecolecon.2010.05.009

Avila-Calero, S. (2017). Contesting energy transitions: Wind power and conflicts in the Isthmus of Tehuantepec. *Journal of Political Ecology, 24*, 992–1012. https://doi.org/10.2458/v24i1.20979

Barry, J., & Proops, J. (1999). Seeking sustainability discourses with Q methodology. *Ecological Economics, 28*, 337–345.

Brondízio, E. S., Gatzweiler, F. W., Zografos, C., Kumar, M., Kadekodi, G. K., McNeely, J. A., Xu, J., & Martinez-Alier, J. (2012). The socio-cultural context of ecosystem and biodiversity valuation. In *The economics of ecosystems and biodiversity: Ecological and economic foundations* (pp. 149–182).

Cardoso, A. (2018). Valuation languages along the coal chain from Colombia to the Netherlands and to Turkey. *Ecological Economics, 146*, 44–59. https://doi.org/10.1016/j.ecolecon.2017.09.012

Del Bene, D. (2018). *Hydropower and ecological conflicts. From resistance to transformations*. PhD thesis. Universitat Autònoma de Barcelona.

Demaria, F. (2010). Shipbreaking at Alang–Sosiya (India): An ecological distribution conflict. *Ecological Economics, Special Section: Ecological Distribution Conflicts, 70*, 250–260. https://doi.org/10.1016/j.ecolecon.2010.09.006

EJOLT. (n.d.). *EJAtlas* [WWW document]. https://ejatlas.org. Accessed 5.13.20.

Funtowicz, S. O., & Ravetz, J. R. (1994). The worth of a songbird: Ecological economics as a post-normal science. *Ecological Economics, 10*, 197–207.

Gerber, J.-F., Veuthey, S., & Martínez-Alier, J. (2009). Linking political ecology with ecological economics in tree plantation conflicts in Cameroon and Ecuador. *Ecological Economics, 68*, 2885–2889. https://doi.org/10.1016/j.ecolecon.2009.06.029

Gsottbauer, E., Logar, I., & van den Bergh, J. (2015). Towards a fair, constructive and consistent criticism of all valuation languages: Comment on Kallis et al. (2013). *Ecological Economics, 112*, 164–169. https://doi.org/10.1016/j.ecolecon.2014.12.014

Himes, A., & Muraca, B. (2018). Relational values: The key to pluralistic valuation of ecosystem services. *Current Opinion in Environmental Sustainability, Sustainability Challenges: Relational Values, 35*, 1–7. https://doi.org/10.1016/j.cosust.2018.09.005

Kallis, G., Gómez-Baggethun, E., & Zografos, C. (2013). To value or not to value? That is not the question. *Ecological Economics, 94*, 97–105. https://doi.org/10.1016/j.ecolecon.2013.07.002

Kapp, K. W. (1975). *The social costs of private enterprise* (2nd printing ed.). Schocken Books.

Kenter, J. O., Hyde, T., Christie, M., & Fazey, I. (2011). The importance of deliberation in valuing ecosystem services in developing countries—Evidence from the Solomon Islands. *Global Environmental Change, 21*, 505–521. https://doi.org/10.1016/j.gloenvcha.2011.01.001

Kothari, A., Salleh, A., Escobar, A., Demaria, F., & Acosta, A. (2019). *Pluriverse: A post-development dictionary*. Tulika Books and Authorsupfront.

Lo, A. Y. (2013). Agreeing to pay under value disagreement: Reconceptualizing preference transformation in terms of pluralism with evidence from small-group deliberations on climate change. *Ecological Economics, 87*, 84–94. https://doi.org/10.1016/j.ecolecon.2012.12.014

Martínez Alier, J. (2020). A global environmental justice movement: Mapping ecological distribution conflicts. *Disjuntiva, 1*, 81. https://doi.org/10.14198/DISJUNTIVA2020.1.2.6

Martinez-Aier, J. (2002). *The environmentalism of the poor*. Edward Elgar Publishing.

Martínez-Alier, J. (2019). Environmental justice. In A. Kothari, A. Salleh, A. Escobar, F. Demaria, & A. Acosta (Eds.), *Pluriverse: A post-development dictionary* (pp. 182–185). Tulika Books and Authorsupfront.

Martinez-Alier, J., Munda, G., & O'Neill, J. (1997). Incommensurability of values in ecological economics. In M. O'Connor & C. Spash (Eds.), *Valuation and the environment–theory, method and practice*. Edward Elgar.

Martinez-Alier, J., Kallis, G., Veuthey, S., Walter, M., & Temper, L. (2010). Social metabolism, ecological distribution conflicts, and valuation languages. *Ecological Economics, 70*, 153–158. https://doi.org/10.1016/j.ecolecon.2010.09.024

Mingorria Martinez, S. (2017). *The nadies waving resistance: Oil palm and sugarcane conflicts in the territory, communities and households of the Q'epchil', PolochicValley, Guatemala*. PhD thesis. Universitat Autònoma de Barcelona.

Munda, G. (1995). *Multicriteria evaluation in a fuzzy environment* (Contributions to economics series). Physica-Verlag.

Nelson, J. A. (2011). *Ethics and the economist: What climate change demands of us*. Ecological Economics.
O'Neill, J. (2007). *Markets, deliberation and environment*. Routledge.
O'Neill, J., & Spash, C. L. (2000). Conceptions of value in environmental decision-making. *Environmental Values, 9*, 521–536. https://doi.org/10.3197/096327100129342191
Rodríguez-Labajos, B., & Martínez-Alier, J. (2013). The economics of ecosystems and biodiversity: Recent instances for debate. *Conservation and Society, 11*, 326–342.
Spash, C. L. (2008). Deliberative monetary valuation: Literature, limitations and perspectives. In C. Zografos & R. B. Howarth (Eds.), *Deliberative Ecological Economics* (pp. 21–49). Oxford University Press.
Spash, C. L., & Hanley, N. (1995). Preferences, information and biodiversity preservation. *Ecological Economics, 12*, 191–208.
Tanasescu, M. (2015). Nature advocacy and the indigenous symbol. *Environmental Values, 24*, 105–122.
Temper, L., Demaria, F., Scheidel, A., Del Bene, D., & Martinez-Alier, J. (2018). The global environmental justice atlas (EJAtlas): Ecological distribution conflicts as forces for sustainability. *Sustainability Science, 13*, 573–584. https://doi.org/10.1007/s11625-018-0563-4
Tlostanova, M. V., & Mignolo, W. (2012). *Learning to unlearn: Decolonial reflections from Eurasia and the Americas*. The Ohio State University Press.
Urkidi, L., & Walter, M. (2011). Dimensions of environmental justice in anti-gold mining movements in Latin America. *Geoforum, 42*, 683–695. https://doi.org/10.1016/j.geoforum.2011.06.003
Vatn, A. (2005). *Institutions and the environment*. Edward Elgar Publishing.
Velicu, I. (2015). Demonizing the sensible and the 'Revolution of our generation' in Rosia Montana. *Globalizations, 12*, 846–858. https://doi.org/10.1080/14747731.2015.1100858
Veuthy, S., & Gerber, J.-F. (2011). *Valuation contests over the commoditisation of the Moabi Tree in South-Eastern Cameroon* [WWW document]. https://doi.org/10.3197/096327111X12997574391805
Willems–Braun, B. (1997). Buried epistemologies: The politics of nature in (post) colonial British Columbia. *Annals of the Association of American Geographers, 87*, 3–31.
Zografos, C. (2007). Rurality discourses and the role of the social enterprise in regenerating rural Scotland. *Journal of Rural Studies, 23*, 38–51.
Zografos, C. (2015a). Value deliberation in ecological economics. In *Handbook of ecological economics*. https://doi.org/10.4337/9781783471416.00008
Zografos, C. (2015b). *Démocratie directe* [Direct Democracy]. D'Alisa, G. Demaria, F., Kallis, G. Décroissance. Vocabulaire pour une nouvelle ère (pp. 187–194).
Zografos, C. (2017). Flows of sediment, flows of insecurity: Climate change adaptation and the social contract in the Ebro Delta, Catalonia. *Geoforum, 80*, 49–60. https://doi.org/10.1016/j.geoforum.2017.01.004
Zografos, C. (2019). Direct democracy. In A. Kothari, A. Salleh, A. Escobar, F. Demaria, & A. Acosta (Eds.), *Pluriverse: A post-development dictionary* (pp. 154–157). Tulika Books and Authorsupfront.
Zografos, C., & Howarth, R. (2008a). *Deliberative ecological economics*. Oxford University Press.
Zografos, C., & Howarth, R. B. (2008b). Towards a deliberative ecological economics. In *Deliberative ecological economics* (pp. 1–20). Oxford University Press.
Zografos, C., & Howarth, R. B. (2010). Deliberative ecological economics for sustainability governance. *Sustainability, 2*, 3399–3417.
Zografos, C., & Martínez-Alier, J. (2009). The politics of landscape value: A case study of wind farm conflict in rural Catalonia. *Environment and Planning A, 41*, 1726–1744.
Zografos, C., & Paavola, J. (2008). Critical perspectives on human action and deliberative ecological economics. In C. Zografos & R. B. Howarth (Eds.), *Deliberative Ecological Economics* (pp. 146–166). Oxford University Press.

Open Access This chapter is licensed under the terms of the Creative Commons Attribution 4.0 International License (http://creativecommons.org/licenses/by/4.0/), which permits use, sharing, adaptation, distribution and reproduction in any medium or format, as long as you give appropriate credit to the original author(s) and the source, provide a link to the Creative Commons license and indicate if changes were made.

The images or other third party material in this chapter are included in the chapter's Creative Commons license, unless indicated otherwise in a credit line to the material. If material is not included in the chapter's Creative Commons license and your intended use is not permitted by statutory regulation or exceeds the permitted use, you will need to obtain permission directly from the copyright holder.

Chapter 6
Post-development: From the Critique of Development to a Pluriverse of Alternatives

Federico Demaria, Ashish Kothari, Ariel Salleh, Arturo Escobar, and Alberto Acosta

6.1 The Conceptualization of (Post-)Development in the Social Sciences, and the Contribution by Joan Martinez Alier

It is impossible to provide a single definition of development. For many, development is the ineluctable strategy by which poor countries need to modernize; for others, it is an imperial imposition by rich capitalist countries on poor ones, and as such, it should be opposed; for yet others, it is a discourse invented by the West for the cultural domination of non-Western societies that need to be denounced as such, beyond its economic effects; finally, for many common people the world over, development has become either a reflection of their aspirations to a dignified life, or an utterly destructive process with which they have to coexist, and not infrequently both at the same time. Taken as a whole, it can be said that development is a complex historical process with social, economic, political and cultural aspects.

Over the past six decades, the conceptualization of development in the social sciences has seen three main moments, corresponding to three contrasting theoretical orientations: modernization theory in the 1950s and 1960s, with its allied

F. Demaria (✉)
Universitat de Barcelona, Barcelona, Spain

A. Kothari
Kalpavriksh (NGO), Pune, Maharashtra, India

A. Salleh
The University of Sydney, Camperdown, Australia

A. Escobar
University of North Carolina Chapel Hill, Chapel Hill, NC, USA

A. Acosta
Facultad Latinoamericana de Ciencias Sociales, Quito, Ecuador

theories of growth; Marxist-inspired dependency theory and related perspectives in the 1960s and 1970s; and critiques of development as a cultural discourse in the 1990s and 2000s. Here we argue that we might have entered into a fourth moment: marked by a focus on a pluriverse of alternatives to development.

The year 2022 marks the thirtieth anniversary of *The Development Dictionary*, edited by Wolfgang (Sachs, 1992). While the *Dictionary* might have fallen short of its intention to write the obituary of development, it did send shockwaves through the activist, policy and scholarly worlds and became an influential text. The relevance and impact of Sachs' book are still felt today. At the same time, there is no dearth of revitalized hegemonic notions, of which 'sustainable development' might be best known, an 'amoeba concept' still floating thanks to its malleability,[1] and indeed given new life in 2015 by the global intergovernmental agreement on Sustainable Development Goals. In this context, we published the book *Pluriverse: A Post-Development Dictionary* (Kothari et al., 2020), which while emulating the spirit of the original *Dictionary* brings both reincarnated worldviews and fresh alternatives to 'development'sharply into view. The starting point is the need to go beyond critique and concentrate on articulating the narratives of those struggling to retain or create diverse ways of life against the homogenising forces of development. There is a need for radical post-development practices, ideas and worldviews to provide an agenda for activists, policy makers and scholars to help in 'transforming our world'. What is needed is an alternative to the 2030 Agenda for Sustainable Development (United Nations, 2013, 2015).

The concept of 'post-development'emerges from the confluence of four main books: first, *The Development Dictionary*, edited by Wolfgang Sachs; second, *Encountering Development* by Arturo Escobar; third, *The History of Development* by Gilbert Rist; fourth, *The Post-Development Reader*, edited by Rahnema and Bawtree (Sachs, 1992; Escobar, 1995; Rist, 2003; Rahnema & Bawtree, 1997). Feminist contributions include Vandana Shiva's *Staying Alive: Women, Ecology and Development* and *The Subsistence Perspective*, authored by Veronika Bennholdt-Thomsen and Maria Mies. Two decades later, our book *Pluriverse: The Post-Development Dictionary* focuses more upon alternatives to, rather than the critique of, development.

In addition to these, the work of activist-scholars such as Gilbert Rist, Helena Norberg-Hodge, Serge Latouche, Majid Rahnema, Wolfgang Sachs, Ashish Nandy, Shiv Visvanathan, Gustavo Esteva (Sachs, 1992), Rajni Kothari, Manfred Max-Neef, François Partant, Bernard Charbonneau and Ivan Illich have gone a long way in drawing the contours of a post-development future. Joan Martinez Alier himself was contributing since the 1980s with an ecological critique of development (Martinez-Alier, 1987), and more recently, wrote a chapter for *Pluriverse* where he

---

[1] Words like 'development' or 'strategy' have been called 'amoeba concepts' or 'plastic words' because of their malleability and the uncanny way in which they are used to fit every circumstance (Poerksen, 2004). Like plastic Lego blocks, they are combinable and interchangeable. In the mouths of experts—politicians, professors, corporate officials and planners—they are used over and over again to explain and justify any type of plans and projects.

shows how a global movement for environmental justice is helping to push society and economy towards environmental sustainability (Martinez-Alier, 2002). By taking seriously the rich knowledge of social movement activists on issues such as environmental justice, resistance to development and alternative forms of valuation, among others, he brought to the fore the fact that activists and even grassroots communities should be considered as theory and knowledge producers in their own right. This is a central insight of post-development theory; it explains why movement-inspired notions such as *Buen Vivir* have become so central to the entire post-development movement. Martinez-Alier's notion of ecological distribution conflicts paved the way for a whole wave of research centered on the relation between environmental destruction and development, thus strengthening calls for post-development. Indeed, a great deal of the web-based project – Environmental Justice Atlas (ejatlas.org) can be seen as a technology for mapping such a relation and as such as contribution to making visible paths towards post-development.

The tendency to attribute all ideas to male scholars dies hard, but in fact, women across the globe pioneered these post-development ideas autonomously from the start. Joan Martinez Alier realized the importance of the feminist contribution early on. Since the 1980s, his work has occasionally overlapped, though usually run parallel with ecofeminists such as Vandana Shiva, who questioned the green revolution in 1988, and Maria Mies, who promoted the subsistence model, in 1999. As Ariel Salleh emphasises, the left ecofeminist position has always offered both a critique of 'development'and advocated post-development alternative livelihoods.

The term post-development can be used to refer either to an era or an approach in which development is no longer the central organizing principle of social life. Even as critiques of development increase in academic spaces, they are equally powerfully arising amongst indigenous peoples, other local communities, womens' rights movements, and other civil society groupings; most prominently amongst the victims of development. Across the world, ancient worldviews resurface alongside new frameworks and visions presenting systemic alternatives for human and planetary well-being. This is forcing the decolonization of knowledge systems and epistemologies, breaking down many of the dualisms that western patriarchal paradigms have engendered between humans and nature.

The idea of post-development is related to at least five other emerging imaginaries:

- Post-capitalism – questioning capitalism's capacity to fully occupy the economy.
- Post- or de-growth – decentring growth from the definition of both economy and social life.
- Post-patriarchy – challenging the primacy of masculinist approaches to political leadership, moral authority, social privilege and control of property.
- Anti-racism – fighting the systemic racism and the oppression of marginalized groups.
- De-coloniality – untangling the production of knowledge from a primarily Eurocentric episteme.

The current mood is 'to search for alternatives in a deeper sense, that is, aiming to break away from the cultural and ideological bases of development, bringing forth other imaginaries, goals, and practices' (Gudynas & Acosta, 2011:75).

The time is ripe to deepen and widen a research, dialogue and action agenda on a variety of worldviews and practices relating to the collective search for an ecologically wise and socially just world. These should be transformative alternatives to the currently dominant processes of globalized development, including its structural roots in modernity,[2] capitalism, state domination, patriarchy, colonialism, racism and more specific phenomena, like casteism, found in some parts of the world. Alternatives should go beyond the false solutions that those in power are proposing in an attempt to 'greenwash' development, including variants of 'sustainable development', market remedies and technofixes. The post-development agenda should investigate the what, how, who and why of all that is transformative and what is not. Equally, though, proponents of post-development need to go beyond a number of weaknesses in their narrative, by acknowledging that development as an idea has not been buried and by sharpening their focus on the structural changes needed to deal with issues of inequity, injustice, deprivation and ecological collapse (Ziai, 2015).

The exploration of alternatives to development already finds concrete expression in a panoply of new or re-emerging concepts and practices such as *buen vivir*, degrowth, ecological *swaraj*, radical feminisms of various kinds, *ubuntu*, commoning, solidarity economy, environmental and climate justice, food and energy sovereignty. These are perhaps the most visible examples of an emergent post-developmentalist epistemic-political field towards a pluriverse.[3] These radical alternatives are becoming not only more visible but, increasingly, genuinely credible and viable. And yet they are still marginal in comparison to the dominant narrative and practice of development. Thus, it is a good moment to make such alternatives more widely known and to facilitate bridges amongst them while respecting their geopolitical and epistemic specificities. It is also critical to build bridges between constructive alternatives and peoples' movements resisting the dominant economic and political systems (Kothari , 2015; Kaul et al. 2022).

The chapter is structured as follows. First, we present a critique of development in its recent reincarnations, like 'sustainable development' (SD), outlining the road from Stockholm 1972 to the Sustainable Development Goals, that is to say, the road from the critique to the defence of economic growth. Second, we introduce the origins and importance of transformative alternative worldviews and practices to development. Third, we outline the purpose and conceptualization of *Pluriverse:*

---

[2] Note that a critique of 'modernity' does not imply a rejection of all that is 'modern', nor an uncritical acceptance of all that is 'traditional'; we are well aware that traditional societies had (and have) many aspects of inequity and injustice, and that elements of what has emerged in contemporary times have been liberating for those previously suppressed. It is the hegemonising, unidirectional, western-centricism of modernity that we are pointing to.

[3] See Walter Mignolo, 'On pluriversality': Available at: http://waltermignolo.com/on-pluriversality/ (Accessed on 09/03/2017).

The Post-Development Dictionary, with a set of questions at the core of the agenda for transformation that we are proposing.

6.2 A Critique of Sustainable Development and Its False Solutions

> Everything must change in order to remain the same. Giuseppe Tomasi di Lampedusa, *The Leopard* (1963)

In 1987, the UN World Commission on Development and the Environment presented the report *Our Common Future,* better known as the Brundtland report, coining the concept of 'sustainable development', then launching it at the Rio Summit on Environment and Development in 1992 – Principle 12 of the Declaration. Within such a framing, the push towards growth and economic liberalization was taken further at subsequent global events relating to sustainable development, though partially concealed behind the rhetoric of environmental sustainability. Compared to the United Nations Conference on Environment and Development in Stockholm 1972, the later conferences involved reframing both the diagnosis and prognosis in relation to the ecological crisis. The focus supposedly became poverty in developing countries, instead of affluence in developed countries, along the lines of the post-materialist thesis of Inglehart. This idea that 'you first need to be rich, in order to be an environmentalist' has been critiqued by Martinez-Alier with his famous concept of environmentalism of the poor (Inglehart, 1990). Economic growth was freed of stigma and redefined as a necessary step towards the solution of environmental problems. (Gómez-Baggethun & Naredo, 2015) This watering down of the initial debates of the 1970s influenced by the *Limits to Growth* report (Meadows et al., 1972) constitutes the core of the 'green economy', a kind of Green Keynesianism with reformist new millennium proposals, such as a Green New Deal, and the 2030 Agenda for Sustainable Development.

At the UN Conference for Sustainable Development in 2012, the so-called Rio + 20 Summit, the 'green economy' concept played a key role as the guiding framework of the multilateral discussions, although resistance from many southern nations meant it was not as central as its proponents may have wished. In preparation for the summit, The United Nations Environmental Programme (UNEP) published a report on 'green economy', defining it 'as one that results in improved human well-being and social equity, while significantly reducing environmental risks and ecological scarcities' (UNEP, 2011). In consonance with the pro-growth approach of sustainable development, the report bypassed any trade-off between economic growth and environmental conservation and conceptualized 'nature'as natural capital, a 'critical economic asset' opening the doors for commodification. In fact, it clearly stated that 'the key aim for a transition to a green economy is to enable economic growth and investment while increasing environmental quality and social inclusiveness (UNEP, 2011).'

This environmental economics approach is based on neoclassical economic theory and a belief that economic growth will de-link or decouple itself from its environmental base through dematerialization and de-pollution because of the improvement in eco-efficiency, viz. increased resource productivity and decreased pollution. In this conceptual framework, market prices are considered the appropriate means for solving environmental issues and exogenous rates of technological progress are expected to counterbalance the effects of resource exhaustion. However, the conflict between a growth-dominated economy and environmental protection cannot be solved with appeals to 'sustainable development', 'eco-efficiency', 'ecological modernisation', 'geo-engineering', 'smart agricultures' or 'cities', 'circular' or 'green economy'. These are false solutions.

The sustainable development approach remains fundamentally flawed on a number of counts. For instance, the final objective for a New Green Deal is the creation of 'resilient low carbon economies, rich in jobs and based on independent sources of energy supply' (UNEP, 2011; NEF, 2008). While on this end there might general agreement, the controversy remains on the means. Among the flaws or weaknesses of the sustainable development approach as articulated thus far in various UN-sponsored documents, including the declaration *Transforming our world: The 2030 Agenda for Sustainable Development*, (UNEP, 2011; United Nations, 2013; SDSN, 2013; United Nations, 2015; United Nations Secretary-General's High-level Panel on Global Sustainability, 2012) are the following (Kothari, 2013):

1. 'Absence of an analysis of the historical and structural roots of poverty, hunger, unsustainability and inequities, which include centralization of state power and capitalist monopolies;
2. Inadequate focus on direct democratic governance (decision-making by citizens and communities in face-to-face settings), beyond the stress on accountability and transparency;
3. Inability to recognize the biophysical limits to economic growth;
4. Continued subservience to private capital, and inability or unwillingness to democratise the economy;
5. Modern science and technology held up as panacea, ignoring their limits and marginalising other forms of knowledge;
6. Culture, ethics and spirituality side-lined;
7. Unbridled consumerism not tackled head-on;
8. Global relations built on localization and self-reliance missing; and,
9. No new architecture of global governance, with a continued reliance on the centrality of nation-states, denying true democratisation'.

These weaknesses outline why and how we consider the solutions that emerge out of sustainable development as false. In the next section, we instead present the alternatives that go beyond development embedding a real potential for transformation.

6.3 From the Critique of Development to Transformative Alternatives

A range of different and complementary notions or worldviews have emerged in various regions of the world that seek to envision and achieve more fundamental transformation than that proposed by the sustainable development approach. Some of these revive long-standing worldviews of indigenous peoples; some have emerged from recent social and environmental movements in but reflect old traditions and philosophies. Arising from different cultural and social contexts, they sometimes differ on the prescription (what shall be done how), but share the main characteristics of the diagnosis (what is the problem and who is responsible for it) as well as similar or equivalent worldviews. *The Post-Development Dictionary* aims to illuminate pathways towards a synergistic articulation of these alternatives to development (Salleh, 1994; Salleh, 1997; Kothari, 2015; Escobar, 2015).

Unlike sustainable development, which is a concept based on false ideological consensus, (Shiva, 1989; Hornborg, 2009) these alternative approaches are irreducible and therefore do not aspire to be adopted as a common goal of governance by the United Nations, the OECD or the African Union. These ideas are born as proposals for radical change from local to global. They reject the current development hegemony, meaning a critique of the homogenisation of cultures due to the widespread adoption of particular technologies and consumption and production models experienced in the global North (Escobar, 1995; Rist, 2003). The Western development model is understood here as an oxymoron (Latouche, 2009); a toxic term to be deconstructed and rejected (Dearden, 2014). In a post-political condition, (Swyngedouw, 2007) pluriversal alternatives affirm dissidence to re-politicise socio-ecological transformation. In short, it is urgent to dissolve the productivist concept of progress as a unidirectional concept, most especially its mechanistic view of economic growth (Kallis, 2015).

Deconstructing development opens up the door for a multiplicity of new and old notions and worldviews, or else a matrix of alternatives (Latouche, 2009). This includes *Buen Vivir*, a culture of life with different names and varieties in various regions of South America; *Ubuntu*, with its emphasis on human mutuality in South Africa and several equivalents in other parts of Africa; *Swaraj*, with a focus on self-reliance and self-governance, in India; and many others (Gudynas, 2011; Kothari, 2014; Metz, 2011). What is important is that while they are ancient, they are re-emerging in modified forms as a part of the narrative of movements that are struggling against development and/or asserting alternative forms of well-being. Ecofeminist arguments represent a further strand in this post-development rainbow (Shiva, 1989; Salleh, 1997).

These worldviews are not a novelty of the twenty-first cntury, but they are rather part of a long practice, ways of living forged in the furnace of humanity's struggle for emancipation and enlightenment *within* rather than outside of nature. In fact, ecofeminists argue that such eco-sufficient knowledge constitutes a vernacular science, learned empirically through labour at the interface with nature (Salleh, 2009).

What is remarkable about these alternative proposals, however, is that they often arise from traditionally marginalized groups. These worldviews are different from dominant Western ones as they emerge from non-capitalist communities or from non-capitalist spaces such as the household sector in the global North (Mies, 1986; Trainer, 1985). They are therefore independent of the anthropocentric and androcentric logic of capitalism, the dominant civilization, as well as with the various state socialist, effectively state capitalist models existing until now. Other approaches emerging from within industrialised countries can also break from dominant logic, such as is the case with degrowth, an example of non-occidentalist West (Sousa Santos, 2009; Demaria et al., 2013; D'Alisa et al., 2014).

Pluriversal alternatives may be distinguished from false solutions in a number of ways. Firstly, they seek to transform the structural roots of global injustice along political, economic, social, cultural and ecological axes. Secondly, they question the core assumptions of the development discourse – growth, material progress, instrumental rationality, the centrality of markets and economy, universality, modernity and its binaries. Thirdly, they encompass a radically different set of ethics and values to those underpinning the current system, including diversity, solidarity, commons, oneness with nature, interconnectedness, simplicity, inclusiveness, equity, non-hierarchy, pluriversality and peace.

At a time when neoliberal governments and rampant extractivism brutalise the everyday life of citizens across the world and in particular the global South, it is crucial that oppositional voices and people's movements engage in a concentrated effort of research, outreach, dialogue and action, informed by and informing grassroots practice. Resistance is crucial, but it is not enough. We need our own narratives. Acts of resistance and regeneration offer hope in the here and now.

6.4 The Post-development Action-Research Agenda: Towards the Pluriverse

This chapter has laid out both a critique of sustainable development as well as the potential of a post-development agenda. It aims to deepen and widen a research, dialogue and action agenda for activists, policy makers and scholars on a variety of worldviews and practices relating to an emerging grassroots collective search for an ecologically wise and socially just world.

The future post-development action-research agenda must investigate the what, how, who and wy of everything that is transformative and also what is not. In particular, what need to be further investigated are:

- What do pluriversal alternatives to development have in common, and how do they differ?
- What potential for tensions and complementarities is there, given that the socio-ecological communities from which they emerge are rooted in specific territories and cultural contexts?

- How to deal with social differentiations of class, race, sex-gender, age, or ability, which are often culturally essentialised?
- What potential for tensions and complementarities is there? At local, national, regional or global level (e.g. Vikalp Sangam, and Global Tapestry of Alternatives).
- How to deal with contradictions within and among alternatives: e.g. pluriversality and universality, without resorting to universal criteria? For instance, how can we deal with those worlds that do not want to relate – ethno-nationalist and imperializing worlds – without going against the principles of the pluriverse?
- Faced with today's global problems, how can the exploration of this pluriverse of alternatives to development, contribute most effectively to transcending the dominant and globalized sociocultural paradigm of industrial civilization?

In conclusion, these alternatives to development practices and worldviews intend to re-politicise the debate on much-needed socio-ecological transformation, affirming dissidence with the current world representations of sustainable development and searching for alternative ones. They highlight the necessity to overcome the modern ontology of one world and expand on the multiplicity of worlds possible. As Escobar argues: 'The modern ontology presumes the existence of One World – a universe. This assumption is undermined by discussions in Transition Discourses, the buen vivir, and the Rights of Nature. In emphasizing the profound relationality of all life, these newer tendencies show that there are indeed relational worldviews or ontologies for which the world is always multiple – a pluriverse. Relational ontologies are those that eschew the divisions between nature and culture, individual and community, and between us and them that are central to the modern ontology. Some of today's struggles could be seen as reflecting the defence and activation of relational communities and worldviews… and as such they could be read as *ontological struggles*; they refer to *a different way of imagining life*, to an other mode of existence. They point towards the pluriverse; in the successful formula of the Zapatista, the pluriverse can be described as "a world where many worlds fit" (Escobar, 2011, 2015).

Joan Martinez-Alier is an incredible observer always attentive to the changes in the world. With all his travels and readings, he has contributed to open new paths and directions towards the pluriverse.

References

D'Alisa, G., Demaria, F., & Kallis, G. (2014). *Degrowth: A vocabulary for a new era*. Routledge.
Dearden, N. (2014, January 22). 'Is development becoming a toxic term?' *The Guardian*. http://www.theguardian.com/global-development-professionals-network/2015/jan/22/development-toxic-term?CMP=share_btn_tw
Demaria, F., Schneider, F., Sekulova, F., & Martinez-Alier, J. (2013). What is degrowth? From an activist slogan to a social movement. *Environmental Values, 22*, 191–215.
Escobar, A. (1995). *Encountering development*. Princeton University Press.
Escobar, A. (2011). Sustainability: Design for the pluriverse. *Development, 54*(2), 137–140.

Escobar, A. (2015). Degrowth, postdevelopment, and transitions: A preliminary conversation. *Sustainability Science, 10*(3), 451–462.

Gómez-Baggethun, E., & Naredo, J. M. (2015). In search of lost time: The rise and fall of limits to growth in international sustainability policy. *Sustainability Science, 10*(3), 385–395.

Gudynas, E. (2011). Buen vivir: today's tomorrow. *Development, 54*(4), 441–447.

Gudynas, E., & Acosta, A. (2011). La renovación de la crítica al desarrollo y el buen vivir como alternativa. *Utopia y Praxis Latinoamerica, 16*(53), 71–83.

Hornborg, A. (2009). Zero-sum world. *International Journal of Comparative Sociology, 50*(3–4), 237–262.

Inglehart, R. (1990). *Culture shift in advanced industrial societies*. Princeton University Press.

Kallis, G. (2015). *The Degrowth alternative*. Great Transition Initiative. http://www.greattransition.org/publication/the-degrowth-alternative

Kaul, S., Akbulut, B., Demaria, F. et al. (2022). Alternatives to sustainable development: what can we learn from the pluriverse in practice?. *Sustainability Science 17*, 1149–1158. https://doi.org/10.1007/s11625-022-01210-2

Kothari, A. (2013). *Missed Opportunity? Comments on two global reports for the post-2015 goals process*. Kalpavriksh and ICCA Consortium. http://www.un-ngls.org/IMG/pdf/Kalpavriksh_and_ICCA_Consortium_-_Post-2015_reports_critique_-_Ashish_Kothari_July_2013.pdf

Kothari, A. (2014). Radical ecological democracy: A way for India and beyond. *Development, 57*(1), 36–45. https://doi.org/10.1057/dev.2014.43

Kothari, A. (2015). Demaria, Federico, and Acosta, Alberto Buen vivir, degrowth and ecological swaraj: Alternatives to development and the green economy. *Development, 57*(3), 362–375.

Kothari, A., Salleh, A., Escobar, A., Demaria, F., & Acosta, A. (2020). *Pluriverse: A post-development dictionary*. Authors Up Front/Tulika/Columbia University Press.

Latouche, S. (2009). *Farewell to growth*. Polity.

Martinez-Alier, J. (1987). *Ecological economics: Energy, environment and society*. Basil Blackwell.

Martinez-Alier, J. (2002). *The environmentalism of the poor: A study of ecological conflicts and valuation*. Edward Elgar.

Meadows, D. H., Meadows, D. L., Randers, J., & Beherns, W. W., III. (1972). *The limits to growth*. Universe Books.

Metz, T. (2011). Ubuntu as a moral theory and human rights in South Africa. *African Human Rights Law Journal, 11*(2), 532–559.

Mies, M. (1986). *Patriarchy and accumulation on a world scale*. Zed.

NEF. (2008). *A Green New Deal: Joined-up policies to solve the triple crunch of the credit crisis, climate change and high oil prices*. Green New Deal Group, New Economics Foundation. Available at http://b.3cdn.net/nefoundation/8f737ea195fe56db2f_xbm6ihwb1.pdf

Poerksen, U. (2004). *Plastic words: The tyranny of a modular language*. Penn State University Press.

Rahnema, M., & Bawtree, V. (1997). *The post-development reader*. Zed Books.

Rist, G. (2003). *The history of development: From Western origins to global faith* (Expanded ed.). Zed Books.

Sachs, W. (1992). *The development dictionary: A guide to knowledge as power*. Zed Books.

Salleh, A. (1994). Naturaleza, mujer, trabajo, capital: la mas profunda contradiccion. *Ecologia Politica, 7*, 35–47.

Salleh, A. (1997). *Ecofeminism as politics: Nature, Marx, and the postmodern*. Zed Books/Palgrave.

Salleh, A. (Ed.). (2009). *Eco-sufficiency & global justice. Women write political ecology*. Pluto Press.

SDSN. (2013). *An action agenda for sustainable development*. Report for the Secretary General, Sustainable Development Solutions Network.

Shiva, V. (1989). *Staying alive: Women, ecology and development*. Zed Books.

Sousa Santos, B. (2009). A non-occidentalist west? Learned ignorance and ecology of knowledge. *Theory, Culture and Society, 26*(7–8), 103–125.

Swyngedouw, E. (2007). Impossible/undesirable sustainability and the post-political condition. In J. R. Krueger & D. Gibbs (Eds.), *The sustainable development paradox* (pp. 13–40). Guilford Press.

Trainer, T. (1985). *Abandon affluence!* Zed.

UNEP. (2011). *Towards a green economy: Pathways to sustainable development and poverty eradication* (A synthesis for policy makers). United Nations Environment Programme. www.unep.org/greeneconomy

United Nations. (2013). *A new global partnership: Eradicate poverty and transform economies through sustainable development* (The report of the high-level panel of eminent persons on the Post-2015 development agenda). United Nations.

United Nations. (2015). *Transforming our world: The 2030 agenda for sustainable development*. United Nations. Available at: https://sustainabledevelopment.un.org/post2015/transformingourworld/publication

United Nations Secretary-General's High-level Panel on Global Sustainability. (2012). *Resilient people, resilient planet: A future worth choosing*. United Nations.

Ziai, A. (2015). Post-development: Pre-mature burials and haunting ghosts. *Development and Change, 46*(4), 833–854.

Open Access This chapter is licensed under the terms of the Creative Commons Attribution 4.0 International License (http://creativecommons.org/licenses/by/4.0/), which permits use, sharing, adaptation, distribution and reproduction in any medium or format, as long as you give appropriate credit to the original author(s) and the source, provide a link to the Creative Commons license and indicate if changes were made.

The images or other third party material in this chapter are included in the chapter's Creative Commons license, unless indicated otherwise in a credit line to the material. If material is not included in the chapter's Creative Commons license and your intended use is not permitted by statutory regulation or exceeds the permitted use, you will need to obtain permission directly from the copyright holder.

Chapter 7
Indigenous and Local Knowledge Contributions to Social-Ecological Systems' Management

Victoria Reyes-García

7.1 Introduction

Social-ecological systems (SES) are complex and adaptive, for which their governance requires holistic understanding of the different components of the system and their relations, capacity to respond to change and uncertainty, and well-functioning institutional frameworks (Janssen & Ostrom, 2006). Indigenous and local knowledge (ILK) systems, or the sophisticated sets of ecological knowledge, management practices, and customary institutions generated by different Indigenous peoples and local communities (IPLC) with long histories of place-based living and time-honored traditions, often entail these characteristics (Berkes, 2017). ILK systems have traditionally guided many social-ecological interactions, resulting in the maintenance of the integrity of many aquatic and terrestrial ecosystems (Cámara-Leret et al., 2019; Kimmerer, 2000). Given that many of the lands and waters that IPLC own or manage are critical for biodiversity conservation and climate change mitigation (Ens et al., 2016; Fa et al., 2020; Porter-Bolland et al., 2012), analyzing how ILK contributes to the governance of complex SES could help in achieving planetary sustainability.

V. Reyes-García (✉)
Institució Catalana de Recerca i Estudis Avançats, Barcelona, Spain

Institut de Ciència i Tecnologia Ambientals, Universitat Autònoma de Barcelona, Cerdanyola del Vallès, Spain

Departament d'Antropologia Social i Cultural, Universitat Autònoma de Barcelona, Cerdanyola del Vallès, Spain
e-mail: victoria.reyes@uab.cat

Indigenous and local knowledge systems use a variety of languages to express multiple and complex values of nature, providing a good example of the need for multi-criteria evaluation proposed by the Barcelona School of Ecological Economics and Political Ecology (Munda, 2008). Many ILK systems capture and reflex human dependence on the interconnected web of life (e.g., Lyver et al., 2017; Reo, 2019). In many of such conceptualizations, humans are viewed as an integral component of nature (Coscieme et al., 2020), and nature is imbued with social, cultural, and spiritual values (Berkes, 2017). Moreover, IPLC conceptualizations of nature often draw on stewardship ethics based on mutual reciprocity between humans and nature, temporary custody for future generations, and health of and attachment to land (Pascual et al., 2017; Reo, 2019). These conceptualizations, which are dynamic and adapt to changes (McMillen et al., 2017), form the basis for land and seascape management (e.g., Joa et al., 2018). Indeed, the defense of the value systems that underpin such conceptualizations has resulted in a myriad of environmental conflicts, particularly when activities based in other valuation systems (e.g., extractive activities with a purely monetary valuation) are imposed on areas managed by IPLC (Scheidel et al., 2020).

In this chapter, I draw on published research to summarize how ILK (1) draws on conceptualizations of nature that contribute to the long-term maintenance of functioning SES, (2) enhances our understanding of complex SES, and (3) articulates resistance to SES degradation and promotes SES restoration. The chapter concludes elaborating on why, although IPLC contributions to complex SES management are growingly recognized, such contributions will not be fully realized unless IPLC are fully acknowledged as equal partners at different levels of environmental governance, as proposed by the post-normal science approach adopted by the Barcelona School of Ecological Economics and Political Ecology.

7.2 Conceptualizations of Nature Embedded in ILK Systems Contribute to Long-Term Maintenance of Complex SES

While acknowledging diversity, many examples show that IPLC conceptualizations of social-ecological relations often build on concepts such as attachment to land, interests in restoration, a powerful stewardship ethics, reciprocity between humans and nature, relational webs – including kinship – with natural elements, and continuity between nature and culture (Díaz et al., 2015; Sterling et al., 2017). These conceptualizations are embedded in customary management practices, such as the protection of sacred forests and fresh or seawater areas or species or taboo enforcement, but also on selective cutting and burning or other biodiversity-enriching small-scale disturbances (Forest Peoples Programm et al., 2016; Guadilla-Sáez et al., 2019; Joa et al., 2018). Such customary management practices extend to the management of coastal ecosystems, including wetlands, mangroves, and seaweed and seagrass beds (Cinner & Aswani, 2007), and of areas particularly sensible to climate change, such as the Arctic (Johnson et al., 2015).

Many examples show that ILK-based SES management arrangements (e.g., traditional agriculture, aquaculture, fishery, and community forestry) contribute to biodiversity maintenance in production landscapes (Chaudhary et al., 2016), ecosystem restoration (Reyes-García et al., 2019a, b), pollution buffering (Fernández-Llamazares et al., 2020), and nutrient cycling (Malley et al., 2016). Examples of these practices include purposive burning to create diversity (Shaffer, 2010; Trauernicht et al., 2015), waste deposition practices resulting in soil carbon enrichment (Solomon et al., 2016), swidden cultivation systems able to maintain forest cover and plant diversity (Takasaki et al., 2022; Wangpakapattanawong et al., 2010), or weeding meadows to maintain grassland productivity and resilience (Babai & Molnár, 2014). Indeed, as a result of the application of these management systems, much of today's world's wild and domesticated biodiversity lies in areas traditionally owned, managed, used, and/or occupied by IPLC (Brondizio & Le Tourneau, 2016; Garnett et al., 2018). Moreover, despite increasing pressures from the expansion of commodity frontiers and resource extraction, biodiversity is declining less rapidly in land and seascapes owned or managed by IPLC than in other ecosystems (IPBES, 2019).

Despite IPLC contributions to the maintenance of global biodiversity, IPLC are often excluded for environmental governance and customary management practices are disappearing (Forest Peoples Programm et al., 2016; Witter et al., 2015). For instance, while more than 40% of government-protected areas overlap with IPLC lands, less than 1% of protected areas are formally governed by IPLC (Garnett et al., 2018). Moreover, in many areas, agricultural expansion, logging, or conservation activities limit or replace customary management practices (Hayes, 2010), with recent proposals to safeguard 30% (Dinerstein et al., 2019) or 50% (Wilson, 2016) of the planet from human use. Researchers have documented that the loss of traditional management systems affects the functionality and stability of the SES previously managed under such rules through landscape homogenization, the increase of invasive species presence, pollution, urbanization, or soil erosion (Fletcher et al., 2020; Fernández-Llamazares et al., 2020; Guadilla-Sáez et al., 2019). This, in turn, has cascading effects on other elements of the SES including reduced abundance and access to culturally valued resources (e.g., food) (Garibaldi, 2009; Kuhnlein, 2014) and deterioration of traditional governance systems and institutions (Oldekop et al., 2012; Sirén, 2017). The erosion of traditional management practices also weakens local conceptualizations of nature (Stocker et al., 2016; Turner et al., 2008), impacting relations with and responsibilities to other-than-human-beings and forces (e.g., Fernández-Llamazares & Virtanen, 2020).

7.3 ILK Enhances Our Understanding of Complex SES

Beyond the actual management of SES, ILK systems encode key information essential for our understanding of complex SES. Thus, ILK systems have been crucial to further scientific understanding of species ecological distribution ranges

(Wilder et al., 2016) and historical population baselines and trends (Bender et al., 2013; Ziembicki et al., 2013), particularly in biologically diverse but little studied regions of the world. For example, ILK has contributed to fisheries science through mapping spawning grounds, understanding seascapes' use and ecology, and documenting fisheries' long-term trends (Lavides et al., 2016; Tesfamichael et al., 2014). In a way, the contributions of ILK systems to the generation of actionable knowledge for sustainability exemplifies the potential of the post-normal science approach for the management of complex issues (Funtowicz & Ravetz, 1993). According to the post-normal science framework which has guided much of the research of the Barcelona School of Ecological Economics and Political Ecology, when facts are uncertain, values in dispute, stakes high, and decisions urgent – as is the case of environmental management – the normal scientific approach is insufficient and new norms of evidence and discourse need to be developed.

In that sense, over the years, ILK systems have contributed not only to provide an enriched picture of biodiversity functioning, but ILK systems have also aid in efforts to sustain nature (Tengö et al., 2014; Wilder et al., 2016). For example, around the world, different place-based, historical land-use practices have been used along with biological data to create more effective national plans to protect biodiversity while supporting local livelihood activities (Diamond & Ansharyani, 2018). Similarly, new knowledge co-produced by scientists and IPLC referring to carbon stocks assessment (Butt et al., 2015), wildlife monitoring (Luzar et al., 2011; Takasaki et al., 2022), or participatory mapping (Herlihy, 2003) has resulted in the development of adaptation strategies to highly variable social-ecological conditions. Moreover, ILK has also contributed to map, monitor, and report changes in SES, including the dynamics of agricultural systems (Coomes et al., 2015), resource over-exploitation (Forest Peoples Programm et al., 2016), invasive species expansion (Bart & Simon, 2013), climate change impacts (Reyes-García et al., 2019a, b), and pollution (Fernández-Llamazares et al., 2020; Orta-Martínez et al., 2007). For example, through a community-based monitoring program which started in 2006 and which builds on their detailed knowledge of the environment, Achuar and Quechua Indigenous Peoples of the Peruvian Amazon have been able to monitor, map and report oil spills impacts on soils, water, wildlife, and their own health (Cartró-Sabaté, 2018; Rosell-Melé et al., 2018; Yusta-García et al., 2017). Their knowledge has uncovered impacts that oil companies had never reported (e.g., concealed oil spills, illegal operations; Orta-Martínez & Finer, 2010), as well as animal geophagia of polluted soils, a behavior unknown to scientists (Cartró-Sabaté, 2018).

7.4 ILK Articulates Resistance to SES Degradation and SES Restoration

IPLC's understandings of the interconnections in nature and of human dependence of the interconnected web of life are at the bases of many environmental conflicts through which Indigenous peoples aim to protect their territories from extractive and industrial development pressures (Scheidel et al., in press; Benyei et al., 2022). Drawing on their understanding of SES functioning and the changes such systems can endure, IPLC have been proactive in implementing innovative strategies to prevent, limit, or stop activities that potentially led to SES degradation (Fernández-Llamazares et al., 2020; Martinez-Alier et al., 2016), sometimes even facing violence for defending the land and resources (Scheidel et al., 2020). Thus, IPLC, through the world, have resisted mining operations, hydrocarbon exploration, infrastructure development, and toxic waste dumping (Martinez-Alier et al., 2016; Orta-Martínez & Finer, 2010; Reyes-García et al., 2020). Some of these actions have been preventive, such as the fight of the Dongria Kondh against bauxite mining in their sacred homelands in India, in which IPLC used their understanding of SES functioning to raise opposition before the activity started to operate (Temper & Martinez-Alier, 2013).

ILK is also at the basis of IPLC activities to restore lands and waters after these areas have been overexploited or degraded by extractive activities (Reyes-García et al., 2019a, b). For instance, some restoration efforts have used insights from local knowledge systems to identify what species to use and which sites to focus on in restoration efforts. ILK can provide baseline ecosystem information on cultural keystone species, i.e., culturally salient species that shape people's identity (Garibaldi & Turner, 2004; Reyes-García et al., 2023), or cultural keystone places, i.e., particular places that are critically important for the flow of ecosystem service and to people's lifeways (Cuerrier et al., 2015). For example, traditional fire management practices have been used to restore overgrown broad-crowned black oak tree stands in California (Long et al., 2003), and in Nepal many local communities contribute to safeguard and restore communal forests and watersheds, thus slowing deforestation, after the Nepali state devolved forests into community control in the 1970s (Paudyal et al., 2015). In some cases, restoration efforts have resulted in a change in the local political context, creating a space for Indigenous spiritual and cultural values to be further reflected in their participation in restoration efforts (Fox et al., 2017).

7.5 Conclusion

Around the world, a myriad of Indigenous and local knowledge systems have achieved the long-term management of functioning ecosystems, informed scientific efforts to maintain ecosystems, and prevented further SES degradation. And, nevertheless, the critical role of IPLC in SES management is not fully recognized in most conservation research, policy, and practice. This is so to the extent that IPLC continue to face challenges of representation in international climate and biodiversity conservation policy processes (Forest Peoples Programm et al., 2016; Witter et al., 2015) and, in some regions of the world, pressures and violence against them are growing (Scheidel et al., 2020). Moreover, ILK systems are globally eroding due to the negative impact of globalization, colonialism, and environmental change (Aswani et al., 2018; Fernández-Llamazares et al., 2021), which endangers not only the very foundations of IPLC' ways of life but also planetary sustainability.

The post-normal science approach proposed by the Barcelona School of Ecological Economics offers principles to bring IPLC to environmental governance. The approach proposes that different groups of interests (beyond scientists) can provide legitimate inputs to the co-production of knowledge for issues affecting them, for which it is first necessary to create the conditions to identify, involve, and engage the relevant communities (Funtowicz & Ravetz, 1993). Drawing in these insights, conservation institutions and governance systems could better reflect ILK contributions and IPLCs crucial roles and rights in planetary sustainability in two complementary ways. On the one side, strengthening frameworks that bridge scientific and Indigenous and local knowledge systems, ensuring collaborative and equitable relations between scientists and IPLC. Such frameworks are vital to the effective co-production of knowledge that enhances conservation strategies, sustainable resource policy and management, and the well-functioning of SES (Tengö et al., 2017; Orlove et al., Accepted). On the other side, these institutions should involve IPLC as partners in governance by promoting inclusive measures that substantially increase the sustainability of land-use practices and the effectiveness of protection (Brooks et al., 2012; Ens et al., 2016). Such measures include safeguarding IPLC knowledge ownership, supporting territorial rights, protecting threatened land defenders, respecting Indigenous laws and principles, and promoting customary management practices.

Acknowledgments The ideas presented here draw in multiple conversations over the years with colleagues and students. I am very grateful of the vivid, critical, and caring research environment created by all of them. Research leading to this chapter has received funding from the European Research Council under an ERC Consolidator Grant (FP7-771056-LICCI). This research contributes to the "María de Maeztu Unit of Excellence" (CEX2019-000940-M).

References

Aswani, S., Lemahieu, A., & Sauer, W. H. H. (2018). Global trends of local ecological knowledge and future implications. *PLoS ONE, 13*(4), e0195440. https://doi.org/10.1371/journal.pone.0195440

Babai, D., & Molnár, Z. (2014). Small-scale traditional management of highly species-rich grasslands in the Carpathians. *Agriculture, Ecosystems and Environment, 182*, 123–130. https://doi.org/10.1016/j.agee.2013.08.018

Bart, D., & Simon, M. (2013). Evaluating local knowledge to develop integrative invasive-species control strategies. *Human Ecology, 41*(5), 779–788. https://doi.org/10.1007/s10745-013-9610-z

Bender, M. G., Floeter, S. R., & Hanazaki, N. (2013). Do traditional fishers recognise reef fish species declines? Shifting environmental baselines in Eastern Brazil. *Fisheries Management and Ecology, 20*(1), 58–67. https://doi.org/10.1111/fme.12006

Benyei, P., Calvet-Mir, L., Reyes-García, V., & Villamayor-Tomas, S. (2022). Indigenous and local knowledge's role in social movement's struggles against threats to community-based natural resource management systems: Insights from a qualitative meta-analysis. *International Journal of the Commons, 16*, 263–277. https://doi.org/10.5334/IJC.1154/

Berkes, F. (2017). *Sacred ecology*. Routledge.

Brondizio, E., & Le Tourneau, F.-M. (2016). Environmental governance for all. *Science, 352*(6291), 1272–1273.

Brooks, J. S., Waylen, K. A., & Mulder, M. B. (2012). How national context, project design, and local community characteristics influence success in community-based conservation projects. *Proceedings of the National Academy of Sciences of the United States of America, 109*(52), 21265. https://doi.org/10.1073/PNAS.1207141110

Butt, N., Epps, K., Overman, H., Iwamura, T., & Fragoso, J. M. V. (2015). Assessing carbon stocks using Indigenous peoples' field measurements in Amazonian Guyana. *Forest Ecology and Management, 338*, 191–199. https://doi.org/10.1016/J.FORECO.2014.11.014

Cámara-Leret, R., Fortuna, M. A., & Bascompte, J. (2019). Indigenous knowledge networks in the face of global change. *Proceedings of the National Academy of Sciences of the United States of America, 116*(20), 9913–9918. https://doi.org/10.1073/pnas.1821843116

Cartró-Sabaté, M. (2018). *Amazo'n'Oil: Exposure to oil and lead for Amazonian wildlife*. PhD thesis. Universitat Autònoma de Barcelona.

Chaudhary, A., Burivalova, Z., Koh, L. P., & Hellweg, S. (2016). Impact of forest management on species richness: Global meta-analysis and economic trade-offs. *Scientific Reports, 6*(1), 23954. https://doi.org/10.1038/srep23954

Cinner, J. E., & Aswani, S. (2007). Integrating customary management into marine conservation. *Biological Conservation, 140*(3–4), 201–216. https://doi.org/10.1016/J.BIOCON.2007.08.008

Coomes, O. T., McGuire, S. J., Garine, E., Caillon, S., McKey, D., Demeulenaere, E., et al. (2015). Farmer seed networks make a limited contribution to agriculture? Four common misconceptions. *Food Policy, 56*, 41–50.

Coscieme, L., da Silva Hyldmo, H., Fernández-Llamazares, Á., Palomo, I., Mwampamba, T. H., Selomane, O., et al. (2020). Multiple conceptualizations of nature are key to inclusivity and legitimacy in global environmental governance. *Environmental Science and Policy, 104*, 36–42. https://doi.org/10.1016/j.envsci.2019.10.018

Cuerrier, A., Turner, N. J., Gomes, T. C., Garibaldi, A., & Downing, A. (2015). Cultural keystone places: Conservation and restoration in cultural landscapes. *Journal of Ethnobiology, 35*(3), 427–448. https://doi.org/10.2993/0278-0771-35.3.427

Diamond, S. K., & Ansharyani, I. (2018). Mismatched priorities, smallholders, and climate adaptation strategies: Landuse scientists, it's time to step up. *Journal of Land Use Science, 13*(4), 447–453. https://doi.org/10.1080/1747423X.2018.1537313

Díaz, S., Demissew, S., Carabias, J., Joly, C., Lonsdale, M., Ash, N., et al. (2015). The IPBES conceptual framework – connecting nature and people. *Current Opinion in Environmental Sustainability, 14*, 1–16. https://doi.org/10.1016/J.COSUST.2014.11.002

Dinerstein, E., Vynne, C., Sala, E., Joshi, A. R., Fernando, S., Lovejoy, T. E., et al. (2019). A global deal for nature: Guiding principles, milestones, and targets. *Science Advances*, eaaw2869. https://doi.org/10.1126/sciadv.aaw2869

Ens, E., Scott, M. L., Rangers, Y. M., Moritz, C., & Pirzl, R. (2016). Putting Indigenous conservation policy into practice delivers biodiversity and cultural benefits. *Biodiversity and Conservation, 25*(14), 2889–2906. https://doi.org/10.1007/s10531-016-1207-6

Fa, J. E., Watson, J. E., Leiper, I., Potapov, P., Evans, T. D., Burgess, N. D., et al. (2020). Importance of Indigenous peoples' lands for the conservation of intact forest landscapes. *Frontiers in Ecology and the Environment, 18*(3), 135–140. https://doi.org/10.1002/fee.2148

Fernández-Llamazares, Á., & Virtanen, P. K. (2020). Game masters and Amazonian Indigenous views on sustainability. *Current Opinion in Environmental Sustainability, 43*, 21–27. https://doi.org/10.1016/j.cosust.2020.01.004

Fernández-Llamazares, Á., Garteizgogeascoa, M., Basu, N., Brondizio, E. S., Cabeza, M., Martínez-Alier, J., et al. (2020). A state-of-the-art review of Indigenous peoples and environmental pollution. *Integrated Environmental Assessment and Management, 16*(3), 324–341. https://doi.org/10.1002/ieam.4239

Fernández-Llamazares, Á., Lepofsky, D., Lertzman, K., Armstrong, C. G., Brondizio, E. S., Gavin, M. C., et al. (2021). Scientists' warning to humanity on threats to Indigenous and local knowledge systems. *Journal of Ethnobiology, 41*(2), 144–169. https://doi.org/10.2993/0278-0771-41.2.144

Fletcher, M. S., Hall, T., & Alexandra, A. N. (2020). The loss of an Indigenous constructed landscape following British invasion of Australia: An insight into the deep human imprint on the Australian landscape. *Ambio, 50*, 138–149.

Forest Peoples Program, International Indigenous Forum on Biodiversity, & Secretariat of the Convention on Biological Diversity. (2016). *Local biodiversity outlooks. Indigenous peoples' and local communities' contributions to the implementation of the strategic plan for biodiversity 2011–2020. A complement to the fourth edition of the Global Biodiversity Outlook*.

Fox, C. A., Reo, N. J., Turner, D. A., Cook, J. A., Dituri, F., Fessell, B., et al. (2017). "The river is us; the river is in our veins": Re-defining river restoration in three Indigenous communities. *Sustainability Science, 12*(4), 521–533. https://doi.org/10.1007/s11625-016-0421-1

Funtowicz, S., & Ravetz, J. (1993). Science for the post-normal age. *Futures, 25*(7), 739–755.

Garibaldi, A. (2009). Moving from model to application: Cultural keystone species and reclamation in Fort McKay, Alberta. *Journal of Ethnobiology, 29*(2), 323–338. https://doi.org/10.2993/0278-0771-29.2.323

Garibaldi, A., & Turner, N. J. (2004). Cultural keystone species: Implications for ecological conservation and restoration. *Ecology and Society, 9*(3), 1. http://www.ecologyandsociety.org/vol9/iss3/art1/

Garnett, S. T., Burgess, N. D., Fa, J. E., Fernández-Llamazares, Á., Molnár, Z., Robinson, C. J., et al. (2018). A spatial overview of the global importance of Indigenous lands for conservation. *Nature Sustainability, 1*(7), 369–374. https://doi.org/10.1038/s41893-018-0100-6

Guadilla-Sáez, S., Pardo-de-Santayana, M., & Reyes-García, V. (2019). The role of traditional management practices in shaping a diverse habitat mosaic in a mountain region of Northern Spain. *Land Use Policy, 89*, 104235. https://doi.org/10.1016/j.landusepol.2019.104235

Hayes, T. M. (2010). A challenge for environmental governance: Institutional change in a traditional common-property forest system. *Policy Sciences, 43*(1), 27–48. https://doi.org/10.1007/s11077-009-9083-5

Herlihy, P. H. (2003). Participatory research mapping of Indigenous lands in Darién, Panama. *Human Organization, 62*(4), 315–331. https://doi.org/10.17730/humo.62.4.fu05tgkbvn2yvk8p

IPBES. (2019). *Summary for policymakers of the global assessment report on biodiversity and ecosystem services of the Intergovernmental Science-Policy Platform on Biodiversity and Ecosystem Services.* (S. Díaz, J. Settele, E. S. Brondízio, H. T. Ngo, M. Guèze, J. Agard, et al., Eds.). IPBES secretariat.

Janssen, M. A., & Ostrom, E. (2006). Governing social-ecological systems. In *Handbook of computational economics* (Vol. 2, pp. 1465–1509). Elsevier. https://doi.org/10.1016/S1574-0021(05)02030-7

Joa, B., Winkel, G., & Primmer, E. (2018). The unknown known – A review of local ecological knowledge in relation to forest biodiversity conservation. *Land Use Policy, 79*, 520–530. https://doi.org/10.1016/J.LANDUSEPOL.2018.09.001

Johnson, N., Alessa, L., Behe, C., Danielsen, F., Gearheard, S., Gofman-Wallingford, V., et al. (2015). The contributions of community-based monitoring and traditional knowledge to Arctic observing networks: Reflections on the state of the field. *Arctic, 68*(5), 28. https://doi.org/10.14430/arctic4447

Kimmerer, R. W. (2000). Native knowledge for native ecosystems. *Journal of Forestry, 98*(8), 4–9. https://doi.org/10.1093/JOF/98.8.4

Kuhnlein, H. V. (2014). How ethnobiology can contribute to food security. *Journal of Ethnobiology, 34*(1), 12–27. https://doi.org/10.2993/0278-0771-34.1.12

Lavides, M. N., Molina, E. P. V., de la Rosa, G. E., Mill, A. C., Rushton, S. P., Stead, S. M., & Polunin, N. V. C. (2016). Patterns of coral-reef finfish species disappearances inferred from fishers' knowledge in global epicentre of marine shorefish diversity. *PLoS ONE, 11*(5), e0155752. https://doi.org/10.1371/journal.pone.0155752

Long, J., Tecle, A., & Burnette, B. (2003). Cultural foundations for ecological restoration on the White Mountain Apache reservation. *Ecology and Society, 8*(1). https://doi.org/10.5751/ES-00591-080104

Luzar, J. B., Silvius, K. M., Overman, H., Giery, S. T., Read, J. M., & Fragoso, J. M. V. (2011). Large-scale environmental monitoring by Indigenous peoples. *BioScience, 61*(10), 771–781. https://doi.org/10.1525/bio.2011.61.10.7

Lyver, P. O. B., Timoti, P., Gormley, A. M., Jones, C. J., Richardson, S. J., Tahi, B. L., & Greenhalgh, S. (2017). Key Māori values strengthen the mapping of forest ecosystem services. *Ecosystem Services, 27*, 92–102. https://doi.org/10.1016/j.ecoser.2017.08.009

Malley, D. F., Watts, P. D., Ulrich, A. E., Malley, D. F., & Watts, P. D. (2016). Lake Winnipeg Basin: Advocacy, challenges and progress for sustainable phosphorus and eutrophication control. *Science of the Total Environment, 542*(Pt B), 1030–1039. https://doi.org/10.1016/j.scitotenv.2015.09.106

Martinez-Alier, J., Temper, L., Del Bene, D., & Scheidel, A. (2016). Is there a global environmental justice movement? *The Journal of Peasant Studies, 43*(3), 731–755. https://doi.org/10.1080/03066150.2016.1141198

McMillen, H., Ticktin, T., & Springer, H. K. (2017). The future is behind us: Traditional ecological knowledge and resilience over time on Hawai'i Island. *Regional Environmental Change, 17*(2), 579–592. https://doi.org/10.1007/s10113-016-1032-1

Munda, G. (2008). *Social multi-criteria evaluation for a sustainable economy*. Springer.

Oldekop, J. A., Bebbington, A. J., Truelove, N. K., Holmes, G., Villamarín, S., & Preziosi, R. F. (2012). Environmental impacts and scarcity perception influence local institutions in Indigenous Amazonian Kichwa communities. *Human Ecology, 40*(1), 101–115. https://doi.org/10.1007/s10745-011-9455-2

Orlove, B., Dawson, N., Sherpa, P., Adelekan, I., Alangui, W., Carmona, R., Coen, D., Nelson, M., Reyes-García, V., Sanago, G., & Wilson, A. (Accepted, pending revisions). Placing diverse knowledge systems at the core of transformative climate research. *Ambio*.

Orta-Martínez, M., & Finer, M. (2010). Oil frontiers and Indigenous resistance in the Peruvian Amazon. *Ecological Economics, 70*(2), 207–218. https://doi.org/10.1016/J.ECOLECON.2010.04.022

Orta-Martínez, M., Napolitano, D. A., MacLennan, G. J., O'Callaghan, C., Ciborowski, S., & Fabregas, X. (2007). Impacts of petroleum activities for the Achuar people of the Peruvian Amazon: Summary of existing evidence and research gaps. *Environmental Research Letters, 2*(4). https://doi.org/10.1088/1748-9326/2/4/045006

Pascual, U., Balvanera, P., Díaz, S., Pataki, G., Roth, E., Stenseke, M., et al. (2017). Valuing nature's contributions to people: The IPBES approach. *Current Opinion in Environmental Sustainability, 26–27*, 7–16. https://doi.org/10.1016/J.COSUST.2016.12.006

Paudyal, K., Baral, H., Burkhard, B., Bhandari, S. P., & Keenan, R. J. (2015). Participatory assessment and mapping of ecosystem services in a data-poor region: Case study of community-managed forests in central Nepal. *Ecosystem Services, 13*, 81–92. https://doi.org/10.1016/j.ecoser.2015.01.007

Porter-Bolland, L., Ellis, E. A., Guariguata, M. R., Ruiz-Mallén, I., Negrete-Yankelevich, S., & Reyes-García, V. (2012). Community managed forests and forest protected areas: An assessment of their conservation effectiveness across the tropics. *Forest Ecology and Management, 268*. https://doi.org/10.1016/j.foreco.2011.05.034

Reo, N. J. (2019). Inawendiwin and relational accountability in Anishnaabeg studies: The crux of the biscuit. *Journal of Ethnobiology, 39*(1), 65. https://doi.org/10.2993/0278-0771-39.1.65

Reyes-García, V., Fernández-Llamazares, Á., McElwee, P., Molnár, Z., Öllerer, K., Wilson, S. J., & Brondizio, E. S. (2019a). The contributions of Indigenous peoples and local communities to ecological restoration. *Restoration Ecology, 27*(1), 3–8. https://doi.org/10.1111/rec.12894

Reyes-García, V., García-del-Amo, D., Benyei, P., Fernández-Llamazares, Á., Gravani, K., Junqueira, A., et al. (2019b). A collaborative approach to bring insights from local indicators of climate change impacts into global climate change research. *Current Opinion in Environment Sustainability, 39*, 1–8. https://doi.org/10.1016/j.cosust.2019.04.007

Reyes-García, V., Fernández-Llamazares, Á., Bauchet, J., & Godoy, R. (2020). Variety of Indigenous peoples' opinions of large infrastructure projects: The TIPNIS road in the Bolivian Amazon. *World Development, 127*, 104751. https://doi.org/10.1016/j.worlddev.2019.104751

Reyes-García, V., Cámara-Leret, R., Renard, D., Zafra-Calvo, N., O'Hara, C., Halpern, B. S., & Díaz, S. (2023). Biocultural vulnerability exposes threats of culturally important species. *Proceedings of the National Academy of Sciences, 120*(2), e2217303120. https://doi.org/10.1073/pnas.2217303120

Rosell-Melé, A., Moraleda-Cibrián, N., Cartró-Sabaté, M., Colomer-Ventura, F., Mayor, P., & Orta-Martínez, M. (2018). Oil pollution in soils and sediments from the Northern Peruvian Amazon. *Science of the Total Environment, 610–611*, 1010–1019. https://doi.org/10.1016/j.scitotenv.2017.07.208

Scheidel, A., Del Bene, D., Liu, J., Navas, G., Mingorría, S., Demaria, F., et al. (2020). Environmental conflicts and defenders: A global overview. *Global Environmental Change, 63*. https://doi.org/10.1016/j.gloenvcha.2020.102104

Scheidel, A., Fernández-Llamazares, Á., Helen Bara, A., Del Bene, D., David-Chavez, D. M., et al. (in press). Global impacts of extractive and industrial development projects on Indigenous peoples' lifeways, lands and rights. *Science Advances*.

Shaffer, L. J. (2010). Indigenous fire use to manage savanna landscapes in southern Mozambique. *Fire Ecology, 6*(2), 43–59. https://doi.org/10.4996/fireecology.0602043

Sirén, A. H. (2017). Changing and partially successful local institutions for harvest of thatch palm leaves. *Ambio, 46*(7), 812–824. https://doi.org/10.1007/s13280-017-0917-7

Solomon, D., Lehmann, J., Fraser, J. A., Leach, M., Amanor, K., Frausin, V., et al. (2016). Indigenous African soil enrichment as a climate-smart sustainable agriculture alternative. *Frontiers in Ecology and the Environment, 14*(2), 71–76. https://doi.org/10.1002/fee.1226

Sterling, E. J., Filardi, C., Toomey, A., Sigouin, A., Betley, E., Gazit, N., et al. (2017). Biocultural approaches to well-being and sustainability indicators across scales. *Nature Ecology & Evolution, 1*(12), 1798–1806. https://doi.org/10.1038/s41559-017-0349-6

Stocker, L., Collard, L., & Rooney, A. (2016). Aboriginal worldviews and colonisation: Implications for coastal sustainability. *Local Environment, 21*(7), 844–865. https://doi.org/10.1080/13549839.2015.1036414

Takasaki, Y., Coomes, O. T., Abizaid, C., & Kalacska, M. (2022). Landscape-scale concordance between local ecological knowledge for tropical wild species and remote sensing of land cover. *Proceedings of the National Academy of Sciences of the United States of America, 119*, e2116446119.

Temper, L., & Martinez-Alier, J. (2013). The god of the mountain and Godavarman: Net Present Value, Indigenous territorial rights and sacredness in a bauxite mining conflict in India. *Ecological Economics, 96*, 79–87. https://doi.org/10.1016/j.ecolecon.2013.09.011

Tengö, M., Brondizio, E. S., Elmqvist, T., Malmer, P., & Spierenburg, M. (2014). Connecting diverse knowledge systems for enhanced ecosystem governance: The multiple evidence base approach. *Ambio, 43*(5), 579–591. https://doi.org/10.1007/s13280-014-0501-3

Tengö, M., Hill, R., Malmer, P., Raymond, C. M., Spierenburg, M., Danielsen, F., et al. (2017). Weaving knowledge systems in IPBES, CBD and beyond – Lessons learned for sustainability. *Current Opinion in Environmental Sustainability, 26–27*, 17–25. https://doi.org/10.1016/J.COSUST.2016.12.005

Tesfamichael, D., Pitcher, T. J., & Pauly, D. (2014). Assessing changes in fisheries using fishers; knowledge to generate long time series of catch rates: A case study from the Red Sea. *Ecology and Society, 19*(1). https://doi.org/10.5751/ES-06151-190118

Trauernicht, C., Brook, B. W., Murphy, B. P., Williamson, G. J., & Bowman, D. M. J. S. (2015). Local and global pyrogeographic evidence that Indigenous fire management creates pyrodiversity. *Ecology and Evolution, 5*(9), 1908–1918. https://doi.org/10.1002/ece3.1494

Turner, N. J., Gregory, R., Brooks, C., Failing, L., & Satterfield, T. (2008). From invisibility to transparency: Identifying the implications. *Ecology and Society, 13*(2). https://doi.org/10.5751/ES-02405-130207

Wangpakapattanawong, P., Kavinchan, N., Vaidhayakarn, C., Schmidt-Vogt, D., & Elliott, S. (2010). Fallow to forest: Applying Indigenous and scientific knowledge of swidden cultivation to tropical forest restoration. *Forest Ecology and Management, 260*(8), 1399–1406. https://doi.org/10.1016/j.foreco.2010.07.042

Wilder, B. T., O'Meara, C., Monti, L., & Nabhan, G. P. (2016). The importance of Indigenous knowledge in curbing the loss of language and biodiversity. *BioScience, 66*(6), 499–509. https://doi.org/10.1093/biosci/biw026

Wilson, E. (2016). Half-Earth: How to save the biosphere. *Half-Earth Our Planet's Fight Life*, 185–188.

Witter, R., Marion Suiseeya, K. R., Gruby, R. L., Hitchner, S., Maclin, E. M., Bourque, M., & Brosius, J. P. (2015). Moments of influence in global environmental governance. *Environmental Politics, 24*(6), 894–912. https://doi.org/10.1080/09644016.2015.1060036

Yusta-García, R., Orta-Martínez, M., Mayor, P., González-Crespo, C., & Rosell-Melé, A. (2017). Water contamination from oil extraction activities in Northern Peruvian Amazonian rivers. *Environmental Pollution, 225*, 370–380. https://doi.org/10.1016/j.envpol.2017.02.063

Ziembicki, M. R., Woinarski, J. C. Z., & Mackey, B. (2013). Evaluating the status of species using Indigenous knowledge: Novel evidence for major native mammal declines in northern Australia. *Biological Conservation, 157*, 78–92. https://doi.org/10.1016/J.BIOCON.2012.07.004

Open Access This chapter is licensed under the terms of the Creative Commons Attribution 4.0 International License (http://creativecommons.org/licenses/by/4.0/), which permits use, sharing, adaptation, distribution and reproduction in any medium or format, as long as you give appropriate credit to the original author(s) and the source, provide a link to the Creative Commons license and indicate if changes were made.

The images or other third party material in this chapter are included in the chapter's Creative Commons license, unless indicated otherwise in a credit line to the material. If material is not included in the chapter's Creative Commons license and your intended use is not permitted by statutory regulation or exceeds the permitted use, you will need to obtain permission directly from the copyright holder.

Chapter 8
Degrowth and the Barcelona School

Giorgos Kallis

8.1 Introduction

Degrowth refers to a radical political and economic reorganization leading to drastically smaller, and much more equitably shared, resource and energy use (Kallis et al., 2018). This chapter, following a personal narrative and recollection (Sect. 8.2), explains how the Barcelona school has shaped theoretically recent degrowth thinking (Sect. 8.3) and then how this new thinking, merged as it is with European/Francophone and Anglo political ecology, renews and transforms the field of ecological economics within which the Barcelona school emerged (Sect. 8.4).

8.2 History of Degrowth and the Barcelona School

I arrived in Barcelona from Berkeley in January 2008. I had encountered the vibrant group of researchers activists coalescing around Joan Martinez Alier in conferences, and I was impressed by the mix of scholarship and dedication to social justice the ICTA team transmitted – and to top it all, located in legendary Barcelona, the city where another world had been, and could still be, possible. In April, everyone I knew from ICTA was heading to a conference in Paris on 'degrowth', which somehow had skipped my attention. I felt like missing a party that no one had invited me to join. I played it cool – well I was working with water and coevolution, what was it for me in a 'degrowth' conference? But I couldn't hold my cool for much longer when people came back from Paris with the contagious excitement of something new being born.

G. Kallis (✉)
ICREA and ICTA, Universitat Autònoma de Barcelona, Bellaterra, Spain

© The Author(s) 2023
S. Villamayor-Tomas, R. Muradian (eds.), *The Barcelona School of Ecological Economics and Political Ecology*, Studies in Ecological Economics 8,
https://doi.org/10.1007/978-3-031-22566-6_8

In my own research on water resource development and urbanization as coevolutionary processes, I had reached the conclusion that the elephant in the room of unsustainable water management was capitalist growth. Powerful growth coalitions between real estate developers, banks, politicians and engineers were organized to make sure water does not limit urban growth, and unless this reality was confronted (politically, but also analytically), all talk about water conservation or demand management was just nice talk. Degrowth was the word I was missing – liberating me, at last, to utter what I stood for and what I saw necessary. Uttering it alongside Joan and our community at ICTA, I could let go of the fear of sounding politically incorrect and committing academic suicide.

The summer after the degrowth conference, Lehman brothers collapsed (no relation) and the greatest economic crisis of our generation unfolded. I remember Joan delivering his first lecture of the Fall semester on 'socially sustainable economic degrowth'. He started by saying that this was the last big crisis he would live to see (wrong – not only he would live much longer, but also crises turned out to be the new normal). He hoped that the global justice movement would grasp the opportunity and turn an undesirable recession into a 'socially sustainable' degrowth. Some would misconstrue his argument as a celebration of the crisis because Gross domestic product (GDP) and emissions fell (a straw man argument that those of us defending degrowth keep encountering in the current pandemic crisis). In the essay that accompanied the lecture, Joan explained clearly to those reading in good faith his thesis: Economic fundamentals were changing, growth encountering its limits. The question then was: Would there be a social force that could make the necessary adaptation sustainable rather than disastrous? What sort of struggles and policies would make this possible (Martinez-Alier, 2009a)?

The Paris conference brought together three streams of thought that together would form 'degrowth': the French school of décroissance/post-development emerging from the radical Continental political ecology of the 1970s with an emphasis on autonomy and conviviality; Tim Jackson and Peter Victor's steady state economics of managing without growth; and the Barcelona school's political ecology that pointed to growth's dependence on unequal exchange and exploitative extraction at the world's 'commodity frontiers'. A historian first and foremost, Joan wrote the first history of degrowth, tracing this early mixing of Anglo ecological economics with Franco political ecology to the translation of an edited volume of Georgescu-Roegen's essays to French in 1979 (entitled 'Demain la décroissance'), unearthing along the way a forgotten advocacy of zero growth in 1972 by EU's president Sicco Mansholt (Martinez-Alier et al., 2010a).

Organizing the second international conference of degrowth in Barcelona in 2010, we formed a group of 20–30 young – and not so young – researchers and activists, which with newcomers and departures has been together ever since (as 'Research & Degrowth'). I am not boasting if I say that our group singlehandedly put degrowth in the academic map, with an avalanche of scientific publications in English. We helped set up the biennial international conferences, out of which a community, research agenda and political discourse emerged (Demaria et al., 2013;

D'Alisa et al., 2014). Today, in a context of climate breakdown and economic crisis, degrowth is prominently covered – positively and negatively – in the pages of the Guardian, the New York Times or the Spectator, attacked by the likes of Stephen Moore, aid and ideological guru of Donald Trump. We kept the discussion around degrowth alive in a period of recession when one would expect it to subside like the 'limits to growth' debate did in the 1970s – and we worked to make it common sense to the extent that a young activist like Greta Thunberg would talk about 'fairy tales of growth' at the podium of the UN.

8.3 Core Concepts of the Barcelona School Informing Degrowth

Behind the claims of degrowth lie core concepts of Joan Martinez-Alier that we have worked with in Barcelona. Let me focus here on three, among many: the energy costs of producing energy; cost-shifting and how it sustains a growing global social metabolism, while causing ecological distribution conflicts; the weak commensurability of different languages of valuation (Martinez-Alier, 2009b).

The insight that to produce energy you have to spend energy, and that hence different energy sources have different 'EROIs' (Energy Returns of Energy Investment) is not Joan Martinez-Alier's of course. Howard Odum and early ecological economists emphasised this aspect of energy production and Stuart Hall among others developed EROI metrics. Martinez-Alier's contribution though was in centering ecological economics as a study of the energetics of economic life. In his history of proto-ecological economists – natural scientists writing in nineteenth and early twentieth century treatises in economics – he showed how the economy can be understood as a process of capturing and distributing useful energy. Economics should be concerned with understanding and calculating these flows and transformations of energy, and not epiphenomenal 'laws' that govern monetary quantities with no necessary connection to physical reality (Martinez-Alier, 1990).

I assume that it was through his research of peasant economies in Andalusia and Peru that Joan appreciated cultural ecology and the arguments of anthropologists about the energy efficiency of peasant societies, and how pre-capitalist production systems yielded more energy out of the energy they invested to capture solar flows. This attention to the energy and resource efficiencies of small-scale and decentralized, sufficiency-oriented production is central in degrowth scholarship, where small is not only beautiful but often better. This connects to the appreciation that the growth machine has been fuelled by high EROI fossil fuels, an input that allowed the expansion of wasteful in terms of resource and energy production systems. This attention to energy returns makes many in the degrowth community sceptical of the possibilities of 'green growth'. Degrowthers postulate that a transition away from fossil fuels, and towards renewable energy sources may slow down the economy (Kallis et al., 2018). Some could go even further and argue that a society powered by solar and wind would need to reorient its economic life around the intermittency of these resources, evolving into a kind of 'new peasant', de-urbanized civilization (Smaje, 2020). This resonates with Georgescu-Roegen's prediction/advocacy of a bioeconomy of solar-powered neo-peasant societies for the future.

Economic production, Martinez-Alier has always argued, is a metabolic process that is entropic. Metabolic means that like our bodies, the economy needs a constant input of energy and resources, which it converts into useful goods and waste. This conversion is entropic, because an amount of energy and materials is always irreversibly lost along the way (Kallis, 2018). Martinez-Alier emphasised how the metabolic requirements of a growth economy meant that a constant influx of materials is extracted from the world's 'commodity frontiers', peripheral territories where untapped resources are found or waste discarded 'out of sight'. Following the work of William Kapp, he conceptualized extraction and waste disposal at these frontiers as a case of 'cost-shifting', core areas and privileged groups profiting at the expense of peripheral areas and disadvantaged people, often exploited along hierarchies of race, ethnicity, gender or class (Martinez-Alier, 2009b). Growth in this reading is not just a matter of technological progress, human capital or a culture of innovation but of unequal exchange, cheapening and exploiting poor people and their natures and securing a low-cost inflow of materials and fossil fuels (Kallis, 2018).

This growth of the global metabolism, Martinez-Alier (2009a) argues, is a source of 'ecological distribution conflicts', conflicts over the distribution of environmental costs and benefits, communities at the commodity frontiers organizing to stop their exploitation. These mobilizations that attempt to put a stop to the engine of growth at its input side, Martinez-Alier claimed, are a force of degrowth (even if not intentionally) – if successful, they will push for a reorientation of the core economies. In that, he sees a natural alliance between the global environmental justice movement that fights against cost-shifting and those who want to see degrowth in Europe (Martinez-Alier, 2009a). Granted, those fighting against specific extractive projects in the periphery might not be against growth as such, and may still welcome some form of economic development in their territory. Here there is a link with the post-development school of thought that highlights the alternative modes of well-being and social organizing emerging from capitalism's peripheries, often in conditions of conflict against extractive, growth-driven projects (Kothari et al., 2019).

Martinez-Alier and out teacn Barcelona studied ethnographically concrete conflicts in commodity frontiers and revealed how the parties involved use different 'languages of valuation' based on different cultural systems of assigning value (Martinez-Alier et al., 2010b). On the one hand, there is the economic language of the market: cost, profit and GDP. On the other, there are valuing systems based on community, sacredness and spirituality that may ascribe intrinsic value to non-human natures or particular ways of being. Different value systems are 'incommensurable', that is there is no common metric – monetary or else – upon which to compare them; but in a true democracy, they could be 'weakly comparable', that is the relative pros and cons of different options for different people according to different value systems deliberated and negotiated. From an analytical perspective, this shifts attention to the power relations through which the economic language of market metrics and exchange values comes to dominate other value systems, and the ways communities can organize and resist market colonization.

This framework speaks directly to original concerns of degrowth scholars, anthropologists like Serge Latouche or Arturo Escobar, who saw a clash between local economic cultures and homogenizing forces of a globalized economy commodifying uncommodified spaces and relations, drawing violently peripheral regions into the circuits of capitalist growth. Escobar (1996) talked of ontological cultural conflicts. And Latouche (2009) developed his theory of degrowth inspired precisely by places he encountered in his fieldwork in South East Asia and Africa, where alternative economic cultures were being crushed by the growth machine. For Latouche, the 'de' of degrowth decolonizes a cultural imaginary in the West saturated by the idea of constant expansion and a singular way of market organization. Degrowth can then be understood as the prioritization of different values and languages of valuation over those of market economies, not least GDP – a 'revaluation' that is part and parcel of slowing down global social metabolism.

8.4 Shaping Ecological Economics

In the previous section, I saw how our ideas in the Barcelona school, inspired by Joan's thought, shaped thinking and research on degrowth. Here I want to show how our thinking on degrowth in turn challenged certain aspects of ecological economics, the interdisciplinary community within which our school was based. In other words, what I argue is that our contribution to degrowth was not simply to bring ecological economics in – rather we have developed a new approach that has changed (or at least aspires to change) both degrowth and ecological economics research.

The origin of ecological economics can intellectually be traced back to the Limits to Growth report, growth in the scale of an economy seen as limited by the external ecosystem that provides material and resources. Ecological economists have distinguished between efficiency and scale – markets may allocate resources efficiently, but still grow the scale of the economy to a level unsustainable by the supporting ecosystems. The particular breed of ecological economics we have developed in Barcelona through our engagement with political economy and ecology and the Francophone school of degrowth points, however, to a very different and more radical type of ecological economics and of understanding society–nature relations.

First, in our work, there is a shift of emphasis from external planetary boundaries to collective processes of self-limitation. The point as I have argued (Kallis, 2019) is not so much whether there are limits to growth and where exactly are they, but instead how to organize effectively to limit growth. Limits in this vein, are part and parcel of what Francophone political ecologists have been calling 'autonomy' – the capacity of collectives to determine their own laws and limits, freed from mythical imperatives of gods, markets or experts. In my book on limits, I take the lead from Martinez-Alier's history of the anarcho-feminists of Emma Goldman who defended birth control not in the name of overpopulation but because they wanted to stop the capitalist and imperial war machines while freeing women to enjoy sex. I argue that degrowth, first and foremost, is such a project of self-limitation – a culture of limits in the pursuit of joy and well-being, not just a defensive strategy of averting disaster and sustaining the current system longer, as one can perhaps interpret the Club of Rome's work.

Second, there is increasing awareness in our writing of the ways under which capitalism structures both the geography and tempo of global social metabolisms. True, non-capitalist industrial systems also pursued growth and had extractive and expansive metabolisms. But accumulation is a capitalist invention, and today capitalism is the only game in town. If we want to understand which resources are extracted, where and when, we need to engage analytically, and politically, with the profit logic of capitalism. Many of the ecological distribution conflicts at the world's frontiers are not only conflicts against the impacts of extraction or disposal but also conflicts against the enclosures that capitalism continues to engender. Cost-shifting in other words is a more general form of appearance of what Marxist scholars have called accumulation by dispossession, the separation of people from their means of production and livelihood, a process that started with the original enclosures, but is constantly repeated in different historical moments as the capitalist growth machine needs to bring new territories and relations into the commodity and accumulation circuit. Degrowth in this sense is part and parcel of decommodification and the social struggles to defend and reclaim the commons.

The limits to growth debate, and to an extent ecological economics were silent on the question of capitalism, but also focussed too much, in my view at least, on the prophetic, 'warning of doom' side, and less so on the affirmation of alternatives, and the politics that can bring them about. And this is our third difference – in our work in Barcelona, we are very much interested in the alternative economic cultures of 'commoning' (of making and defending commons) that different communities, both within core cities or peripheral frontiers, juxtapose to capitalism. We are interested in alternatives – from cooperatives and community currencies to the agroecology of the Via Campesina movement – because of the embryonic forms of an alternative (post-growth, post-capitalism) economy they represent, but also because we seek to understand how these alternatives can organize politically and evolve into a bigger force that can bring systemic change (Kallis et al., 2020).

Our work in Barcelona does not just lament the power of an ever-encroaching capitalist growth machine but seeks to elevate and celebrate social opposition against the logics of capitalism – the myriad local movements of opposition to extractivism and growth (Scheidel et al., 2020). It is the intertwining of environmental justice movements with agrarian justice movements, indigenous claims, gender struggles, and even sometimes actual or potential working-class movements that can forge a transition towards economies that do put limits to growth. This transition is being helped by the successful episodes in which oil is left in the soil, coal in the hole and copper or bauxite extraction is stopped by grassroots movements. One main task of us as researchers in the transition is to be active in such movements of environmental justice (e.g. Ende Gelände in Germany and thousands of other movements in all countries, very often repressed by force) and to learn and support the vocabularies and repertoires of action that they develop against the logic of capitalist growth.

And this is the main legacy of Joan Martinez-Alier and the school of thought he helped build in Barcelona: an activism-oriented research that seeks environmental and social justice and which puts first and foremost the ideas and the alternatives that emerge from the grassroots.

References

D'Alisa, G., Demaria, F., & Kallis, G. (Eds.). (2014). *Degrowth: A vocabulary for a new era*. Routledge.

Demaria, F., Schneider, F., Sekulova, F., & Martinez-Alier, J. (2013). What is degrowth? From an activist slogan to a social movement. *Environmental Values, 22*(2), 191–215.

Escobar, A. (1996). Construction nature: Elements for a post-structuralist political ecology. *Futures, 28*(4), 325–343.

Kallis, G. (2018). *Degrowth*. Agenda Publishing.

Kallis, G. (2019). *Limits. Why Malthus was wrong and why environmentalists should care*. Stanford University Press.

Kallis, G., Kostakis, V., Lange, S., Muraca, B., Paulson, S., & Schmelzer, M. (2018). Research on degrowth. *Annual Review of Environment and Resources, 43*, 291–316.

Kallis, G., Paulson, S., D'Alisa, G., & Demaria, F. (2020). *The case for degrowth*. Polity Press.

Kothari, A., Salleh, A., Escobar, A., Demaria, F., & Acosta, A. (Eds.). (2019). *Pluriverse: A post-development dictionary*. Tulika Books and Authorsupfront.

Latouche, S. (2009). *Farewell to growth*. Polity Press.

Martinez-Alier, J. (1990). *Ecological economics: Energy, environment and society*. Blackwell.

Martinez-Alier, J. M. (2009a). Socially sustainable economic de-growth. *Development and Change, 40*(6), 1099–1119.

Martinez-Alier, J. M. (2009b). Social metabolism, ecological distribution conflicts, and languages of valuation. *Capitalism Nature Socialism, 20*(1), 58–87.

Martínez-Alier, J., Pascual, U., Vivien, F. D., & Zaccai, E. (2010a). Sustainable de-growth: Mapping the context, criticisms and future prospects of an emergent paradigm. *Ecological Economics, 69*(9), 1741–1747.

Martínez-Alier, J., Kallis, G., Veuthey, S., Walter, M., & Temper, L. (2010b). Social metabolism, ecological distribution conflicts, and valuation languages. *Ecological Economics, 70*(2), 153–158.

Scheidel, A., Del Bene, D., Liu, J., Navas, G., Mingorría, S., Demaria, F., Avila, S., Roy, B., Ertör, I., Temper, L., & Martínez-Alier, J. (2020). Environmental conflicts and defenders: A global overview. *Global Environmental Change, 63*, 102104.

Smaje, C. (2020). *A small farm future*. Chelsea Green Publishing.

Open Access This chapter is licensed under the terms of the Creative Commons Attribution 4.0 International License (http://creativecommons.org/licenses/by/4.0/), which permits use, sharing, adaptation, distribution and reproduction in any medium or format, as long as you give appropriate credit to the original author(s) and the source, provide a link to the Creative Commons license and indicate if changes were made.

The images or other third party material in this chapter are included in the chapter's Creative Commons license, unless indicated otherwise in a credit line to the material. If material is not included in the chapter's Creative Commons license and your intended use is not permitted by statutory regulation or exceeds the permitted use, you will need to obtain permission directly from the copyright holder.

Part III
Social Metabolism

Chapter 9
Agrarian Metabolism and Socio-ecological Transitions to Agroecology Landscapes

Enric Tello and Manuel González de Molina

9.1 From Land Reform and Agrarian Capitalism to Energy Accounting of Agriculture

After studying law and economics at the University of Barcelona, Joan Martínez Alier started doing research in agricultural economics at St Antony's College in Oxford from 1966 to 1973, publishing *La estabilidad del latifundismo* and an upgraded English version of this same book, which dismissed the hypothesis of the supposedly backward 'feudal' character of Andalusia's large states (Martínez Alier, 1968, 1971). His work also rejected the 'primitive' character of the labourers' resistance and rebellions claimed by some Marxists (Hobsbawm, 1959) and the Spanish Communist Party, which was then leading the resistance to Franco's dictatorship. He then went to Cuba and Perú to study, with Verena Stolcke, how surplus was extracted from those who worked the land either through hired labour or different agrarian systems of tenancy, treating these peasant economies in ways that went beyond orthodox neoclassical economics (Martínez Alier, 1977, 1978a). This research connected him with the views of the Russian Narodnik Aleksandr V. Chayanov (1966 [1925]; Martínez Alier, 1978b), then being revived by Theodor Shanin (1971, 1973, 1974) and others. His contributions had a broad impact on international debates on the agrarian question of the time (Kay, 1974, 1977), together with the books and articles on Spanish agriculture of his friend José Manuel Naredo (1971, 1978a, b, and Naredo et al., 1975).

E. Tello (✉)
Universitat de Barcelona, Barcelona, Spain
e-mail: tello@ub.edu

M. González de Molina
Universidad Pablo Olavide, Seville, Spain

The two friends also collaborated in editing the political journal *Cuadernos de Ruedo Ibérico*, issued in Paris (1965–1979) with a critical stance towards the policies adopted by the Spanish left opposition parties at the end of the Francoist dictatorship (Naredo & Martínez Alier, 1976) and making an early connection with the awakening of environmentalism, which also started in Spain with the opposition to nuclear plants and big new dams (Naredo, 1973, 1981; Gaviria et al., 1978). This drew their attention to the energy question (Naredo & Martínez Alier, 1979; Martínez Alier, 1980, 1982, 2019: 213–220). Martínez Alier had already made energy accounts of food baskets in his studies of agrarian economics and went deeper into the subject after reading Roy Rappaport's *Pigs for the Ancestors* (1968). Together with Pablo Campos, José Manuel Naredo calculated the first energy balance for Spanish agriculture and its declining rates of return with the Green Revolution, one of the world's first examples of agricultural EROI calculation (Campos & Naredo, 1980; Naredo & Campos, 1980).

The two authors also started reviving the work on energy economics of Sergei Podolinsky, another socialist-oriented Ukrainian Narodnik, pointing out the failed opportunity to start an ecosocialist current a century earlier than it took place, owing to Marx's doubts and death, and Engels' final refusal (Martínez Alier & Naredo, 1982; Martínez Alier, 1995). Podolinsky's attempt was praised by Valdimir Vernadsky (2007: 212) in the Soviet Union in 1924, when such an event entailed great risk, like the trial and death of Chayanov in the 1930s. However, some recent ecological Marxists still try to dismiss his forerunner essay (Foster & Burkett, 2008). They are wrong, and Martínez Alier and Naredo were right in putting the finger on the sore of the Marxist tradition. The only 'obituary' that deserves to be explained is why, since 1883, so many Marxists ignored Marx's ecological and energy-related inklings, with only a very few exceptions (Sacristán, 1992; Tello, 2016).

9.2 Growing Up as Historians in the Debates over the Agrarian Question in the 1970s

The influence of Martínez Alier on the authors of this chapter began during the last years of the Franco dictatorship and at the beginning of the disappointing transition to the current Spanish parliamentary monarchy. At that time, most of the intellectuals who were linked to the Communist Party denied that a bourgeois revolution had taken place in Spain and considered the *latifundio* and *minifundio* to be 'feudal remnants' of an agrarian system that was not yet fully capitalist. According to Martínez Alier and Naredo, the idea that the *latifundios* were largely inefficient could no longer be sustained. The large estates were among the protagonists of the process of 'agrarian modernization' that was then already taking place. The need for agrarian reform could no longer be argued on the grounds of promoting economic development. The only valid reason in its favour was ethical, not economic.

Industrial agriculture, which was then being championed by left and right political parties alike, had profound ecological impacts, which Naredo and Martínez Alier were pioneers in denouncing. However, the productivism of this 'agricultural modernization' still inspired the Agrarian Reform Law approved by the Andalusian Parliament in 1984, when Spain's imminent entry into the European Common Market diverted political concerns towards production surpluses rather than more intensive agriculture (Naredo & González de Molina, 2002).

This agrarian debate framed our training as historians when we graduated and began our doctoral theses (González de Molina, 2020a, b; Tello, 2020). At this time, a vigorous labour movement had re-emerged in Andalusia in pursuit of land reform from below, making the works of Martínez Alier and Naredo even more relevant in political and historiographical terms. Manuel González de Molina got in touch with them for the first time in a conference on agrarian reform in the University of Granada, initiating a relationship that has lasted until today. That collaboration between academics and members of the *Sindicato de Obreros de Campo* (SOC) continued throughout the 1980s, and the ecological component gained more and more momentum. It gave rise to an original socioenvironmental synthesis around the so-called Andalusian Pact for Nature, a movement with a peasant base that progressively assumed clear environmentalist approaches (Herrera et al., 2010).

Joan Martínez Alier participated in the seminars then organized in Córdoba, Baeza (Jaén) and La Rábida (Huelva) by Manuel González de Molina and Eduardo Sevilla Guzmán, seminars linked to the first peasant experiences with organic agriculture in the context of the SOC cooperatives (González de Molina & Guzmán, 2017). That was the true birth of agroecology in Spain, opening a fruitful debate on the role of the peasantry in social and political change. From these experiences and debates emerged a proposal for an ecological neo-populism, which was actively spread by Martínez Alier (1985a, b, 1987b, 1988, and Flores Galindo & Martínez Alier, 1988). Eduardo Sevilla Guzmán and Manuel González de Molina coordinated a book on *Ecología, Campesinado e Historia* (1993), including a chapter by Víctor Toledo (1993), emphasizing the ecological rationality of peasant production that laid the socioecological foundations for neo-populist proposal from within agroecology.

Enric Tello met Joan Martínez Alier for the first time in 1979 at a *Seminar on the energy crisis in a capitalist society* at the University of Barcelona, organized by the Catalan Antinuclear Committee, of which he was member. It opened with a lecture by Martínez Alier on *Energy and agrarian economy* (Martínez Alier, 1980) and closed with another by the Marxist Manuel Sacristán (1980) on *Why does the environmentalist movement lack economists?* Martínez Alier was then writing his well-known book *Ecological Economics* (1987a), first issued in Catalan (Martínez Alier, 1984). In 1980, Sacristán founded the first ecosocialist journal in the world, called *Mientras Tanto* (the second, *Capitalism Nature Socialism*, only appeared in 1992; Tello, 2003, 2016). Enric Tello joined its editorial board in 1982, the agrarian historian Ramon Garrabou already being a member. At that time this red, green and violet journal paid attention to the birth of the German Greens and published the first European Ecosocialist Manifesto (Antunes et al., 1990). However, Sacristan died in 1985. When Víctor Toledo, Manuel González de Molina and others sent a co-authored article to *Minestras Tanto* claiming the political importance of new peasant organizations like Via *Campesina* as a transformational subject of social change, a group of neo-orthodox Marxists on the editorial board imposed an aggressive veto against publishing it. It was the beginning of the end of this journal's interest in views from political ecology, which were spread instead by the journal *Ecología Política*, founded by Martínez Alier in 1991.

In 1992, with Miguel Altieri, Stephan Gliessman and Víctor Toledo, Martínez Alier also participated in the first agroecology course taught in Spain in Baeza and has continued participating in it and in the doctorate and master's courses at La Rábida ever since. Besides explaining the new field of ecological economics (Martínez Alier, 1993), he advanced from the 'ecological neo-Narodnism' (Martínez Alier, 1985a, b) towards what later became the *Environmentalism of the poor* (Martínez Alier, 2002). He also began proposing the development of a socioecological history (Martínez Alier, 1990, 1993). All these approaches encouraged new research on Andalusian peasant conflicts in the nineteenth century (Cobo et al., 1992). Environmental history emerged in Spain at the beginning of the 1990s in the recently founded Spanish Society for Agrarian History (SEHA). After some debates on forest history and common pool resources (González de Molina, 2000), Ramón Garrabou organized the first Spanish environmental history encounter in Girona, which led to the first special issue published on the subject edited by González de Molina and Martínez Alier (1993).

9.3 From Agrarian History to the Environmental History of Agroecosystems

The First Spanish Environmental History Conference was held in Andújar (Jaen) in 1999, with an opening lecture by Rolf Peter Sieferle. The best contributions were published in a volume edited by Martínez Alier and González de Molina (2011). The Second Spanish Environmental History Conference took place in Huesca in 2001 with more than 200 registered participants and an opening lecture by John McNeill, leading to another book (Sabio & Iriarte, 2003). Martínez Alier also encouraged the creation of the European Society for Environmental History (ESEH) and put Manuel González de Molina in touch with the Social Ecology Institute (SEC) in Vienna. In 1999, he presented a first paper there considering the use of environmental variables to explain the contemporary history of Spanish agriculture, later published in a collective book in Spanish and in *Ecological Economics* (González de Molina, 2001, 2002). Through these and other channels he was in touch with approaches from social metabolism (Ayres & Simonis, 1994; Opschoor, 1997; Fischer-Kowalski, 1998; Fischer-Kowalski & Hüttler, 1998; Carpintero, 2005). At the same time, Enric Tello started calculating the energy and water balances of the social metabolism of Barcelona city by participating in several environmental campaigns (Tello, 2002, 2005; Puig et al., 2003; Roca et al., 2006).

Enric Tello and Manuel González de Molina met each other in the SEHA, founded in 1990. In 1994, they both attended the first meeting to bring together agronomists, soil scientists and biologists with agrarian historians, launched at the SEHA by Ramon Garrabou and José Manuel Naredo, which gave rise to a first book on soil fertility management (Garrabou & Naredo, 1996). The experience was very successful and led to other meetings held in 1996 devoted to soil water uses and balances, giving rise to a second book (Garrabou & Naredo, 1999). A last meeting held in 1999 focused on agricultural landscapes (Tello, 1999). It also produced a third book (Garrabou & Naredo, 2008), which took a step forward due to the methodology proposed by José Manuel Naredo to address it. That is, using the energy and material flow accounting of agricultural metabolism to study cultural landscapes as a territorial 'imprint' of the biophysical flows driven by different types of farming throughout history. This meant recalculating the energy balances of Spanish agriculture made in the early 1980s. But 20 years later, material and energy flow accounting (MEFA) had been developed in ecological economics. Martínez Alier and others had created the ICTA at the Autonomous University of Barcelona in 1997 (Martínez Alier, 2019: 61–64) and were joined by Mario Giampietro in 2007. His articles and books on the energy analysis of agroecosystems became key references on the subject (Giampietro & Pimentel, 1991; Giampietro et al., 1994; Giampietro, 1997, 2004). To apply a socio-metabolic approach to agroecosystems required novel MEFA accountancies.

Fascinated by the analytical potential of this biophysical analysis for agroecology and environmental history, Manuel González de Molina began to collaborate with Víctor Toledo at the UNAM Institute of Ecology in Morelia (Mexico). Their first joint contribution on social metabolism was presented at the International Symposium on Environmental History held in Xalapa (Mexico) in 2001, in which Joan Martínez Alier also participated (Toledo & González de Molina, 2007). The agroecologists Gloria Guzmán and Manuel González de Molina calculated and published their first historical energy balances of an Andalusian farm system in the same issue of *Historia Agraria* in which Ramon Garrabou, Enric Tello, Xavier Cussó and the agronomist José Ramón Olarieta had done this in a Catalan case study, also published in *Ecological Economics* (González de Molina & Guzmán, 2006; Guzmán & González de Molina, 2006; Cussó et al., 2006a, b; Tello et al., 2006). Both teams started applying the same type of energy accounting that had already been used by Naredo and Campos (1980), but that first attempt led to a methodological disagreement. Pablo Campos criticised Enric Tello on the grounds that the Catalan energy balance failed to treat the internal provision of services, like animal traction and manure obtained reusing internal biomass flows, as a cost. Pablo Campos and Javier López-Linage did so in working out the energy balances of the Spanish wood pasture or *dehesa* and other case studies (Campos & López-Linage, 1997; Campos & Casado, 2004; López-Linage, 2007). The Catalan research group adopted it, but the Andalusian researchers did not.

In a debate between the two teams, together with Naredo, Campos and López-Linage, held in Madrid in 2007, González de Molina criticized the fact that Campos' calculation of agricultural energy balances only focused on the farmers' viewpoint from the perspective of environmental economics. Tello replied that the importance of these internal reuses of biomass and of internal services provided by the multiple uses of livestock were a hallmark of traditional organic agriculture compared to the linear industrial ways of farming. Gloria Guzmán pointed out that farming biomass reuses played very important roles from an agroecology standpoint other than provisioning internal economic services to farmers. These conceptual and methodological difficulties made apparent, once again, the lack of a socio-metabolic energy-flow analysis conducted from an agroecological point of view. Clearly, new types of concepts, models and indicators were required to study agrarian metabolism from a long-term historical perspective.

9.4 Advances in the Study of Agrarian Metabolism as a Tool for the New Agroecological Transition

During the First World Conference on Environmental History, held in Copenhagen in 2009, the North American environmental historian Geoff Cunfer and Fridolin Krausmann from the Vienna SEC proposed to join with Manuel González de Molina, Enric Tello and Stefania Gallini to draw up a proposal to study the historical socioecological transitions of agricultural systems using socio-metabolic calculations, to be submitted for a Partnership Grant to the Social Sciences and Humanities Research Council of Canada. The research project *Sustainable Farm Systems: Long-term Socioecological Metabolism of Western Agriculture* (SFS) was approved in 2012 and allowed the assembly, up to 2018, of a network of teams in Canada (Saskatchewan and Edward Prince Island), Austria (Vienna), Spain (Seville and Barcelona), Colombia (Bogotá and Cali) and Cuba, later enlarged with collaborators in the United States, the Czech Republic and Costa Rica. To date, 82 energy balances of past and present farm systems have been calculated using a novel agricultural energy analysis. Many of them were published in a special issue of *Regional Environmental Change* in 2018 (Gingrich et al., 2018a, b), while others are still being published. This is the largest dataset of energy and soil nutrient balances of farm systems to use the same methods of calculation from a long-term historical perspective thus far. It is considered the most circular of all the current different methods used in the field by the latest review article on the subject (Hercher-Pasteur et al., 2020).

So as to provide comparable results, compiling these case studies required solving the above-mentioned debate on calculating energy. To consider the important role of farmers' internal reuses of biomass, Enric Tello and other SFS researchers started to draw increasingly circular flow diagrams of the socio-metabolic interaction among the fund components of agroecosystems (i.e., the different land uses of farmland, the livestock, the farming community and the rest of society) during the sojourn of Geoff Cunfer as visiting researcher to the University of Barcelona in 2012. This led them to realize that using a single EROI indicator of the energy performance of a complex agroecosystem became a straitjacket for this circular approach. The conceptual barrier started to be overcome in 2013 while Enric Tello was a visiting researcher in the Vienna SEC, working with Fridolin Krausmann and Simone Gingrich. The energy return to all inputs invested, including internal biomass reuses, was called Final EROI. It was then decomposed into an Internal Final EROI and an External Final EROI (the latter being the only agricultural EROI formerly calculated in the literature). This multi-EROI analytical approach avoided the internal flows of agroecosystems driven by farmers remaining into a black box (Tello et al., 2015). The three agricultural EROIs are related, so that the Final EROI equals the product of IFEROI and EFEROI divided by their sum. This equation allows the performance of optimality analyses to be contrasted with the actual historical evolution of these energy returns (Tello et al., 2016).

Two main general results using this novel multi-EROI energy analysis have been obtained thus far for North America and Europe. First, throughout the socioecological transition from past organic to industrial agriculture, farm systems have fallen into an energy trap of much lower EFEROI values due to the sharp increase in the consumption of fossil-based external inputs above the growth in final produce. This confirms the diagnosis made by the first agricultural energy balances half a century ago (Leach, 1976; Pimentel & Pimentel, 1979), despite the recent small EFEROI increases observed in the last two decades (Pellegrini & Fernández, 2018). Second, although the changes in IFEROI and FEROI values have been less significant (Gingrich et al., 2018b), the relative maintenance of their values over time conceals deep structural changes in the composition and integration of agroecosystems among live funds, mainly driven by the dietary transition (which increased the proportion of cereals diverted towards animal feed at the expense of crop by-products and of poor natural pastures being increasingly wasted), and by the forest transition (which reduced forestry in the Global North, lowering the more energy-dense component in the output obtained with lower energy inputs). Behind the energy trap of industrial farming lies a loss in the complexity and circularity of agroecosystems and the integration of their fund-flow patterns.

Gloria Guzmán, Manuel González de Molina and their Andalusian colleagues in the Agroecosystem History Lab agreed on this novel ecological economic multi-EROI, which was adopted by the whole SFS research project, but stressed the agroecological importance of the total photosynthetic Net Primary Production of agroecosystems, and the part of it that remains unharvested and feeds the food chains of non-colonized species either above ground in the landscape or below ground in the soil biota. This perspective linking agrarian metabolism to farm-associated biodiversity and related ecosystem services was incorporated by the Spanish SFS teams, but has not yet been adopted by others, leading to other types of EROI and other methodological advances being made by the Andalusian team (Aguilera et al., 2015). Gloria Guzmán and Manuel González de Molina took the lead in developing this agroecology energy analysis (Guzmán et al., 2014, 2018; Soto et al., 2016; Guzmán & González de Molina, 2015, 2017; González de Molina, 2020b), relying on the conceptual approach to agrarian metabolism put forward by González de Molina and Victor Toledo (2011, 2014). This led to two different but compatible versions of multi-EROI energy analysis: agroecological and bioeconomic.

At the same time, the SFS Catalan team took over Ramon Margalef's hypothesis on the capacity of a mosaic pattern of complex agricultural landscapes to host biodiversity. The whole energy turnover of an agroecosystem, starting from the solar Net Primary Production (NPP) up to the decomposition of food chains of soils, were calculated using graph modelling. This graph avoids double-counting energy flows by replacing the addition of their energy content with the proportions of each flow, which are split into two at each node, one looping inside the agroecosystem, the other either going outside it or coming into it. This new analytical approach allowed an Energy-Landscape Integrated Analysis (ELIA) using as indicators the share of NPP that remain temporarily stored within the agroecosystem; the Shannon index of how evenly NPP energy flows circulate across all possible paths of the graph, taken as an indicator of the information complexity of the agroecosystem (Sherwin & Prat-i-Fornells, 2019); and the land cover diversity of the landscape, which corresponds to this pattern of energy flows, calculated by means of another Shannon index through GIS in digital maps (Marull et al., 2016, 2019a; Font et al., 2020).

These three ELIA indicators (energy storage, information complexity and landscape heterogeneity) can be assessed in each cell of a grid applied through GIS to the digital map of any landscape to evaluate across gradients how heterogeneous land covers are (L), how complex the energy fund-flow pattern is (I) and how dissipative the socio-metabolic energy flows become (E). The latter E indicator means monitoring what Martínez Alier and Naredo (1982) called the Podolinsky Principle in agricultural landscapes. The former two (I and L) build on the proposition of Ramon Margalef (2006 [1973]) that a differentiated pattern in the spatial distribution of external energy flows turns ecosystems into agroecosystems, giving rise to complex landscape mosaics with heterogeneous land covers which offer differentiated habitats to non-domesticated species (Loreau et al., 2010; Jackson et al., 2011). Applied to Mediterranean agroecosystems, and tested with actual data on biodiversity locations, the ELIA model has proved to have a good predictive capacity (Marull et al., 2019b). Further ongoing improvements proposed by Manuel González de Molina and Gloria Guzmán aim at differentiating the agroecological importance of the diverse paths taken by matter-energy flows across the graph.

Another development of our agrarian metabolism approach is SAFRA, a prospective nonlinear programming model used to forecast the different landscape configurations that will arise from optimizing the uses of land, labour and other resources from a set of site-specific biophysical restrictions and capacities of agroecosystems, according to different social aims, including the diet to be provided (Padró et al., 2019, 2020). The model is devised to improve deliberative processes of agroecology transitions towards more sustainable agri-food systems and territories (González de Molina et al., 2019; González de Molina & Lopez-Garcia, 2021).

Other advances have started calculating social and gender inequalities from a socio-metabolic point of view. Manuel González de Molina and Víctor Toledo (2011, 2014) began this task in their books, and some SFS researchers have followed this approach (Gizicki-Neundlinger et al., 2017; Gizicki-Neundlinger & Güldner, 2017; Marco et al., 2020a, b). Finally, the socio-metabolic approach developed so far has started to combine environmental, social and economic dimensions in the recent book *The Social Metabolism of Spanish Agriculture* (González de Molina et al., 2020). There is a long way to go in making socio-metabolic calculations from these social and gender perspectives – an avenue for young scholars to advance.

References

Aguilera, E., Guzmán, G. I., Infante-Amate, J., Soto, D., García-Ruiz, R., Herrera, A., Villa, I., Torremocha, E., Carranza, G., & González de Molina, M. (2015). *Embodied energy in agricultural inputs. Incorporating a historical perspective*. Documentos de Trabajo de la Sociedad Española de Historia Agraria 1507, Sociedad Española de Historia Agraria. Available online at: https://ideas.repec.org/p/seh/wpaper/1507.html

Antunes, C., Juquin, P., Kemp, P., Stengers, I., Telkämper, W., & Wolf, F. O. (1990). Manifiesto ecosocialista. *Mientras Tanto, 41*, 59–171. https://www.jstor.org/stable/27819829

Ayres, R. U., & Simonis, U. E. (Eds.). (1994). *Industrial metabolism: Restructuring for sustainable development*. United Nations University Press. https://library.wur.nl/WebQuery/titel/905628

Campos, P., & Casado, J. M. (2004). *Cuentas ambientales y actividad económica*. Colegio de Economistas.

Campos, P., & López-Linage, J. (1997). *Renta y naturaleza en Doñana*. Icaria.

Campos, P., & Naredo, J. M. (1980). La energía en los sistemas agrarios. *Agricultura y Sociedad, 15*, 17–113. http://www.mapa.es/ministerio/pags/biblioteca/revistas/pdf_ays/a015_02.pdf

Carpintero, O. (2005). *El metabolismo de la economía española: Recursos naturales y huella ecológica (1955–2000)*. Fundación César Manrique. https://fcmanrique.org/fcm-publicacion/el-metabolismo-de-la-economia-espanola-2/?cpg=2&lang=es

Chayanov, A. V. (1966 [1925]). *The theory of peasant economy*. The University of Wisconsin Press.

Cobo, F., Cruz, S., & González de Molina, M. (1992). Privatización del monte y protesta campesina en Andalucía Oriental (1836–1920). *Agricultura y Sociedad, 65*, 253–302. https://www.mapa.gob.es/ministerio/pags/biblioteca/revistas/pdf_ays/a065_07.pdf

Cussó, X., Garrabou, R., Olarieta, J. R., & Telllo, E. (2006a). Balances energéticos y usos del suelo en la agricultura catalana: una comparación entre mediados del siglo XIX y finales del siglo XX. *Historia Agraria, 40*, 471–500. https://www.historiaagraria.com/FILE/articulos/xavier-balances.pdf

Cussó, X., Garrabou, R., & Tello, E. (2006b). Social metabolism in an agrarian region of Catalonia (Spain) in 1860–1870: Flows, energy balance and land use. *Ecological Economics, 58*(1), 49–65. https://doi.org/10.1016/j.ecolecon.2005.05.026

Fisher-Kowalski, M. (1998). Society's metabolism: The intellectual history of materials flow analysis, part I, 1860–1970. *Journal of Industrial Ecology, 2*(1), 61–77. https://doi.org/10.1162/jiec.1998.2.1.61

Fisher-Kowalski, M., & Hüttler, W. (1998). Society's metabolism: The intellectual history of materials flow analysis, part II, 1970–1998. *Journal of Industrial Ecology, 2*(4), 107–136. https://doi.org/10.1162/jiec.1998.2.4.107

Flores Galindo, A. & Martínez Alier, J. (1988). Agricultura, alimentación y medio ambiente en Perú. *Mientras Tanto, 34*, 79–89. https://www.jstor.org/stable/27819708

Font, C., Padró, R., Cattaneo, C., Marull, J., Tello, E., Alabert, A., & Farré, M. (2020). How farmers shape cultural landscapes: Dealing with information in farm systems (Vallès County, Catalonia, 1860). *Ecological Indicators, 112*, 106104. https://doi.org/10.1016/j.ecolind.2020.106104

Foster, J. B., & Burkett, P. (2008). The Podolinsky myth: An obituary. Introduction to 'Human Labour and Unity of Force' by Sergei Podolinsky. *Historical Materialism, 16*, 115–161. https://johnbellamyfoster.org/wp-content/uploads/2014/07/The-Podolinsky-Myth-An-Obituary-Introduction-to-Human-Labour-and-Unity-of-Force-by-Sergei-Podolinsky.pdf

Garrabou, R., & Naredo, J. M. (Eds.). (1996). *La fertilización en los sistemas agrarios: Una perspectiva histórica*. Fundación Argentaria/Visor. http://www.fcmanrique.org/recursos/publicacion/4d38453a1a%20fertilizacion%20en%20los%20sistemas%20agrarios%20(parte%201).pdf

Garrabou, R., & Naredo, J. M. (Eds.). (1999). *El agua en los sistemas agrarios: Una perspectiva histórica*. Fundación Argentaria/Visor. http://fcmanrique.org/fcm-publicacion/el-agua-en-los-sistemas-agrarios-una-perspectiva-historica-2/?lang=es

Garrabou, R., & Naredo, J. M. (Eds.). (2008). *El paisaje en perspectiva histórica: Formación y transformación del paisaje en el mundo mediterráneo*. SEHA-PUZ. http://seha.info/FILE/monografias/seha_monografia_6_ramon_garrabou_y_jose_manuel_naredo-el_paisaje_en_perspectiva_historica_formacion_y_transformacion_del_paisaje_en_el_mundo_mediterraneo.pdf

Gaviria, M., Naredo, J. M., & Serna, J. (coords.) (1978). *Extremadura saqueada: Recursos naturales y autonomía regional*. Ruedo Ibérico—Ibérica de Ediciones y Publicaciones. http://www.elrincondenaredo.org/Biblio-1978-Extremadura_saqueada.pdf

Giampietro, M. (1997). Socioeconomic constraints to farming with biodiversity. *Agriculture, Ecosystems and Environment, 62*(2–3), 145–167. https://doi.org/10.1016/S0167-8809(96)01137-1

Giampietro, M. (2004). *Multi-scale integrated analysis of agroecosystems*. CRC Press.

Giampietro, M., & Pimentel, D. (1991). Energy efficiency: Assessing the interaction between humans and their environment. *Ecological Economics, 4*, 117–144. https://doi.org/10.1016/0921-8009(91)90025-A

Giampietro, M., Bukkens, S. G. F., & Pimentel, D. (1994). Models of energy analysis to assess the performance of food systems. *Agricultural Systems, 45*, 19–41. https://doi.org/10.1016/S0308-521X(94)90278-X

Gingrich, S., Cunfer, G., & Aguilera, E. (2018a). Agroecosystem energy transitions: Exploring the energy-land nexus in the course of industrialization. *Regional Environmental Change, 18*, 929–936. https://doi.org/10.1007/s10113-018-1322-x

Gingrich, S., Marco, I., Aguilera, E., Padró, R., Cattaneo, C., Cunfer, G., Guzmán, G. I., MacFadyen, J., & Watson, A. (2018b). Agroecosystem energy transitions in the old and new worlds: Trajectories and determinants at the regional scale. *Regional Environmental Change, 18*, 1089–1101. https://doi.org/10.1007/s10113-017-1261-y

Gizicki-Neundlinger, M., & Güldner, D. (2017). Surplus, scarcity and soil fertility in pre-industrial Austrian agriculture: The sustainability costs of inequality. *Sustainability, 9*(2), 265. https://doi.org/10.3390/su9020265

Gizicki-Neundlinger, M., Gingrich, S., Güldner, D., Krausmann, F., & Tello, E. (2017). Land, food and labour in pre-industrial agro-ecosystems: A socio-ecological perspective on early 19th century seigneurial systems. *Historia Agraria, 71*, 37–78. https://www.historiaagraria.com/FILE/articulos/HA71_web_Gizicki-Neundlinger_etal.pdf

González de Molina, M. (2000). De la "cuestión agraria" a la "cuestión ambiental" en la historia agraria de los noventa. *Historia Agraria, 22*, 19–36. https://www.historiaagraria.com/FILE/articulos/HA22_gonzalez.pdf

González de Molina, M. (2001). Condicionamientos ambientales del crecimiento agrario español (siglos XIX y XX). In J. Pujol et al. (Eds.), *El pozo de todos los males: Sobre el atraso en la agricultura española contemporánea* (pp. 43–94). Crítica.

González de Molina, M. (2002). Environmental constraints on agricultural growth in 19th century Granada (Southern Spain). *Ecological Economics, 41*(2), 257–270. https://doi.org/10.1016/S0921-8009(02)00030-7

González de Molina, M. (2020a). La «cuestión agraria» alrededor de 1975. In A. Díaz Geada & L. Gernández Prieto (coords.) (Eds.), *Senderos de la historia: Miradas y actores en medio siglo de historia rural* (pp. 45–58). Comares.

González de Molina, M. (2020b). Strategies for scaling up agroecological experiences in the European Union. *International Journal of Agriculture and Natural Resources, 47*(3), 187–203. https://doi.org/10.7764/ijanr.v47i3.2257

González de Molina, M., & Guzmán, G. I. (2017). On the Andalusian origins of agroecology in Spain and its contribution to shaping agroecological thought. *Agroecology and Sustainable Food Systems, 41*(3–4), 256–275. https://doi.org/10.1080/21683565.2017.1280111

González de Molina, M., & Lopez-Garcia, D. (2021). Principles for designing agroecology-based local (territorial) agri-food systems: A critical revision. *Agroecology and Sustainable Food Systems, 45*(7), 1050–1082. https://doi.org/10.1080/21683565.2021.1913690

González de Molina, M., & Martínez Alier, J. (Eds.). (1993). *Historia y ecología. Ayer,* 11. Marcial Pons.

González de Molina, M., & Toledo, V. M. (2011). *Metabolismos, naturaleza e historia: Hacia una teoría de las transformaciones socioecológicas.* Icaria.

González de Molina, M., & Toledo, V. M. (2014). *The social metabolism: A socio-ecological theory of historical change.* Springer [a second extended edition is scheduled for 2023].

González de Molina, M., Petersen, P. F., Garrido Peña, F., & Caporal, F. R. (2019). *Political agroecology: Advancing the transition to sustainable food systems.* CRC Press. https://doi.org/10.1201/9780429428821

González de Molina, M., Soto Fernández, D., Guzmán Casado, G., Infante-Amate, J., Aguilera Fernández, E., Vila Traver, J., & García Ruiz, R. (2020). *The social metabolism of Spanish agriculture, 1900–2008: The Mediterranean way towards industrialization.* Springer Open Access. https://www.springer.com/gp/book/9783030208998

Guzmán, G. I., & González de Molina, M. (2006). Sobre las posibilidades del crecimiento agrario en los siglos XVIII, XIX y XX: Un estudio de caso desde la perspectiva energética. *Historia Agraria, 40*, 437–470. https://www.historiaagraria.com/FILE/articulos/gloriasobre.pdf

Guzmán, G. I., & González de Molina, M. (2015). Energy efficiency in agrarian systems from an agroecological perspective. *Agroecology and Sustainable Food Systems, 39*(8), 924–952. https://doi.org/10.1080/21683565.2015.1053587

Guzmán, G. I., & González de Molina, M. (Eds.). (2017). *Energy in agroecosystems: A tool for assessing sustainability.* CRC Press.

Guzmán, G. I., Aguilera, E., Soto, D., Cid, A., Infante, J., García-Ruiz, R., Herrera, A., Villa, I., & González de Molina, M. (2014). *Methodology and conversion factors to estimate the net primary productivity of historical and contemporary agroecosystems.* Documentos de Trabajo de la Sociedad Española de Historia Agraria 1407, Sociedad Española de Historia Agraria. https://ideas.repec.org/cgi-bin/htsearch?ul=seh%2Fwpaper&q=Guzm%C3%A1n

Guzmán, G. I., González de Molina, M., Soto Fernández, D., Infante-Amate, J., & Aguilera, E. (2018). Spanish agriculture from 1900 to 2008: A long-term perspective on agroecosystem energy from an agroecological approach. *Regional Environmental Change, 18*, 995–1008. https://doi.org/10.1007/s10113-017-1136-2

Hercher-Pasteur, J., Loiseau, E., Sinfort, C., & Hélias, A. (2020). Energetic assessment of the agricultural production system. A review. *Agronomy for Sustainable Development, 40*, 29. https://doi.org/10.1007/s13593-020-00627-2

Herrera, A., González de Molina, M., & Soto, D. (2010). «El Pacto Andaluz por la Naturaleza» (1985): La confluencia del movimiento campesino y el movimiento ecologista. *Historia Agraria, 50*, 121–147. https://www.historiaagraria.com/FILE/articulos/ha50_pacto_andaluz.pdf

Hobsbawm, E. J. (1959). *Primitive rebels: Studies in archaic forms of social movement in the 19th and 20th centuries.* Manchester University Press.

Jackson, L. E., Pulleman, M. M., Brussaard, L., Bawa, K. S., Brown, G. G., Cardoso, I. M., de Ruiter, P. C., García-Barrios, L., Hollander, A. D., Lavelle, P., Ouédraogo, E., Pascual, U., Setty, S., Smukler, S. M., Tscharntke, T., & Van Noordwijk, M. (2011). Social-ecological and regional adaptation of agrobiodiversity management across a global set of research regions. *Global Environmental Change, 22*(3), 623–639. https://doi.org/10.1016/j.gloenvcha.2012.05.002

Kay, C. (1974). Comparative development of the European manorial system and the Latin American hacienda system. *Journal of Peasant Studies, 2*(1), 69–98. https://doi.org/10.1080/03066157408437916

Kay, C. (1977). The Latinamerican hacienda system: Feudal or capitalist? *Jahrbuch für Geschichte Lateinamerikas—Anuario de Historia de América Latina, 14*, 369–377. https://journals.sub.uni-hamburg.de/hup1/index.php/jbla/degruyter

Leach, G. (1976). *Energy and food production*. IPC Science and Technology Press.

López-Linage, J. (2007). *Modelo productivo y población campesina del occidente asturiano, 1940–1975*. Ministerio de Agricultura [with a foreword of Enric Tello].

Loreau, M., Mouquet, N., & Gonzalez, A. (2010). Biodiversity as spatial insurance in heterogeneous landscapes. *Proceedings of the National Academy of Sciences, 100*(22), 12765–12770. https://doi.org/10.1073/pnas.2235465100

Marco, I., Padró, R., & Tello, E. (2020a). Dialogues on nature, class and gender: Revisiting socio-ecological reproduction in past organic advanced agriculture (Sentmenat, Catalonia, 1850). *Ecological Economics, 169*, 106395. https://doi.org/10.1016/j.ecolecon.2019.106395

Marco, I., Padró, R., & Tello, E. (2020b). Labour, nature, and exploitation: Social metabolism and inequality in a farming community in mid-19th century Catalonia. *Journal of Agrarian Change, 20*(3), 408–436. https://doi.org/10.1111/joac.12359

Margalef, R. (2006 [1973]). Ecological theory and prediction in the study of the interaction between man and the rest of the biosphere. In H. Sicht (Ed.), *Ökologie und Lebensschutz in internationaler Sicht*. Rombach (pp. 307–353) [republished and translated into Catalan, Spanish, and English in the journal *Medi Ambient. Tecnologia i Cultura, 38*, 39–61, 82–94, 114–125. http://www.gencat.cat/mediamb/publicacions/Memories/Revista/cat38.pdf]

Martínez Alier, J. (1968). *La estabilidad del latifundismo*. Ruedo Ibérico.

Martínez Alier, J. (1971). *Labourers and landowners in Southern Spain*. George Allen and Unwin.

Martínez Alier, J. (1977). *Haciendas, plantations and collective farms*. Frank Cass.

Martínez Alier, J. (1978a). La actualidad de la reforma agraria. *Agricultura y Sociedad, 7*, 223–243. https://www.mapa.gob.es/ministerio/pags/biblioteca/revistas/pdf_ays/a007_09.pdf

Martínez Alier, J. (1978b). Renda de la terra, explotació i excedent. *Estudis d'Història Agrària, 1*, 38–63. https://raco.cat/index.php/EHA/article/download/99499/145491

Martínez Alier, J. (1980). La crisis energética y la agricultura moderna. *Boletín de Información sobre Energía Nuclear, 11-13*, 11–16.

Martínez Alier, J. (1982). L'anàlisi energètica i la ciència econòmica. *Mientras Tanto, 12*, 47–57. https://www.jstor.org/stable/27819337

Martínez Alier, J. (1984). *L'ecologisme i l'economia*. Edicions 62.

Martínez Alier, J. (1985a). Réplica a mis críticos. *Mientras Tanto, 23*, 37–43. https://www.jstor.org/stable/27819505

Martínez Alier, J. (1985b). La base social del ecologismo de izquierda: ¿un neopopulismo ecológico? *Mientras Tanto, 25*, 21–28. https://www.jstor.org/stable/27819544

Martínez Alier, J. (1987a). *Ecological economics: Energy, society and environment*. Blackwell.

Martínez Alier, J. (1987b). Utopismo ecológico: Popper-Lynkeus y Ballod-Atlanticus. *Mientras Tanto, 33*, 71–85. https://www.jstor.org/stable/27819689

Martínez Alier, J. (1988). El marxismo y la economía ecológica. *Mientras Tanto, 35*, 127–147. https://www.jstor.org/stable/27819730

Martínez Alier, J. (1990). La interpretación ecologista de la historia socioeconómica: algunos ejemplos andinos. *Historia Social, 7*, 137–162. https://dspace.unia.es/bitstream/handle/10334/597/12JVIII.pdf?sequence=1

Martínez Alier, J. (1993). Hacia una historia socioecológica: algunos ejemplos andinos. In: Sevilla Guzmán, E., González de Molina, M. (eds.) (1993). *Ecología, campesinado e historia*. Ediciones La Piqueta, pp. 219–256.

Martínez Alier, J. (Ed.). (1995). *Los principios de la Economía Ecológica: Textos de P. Geddes, S. A. Podolinsky y F. Soddy*. Fundación Argentaria—Visor. http://fcmanrique.org/fcm-publicacion/los-principios-de-la-economia-ecologica-2/?lang=es

Martínez Alier, J. (2002). *The environmentalism of the poor: A study of ecological conflicts and valuation*. Edward Elgar (Spanish version in Barcelona: Icaria, 2005).

Martínez Alier, J. (2019). *Demà serà un altre dia: Una vida fent economía ecològica i ecología política*. Icaria.

Martínez Alier, J., & González de Molina, M. (Eds.). (2011). *Naturaleza transformada*. Icaria.

Martínez Alier, J., & Naredo, J. M. (1982). A Marxist precursor of energy economics: Podolinsky. *Journal of Peasant Studies, 9*(2), 207–224. https://doi.org/10.1080/03066158208438162

Marull, J., Font, C., Padró, R., Tello, E., & Panazzolo, A. (2016). Energy–landscape integrated analysis: A proposal for measuring complexity in internal agroecosystem processes (Barcelona Metropolitan Region, 1860–2000). *Ecological Indicators, 66*, 30–46. https://doi.org/10.1016/j.ecolind.2016.01.015

Marull, J., Cattaneo, C., Gingrich, S., González de Molilna, M., Guzmán, G. I., Watson, A., MacFadyen, J., Pons, M., & Tello, E. (2019a). Comparative Energy-Landscape Integrated Analysis (ELIA) of past and present agroecosystems in North America and Europe from the 1830s to the 2010s. *Agricultural Systems, 175*, 46–57. https://doi.org/10.1016/j.agsy.2019.05.011

Marull, J., Herrando, S., Brotons, L., Melero, Y., Pino, J., Cattaneo, C., Pons, M., Lobet, J., & Tello, E. (2019b). Building on Margalef: Testing the links between landscape structure, energy and information flows driven by farming and biodiversity. *Science of the Total Environment, 674*, 603–614. https://doi.org/10.1016/j.scitotenv.2019.04.129

Naredo, J. M. (1971). *La evolución de la agricultura en España. Desarrollo capitalista y crisis de las formas de producción tradicionales*. Estela [3ª ed. corregida, Barcelona, Laia, 1978, 4ª edición, 1940–1990 corregida y ampliada, Universidad de Granada, 1996].

Naredo, J. M. (1973, February 24). El mito del crecimiento. *Cambio 16*, 59.

Naredo, J. M. (1978a). Ideología y realidad en el campo de la reforma agraria. *Agricultura y Sociedad, 7*, 199–221. http://www.elrincondenaredo.org/Biblio-AYS-1978-n07.pdf

Naredo, J. M. (1978b). La visión tradicional del problema del latifundio y sus limitaciones. En: Anes Álvarez, G. et al. (ed.), *La economía agraria en la historia de España: Propiedad, Explotación, comercialización, rentas*. Alfaguara, pp. 237–244.

Naredo, J. M. (1981). La estafa nuclear y sus propagandistas. *Revista Bicicleta, 41-42*, 49–52. http://www.elrincondenaredo.org/Biblio-CRI-1981-La%20estafa%20nuclear%20y%20sus%20propagandistas.pdf

Naredo, J. M., & González de Molina, M. (2002). Reforma Agraria y desarrollo económico en la Andalucía del siglo XX. In M. González de Molina (Ed.), *La Historia de Andalucía a debate (volume II). El campo andaluz* (pp. 88–116). Editorial Anthropos. https://www.centrodeestudiosandaluces.es

Naredo, J. M., & Martínez Alier, J. (1976). Los dos primeros gobiernos de la Monarquía y sus relaciones con el poder económico. *Cuadernos de Ruedo Ibérico, 51–53*, 103–116. http://www.elrincondenaredo.org/Biblio-CRI-1976-51-103.pdf

Naredo, J. M., & Martínez Alier, J. (1979, May–December). La noción de 'fuerzas productivas' y la cuestión de energía. *Cuadernos de Ruedo Ibérico*, 63–66.

Naredo, J. M., Leal, J. L., Leguina, J., & Terrafeta, L. (1975). *La agricultura en el desarrollo capitalista español (1940–1970)*. Siglo XXI [3ª edición actualizada, 1987].

Naredo, J. M., & Campos, P. (1980). Los balances energéticos de la agricultura española. *Agricultura y Sociedad, 15*, 163–255. http://www.mapa.es/ministerio/pags/biblioteca/revistas/pdf_ays/a015_04.pdf

Opschoor, J. B. (1997). Industrial metabolism, economic growth and institutional change. In M. Redclift & G. Woodgate (Eds.), *The international handbook of environmental sociology* (pp. 274–287). Edward Elgar.

Padró, R., Marco, I., Font, C., & Tello. (2019). Beyond Chayanov: A sustainable agroecological farm reproductive analysis of peasant domestic units and rural communities (Sentmenat; Catalonia, 1860). *Ecological Economics, 160*, 227–239. https://doi.org/10.1016/j.ecolecon.2019.02.009

Padró, R., Tello, E., Marco, I., Olarieta, J. R., Grasa, M., & Font, C. (2020). Modelling the scaling up of sustainable farming into agroecology territories: Potentials and bottlenecks at the landscape level in a Mediterranean case study. *Journal of Cleaner Production, 275*, 124043. https://doi.org/10.1016/j.jclepro.2020.124043

Pellegrini, P., & Fernández, R. J. (2018). Crop intensification, land use, and on-farm energy-use efficiency during the worldwide spread of the green revolution. *Proceedings of the National Academy of Sciences, 115*(10), 2335–2340. www.pnas.org/cgi/doi/10.1073/pnas.1717072115

Pimentel, D., & Pimentel, M. (1979). *Food, energy, and society*. Edward Arnold.

Puig, I., Roca, J., Tello, E., & Esquerrà, J. (2003). Greening local taxes in Barcelona to Foster energy efficiency. In S. Ulgiati, M. T. Brown, M. Giampietro, R. A. Herendeen, & K. Mayumi (Eds.), *Reconsidering the importance of energy* (pp. 535–539). SGE Editoriali.

Rappaport, R. A. (1968). *Pigs for the ancestors: Ritual in the ecology of a New Guinea people*. Yale University Press.

Roca, J., Tello, E., & Padilla, E. (2006). Ahorro de agua y tarifas domésticas. In A. Estevan & N. Prat (coords.) (Eds.), *Alternativas para la gestión del agua en Cataluña: Una visión desde la perspectiva de la Nueva Cultura del Agua* (pp. 131–188). Bakeaz.

Sabio, A., & Iriarte, I. (Eds.). (2003). *La construcción histórica del paisaje agrario en España y Cuba*. La Catarata.

Sacristán, M. (1980). ¿Por qué faltan economistas en el movimiento ecologista? In *Boletín de Información sobre Energía Nuclear* (pp. 11–13, 63–67) [reprinted in Sacristán, M. (1987). *Pacifismo, ecología y política alternativa* (pp. 48–63). Icaria]. https://espai-marx.net/?p=12115

Sacristán, M. (1992). Political ecological considerations in Marx. *Capitalism Nature Socialism, 3*(1), 37–48. https://doi.org/10.1080/10455759209358472

Sevilla Guzmán, E., & González de Molina, M. (Eds.). (1993). *Ecología, campesinado e historia*. Ediciones La Piqueta.

Shanin, T. (Ed.). (1971). *Peasants and peasant societies: Selected readings*. Penguin Books.

Shanin, T. (1973). The nature and logic of the peasant economy—I: A generalization. *Journal of Peasant Studies, 1*(1), 63–80. https://doi.org/10.1080/03066157308437872

Shanin, T. (1974). The nature and logic of the peasant economy—II: Diversity and change: III. Policy and intervention. *Journal of Peasant Studies, 1*(2), 186–206. https://doi.org/10.1080/03066157408437883

Sherwin, W. B., & Prat-i-Fornells, N. (2019). The introduction of entropy and information methods to ecology by Ramon Margalef. *Entropy, 21*, 794. https://doi.org/10.3390/e21080794

Soto, D., Infante-Amate, J., Guzmán, G. I., Cid, A., Aguilera, E., García, R., & González de Molina, M. (2016). The social metabolism of biomass in Spain, 1900–2008: From food to feed-oriented changes in the agro-ecosystems. *Ecological Economics, 128*, 130–138. https://doi.org/10.1016/j.ecolecon.2016.04.017

Tello, E. (1999). La formación histórica de los paisajes agrarios mediterráneos: una perspectiva coevolutiva. *Historia Agraria, 19*, 195–212. https://historiaagraria.com/FILE/articulos/HA19_tello.pdf

Tello, E. (2002). La 'guerra del agua y la Tasa Ambiental de Gestión de Residuos Municipales en el área metropolitana de Barcelona. Experiencias y lecciones para el desarrollo de la fiscalidad ecológica. In D. Romano & P. Barrenchea (Eds.), *Instrumentos económicos para la prevención y reciclaje de los residuos urbanos* (pp. 119–132). Bakeaz.

Tello, E. (2003). Leer a Sacristán en el crisol de un nuevo comienzo. In S. López Arnal (Ed.), *Manuel Sacristán Luzón. M.A.R.X.—Máximas, aforismos y reflexiones con algunas variables libres (con prólogo de Jorge Riechmann y epílogo de Enric Tello)* (pp. 457–503). El Viejo Topo.

Tello, E. (2005). Changing course? Principles and tools for local sustainability. In T. Marshall (Ed.), *Transforming Barcelona* (pp. 110–225). Routledge.

Tello, E. (2016). Manuel Sacristán at the onset of ecological Marxism after Stalinism. *Capitalism Nature Socialism, 27*(2), 32–50. https://doi.org/10.1080/10455752.2016.1165272

Tello, E. (2020). La cuestión agraria en perspectiva histórica. Mirando hacia atrás para seguir adelante. In A. Díaz Geada & L. Gernández Prieto (coords.) (Eds.), *Senderos de la historia: Miradas y actores en medio siglo de historia rural* (pp. 111–125). Comares.

Tello, E., Garrabou, R., & Cussó, X. (2006). Energy balance and land use: The making of an agrarian landscape from the vantage point of social metabolism (the Catalan Vallès County in 1860/1870). In M. Agnoletti (Ed.), *The conservation of cultural landscapes* (pp. 42–56). CABI International.

Tello, E., Galán, E., Cunfer, G., Guzmán, G. I., González de Molina, M., Krausmann, F., Gingrich, S., Marco, I., Padró, R., Sacristán, V., & Moreno-Delgado, D. (2015). *A proposal for a workable analysis of Energy Return on Investment (EROI) in agroecosystems. Part I: Analytical approach* (Social ecology working paper 156). Social Ecology Institute. Retrieved from URL: https://www.uni-klu.ac.at/socec/downloads/WP156_web%282%29.pdf.

Tello, E., Galán, E., Sacristán, V., Cunfer, G., Guzmán, G. I., González de Molina, M., Krausmann, F., Gingrich, S., Padró, R., Marco, I., & Moreno-Delgado, D. (2016). Opening the black box of energy throughputs in farm systems: A decomposition analysis between the energy returns to external inputs, internal biomass reuses and total inputs consumed (the Valles County, Catalonia, c.1860 and 1999). *Ecological Economics, 121*, 160–174. https://doi.org/10.1016/j.ecolecon.2015.11.012

Toledo, V. M. (1993). La racionalidad ecológica de la producción campesina. In: Sevilla Guzmán, E., González de Molina, M. (eds.) (1993). *Ecología, campesinado e historia*. Ediciones La Piqueta, pp. 197–218.

Toledo, V. M., & González de Molina, M. (2007). El metabolismo social las relaciones entre la sociedad y la naturaleza. In F. Garrido Peña, M. González de Molina, J. L. Serrano Moreno, & J. L. Solana Ruiz (coords.) (Eds.), *El paradigma ecológico en las ciencias sociales* (pp. 85–112). Icaria.

Vernadsky, V. (2007). *Geochemistry and the biosphere*. Synergetic Press.

Open Access This chapter is licensed under the terms of the Creative Commons Attribution 4.0 International License (http://creativecommons.org/licenses/by/4.0/), which permits use, sharing, adaptation, distribution and reproduction in any medium or format, as long as you give appropriate credit to the original author(s) and the source, provide a link to the Creative Commons license and indicate if changes were made.

The images or other third party material in this chapter are included in the chapter's Creative Commons license, unless indicated otherwise in a credit line to the material. If material is not included in the chapter's Creative Commons license and your intended use is not permitted by statutory regulation or exceeds the permitted use, you will need to obtain permission directly from the copyright holder.

Chapter 10
Multi-scale Integrated Analysis of Societal and Ecosystem Metabolism

Mario Giampietro

10.1 Introduction

The field of ecological economics was born out of the perceived need for a novel framing of the sustainability discussion, breaking away from the obsolete narratives of orthodox economics and the pitfalls of reductionism (Christensen, 1989; Martínez-Alier and Schlüpmann, 1987; Røpke, 2004). Early proponents emphasized the importance of the biophysical dimension of the economic system and the key role of non-equilibrium thermodynamics and systems ecology in shaping societal functioning and development (Georgescu-Roegen, 1971, 1975, 1977; Hall et al., 1986; Odum, 1971; Prigogine, 1980; White, 1943; Zipf, 1941). Multi-Scale Integrated Analysis of Societal and Ecosystem Metabolism (MuSIASEM) was developed precisely to respond to these fundamental challenges. Indeed, MuSIASEM can be aptly described as a semantically open accounting framework for quantifying the complex biophysical interactions between human society and its natural environment that shape the metabolic pattern of social-ecological systems.

MuSIASEM is firmly grounded on the main conceptual pillars that drove the foundation of the field ecological economics:

1. Hierarchy theory provides the recognition that a complex system (such as social-ecological systems) can only be studied by integrating representations referring to different scales of analysis (Allen & Starr, 1982; Giampietro et al., 2006).

M. Giampietro (✉)
Institut de Ciència i Tecnologia Ambientals, Universitat Autònoma de Barcelona, Bellaterra, Spain

Institució Catalana de Recerca i Estudis Avançats (ICREA), Barcelona, Spain
e-mail: Mario.Giampietro@uab.cat

2. Non-equilibrium thermodynamics provides the systemic definition of the state-pressure relation in the representation of the metabolic pattern across different levels of analysis (Prigogine, 1980).
3. Theoretical ecology provides the definition and quantification of concepts such as adaptability and evolution of the identity of social-ecological systems (Odum, 1971).
4. Georgescu-Roegen's biophysical economics, and in particular his flow-fund model, provides for the assessment of the inputs in the extended biophysical production function (Georgescu-Roegen, 1971).

MuSIASEM is a unique approach, both in terms of its process of quantification—based on relational analysis describing impredicative entanglement across scales (Louie, 2009; Rosen, 1991)—and in terms of its applications—the production and use of quantitative information for governance in the form of quantitative storytelling.

In the remainder of this chapter, I first provide a brief history of MuSIASEM. Following, I present an overview of its theoretical foundations and its working. I then cite examples of recent applications. The chapter concludes with a reflection on the current impasse in the field of ecological economics and the potential role of the Barcelona School of Ecological Economics in moving forward.

10.2 Brief History of MuSIASEM

MuSIASEM was first put forward as a multi-scale quantitative accounting framework by Mario Giampietro and Kozo Mayumi in 2000 (Giampietro & Mayumi, 2000a, b). Fruit of previous work on the subject (Giampietro, 1994, 1997; Giampietro et al., 1997; Giampietro & Mayumi, 1997), it aimed to provide a coherent quantification of the biophysical factors determining the feasibility, viability, and desirability of the metabolism of social-ecological systems. Early applications of MuSIASEM focused specifically on the metabolic pattern of food production (Giampietro, 2003) and energy (Giampietro et al., 2009, 2012, 2013b; Ramos-Martín et al., 2007, 2009). Pioneering work on integrating water metabolism into MuSIASEM soon followed (Madrid et al., 2013; Madrid-López & Giampietro, 2015). In 2012–13, a first attempt was made to extend its application to the complex water-energy-food nexus (Giampietro et al., 2013a, 2014), thus paving the way for the development of a comprehensive biophysical picture of the metabolic pattern.

In the EU-funded Horizon2020 project "Moving towards Adaptive Governance in Complexity: Informing Nexus Security" (MAGIC) (2016–2020), a series of important conceptual breakthroughs were made that solved earlier limitations of MuSIASEM related to the openness of social-ecological systems (e.g., trade in a globalized economy) and the quantification of the metabolic pattern across scales. Notably, the use of "metabolic processors," an instrumental tool developed in relational biology, enabled the analysis of impredicative relations over structural and functional elements across different levels of the metabolic pattern of the social-ecological system (Giampietro et al., 2021). This progress resulted in the development of the "MuSIASEM toolkit," a set of defined tools based on the MuSIASEM accounting framework that, used in combination, effectively organizes a multi-level and multi-dimensional quantitative characterization of a metabolic pattern in relation to four criteria of performance: feasibility, viability, desirability, and security (openness) (see, Giampietro et al., 2021).

MuSIASEM became part of the Barcelona School of Ecological Economics in 2007. Nonetheless, being firmly based in complexity theory, it retained its own identity, clearly differentiating itself from the more conventional applications of material (and energy) flow analysis.

10.3 Theoretical Foundation

MuSIASEM explicitly recognizes that "production" and "consumption" are intricately linked and cannot be studied separately—something that is, unfortunately, still routinely done in conventional economic and biophysical analyses. The way a society produces secondary inputs invariably affects the way it consumes secondary inputs in an *impredicative relation*. Hence, as argued by Zipf as early as 1941 (Zipf, 1941), human societies must be studied as complex sociobiological systems. MuSIASEM, therefore, starts out from the basic premise that the metabolism of social-ecological systems is not a mere metaphor; social-ecological systems *are* complex metabolic networks.

Indeed, in MuSIASEM, the external referent for the entanglement between biophysical flows and funds is the metabolic pattern of social-ecological systems (Giampietro, 2018). This means that the various conversions of energy and material inputs taking place in the system—needed to reproduce its structural elements and express the expected functions—impose a set of relations over the profiles of the inputs and the outputs of the system: energy, water, food, other materials on the input side and waste, and emissions and other pollutants on the sink side. The relation between the expression of a given metabolic pattern inside the system (inputs and outputs of the parts), on the one hand, and the requirement of a specific profile of inputs and outputs of the whole exchanged with the environment, on the other hand, is governed by the *state-pressure relation* typical of dissipative structures and well defined in the field of non-equilibrium thermodynamics (Giampietro & Renner, 2021; Glansdorff & Prigogine, 1971; Prigogine, 1980). The state-pressure conceptualization is crucial as it allows a definition and quantification of the DPSIR framework (driving forces, pressures, states, impacts, responses) across different levels and dimensions of analysis.

In the operationalization adopted in MuSIASEM, social-ecological systems are defined as "metabolic networks in which constituent components stabilize each other in an impredicative (self-referential) set of relations in presence of favorable boundary conditions" (Renner et al., 2021). The characteristics of the stabilized metabolic network represent the *state of the system* and the characteristics of the favorable boundary conditions the *associated pressure* exerted on the environment. Such a representation of the state allows for the analysis of characteristics such as the relative size of constituent components (parts), their metabolic rates, and, more in general, a definition of societal identity (Renner et al., 2021). More specifically, the state of a social-ecological system is defined as the specific combination of relations among structural and functional elements (parts). These relations must guarantee an internal dynamic equilibrium between the aggregate requirement of goods and services needed to support the metabolism of the various constituent components of the system and the ability of the internal structural and functional elements to produce and/or import these goods and services.

In the analysis of the state of the system (viability seen from the inside), only the metabolism of secondary flows is observed (i.e., energy carriers, food products, and other material goods consumed in the economic process). On the contrary, in the analysis of the interaction of the system with its local environment (feasibility seen from the outside), the state of the system (what is going on inside the socio-economic system) is coupled with the set of environmental pressures resulting from the interaction of the system with the environment (what is going on in the ecological systems). The analysis of viability and feasibility requires the adoption of two different descriptive domains and different metrics. (For example, in a metabolic analysis, 1 joule of coal—a primary energy source produced by the biosphere—is not "the same" as 1 joule of electricity—an energy carrier or secondary energy produced and consumed in the technosphere). The study of the state-pressure relation demands an accounting system that can track primary flows: flows crossing the interface between the socio-economic and the ecological system, either extracted from primary sources (e.g., coal mines, wind, aquifers, rain) or discharged into primary sinks (e.g., GHG emissions into the atmosphere). Thus, the analysis of primary flows is prerequisite to study the environmental pressures and related environmental impacts (feasibility).

The openness of modern economies through international trade represents a major challenge for the analysis of the state-pressure relation across hierarchical levels/scales of analysis. We can define local feasibility in relation to the state-pressure relation observed inside the boundaries of the social-ecological system. However, we also have to consider the set of externalized pressures to the biosphere outside of the borders of the system. Openness is, therefore, a crucial piece of information for deliberating about resource security and burden shifting. In order to account for the openness of the system in the analytical framework, MuSIASEM distinguishes between:

1. Internal or domestic supply systems operating inside the system, for which we can calculate both feasibility and viability from observed data.
2. Externalized or "virtual" supply systems that are embodied in the imported commodities.

For the latter, we can only define notional representations of the inputs required for producing the imported commodities (relevant for feasibility and viability). These can be assessed by measuring the flows of imported commodities (food, energy, and products), a task that requires the use of yet another metric. The relative share of the internal consumption that is produced internally provides an indication of the "metabolic security" of a society (and hence its degree of biophysical resource security) as well as the burden-shifting outside of its boundaries.

10.4 How Does MuSIASEM Work in Practice?

MuSIASEM has three main practical goals:

1. To generate a useful quantitative characterization of the *state* of the system. This includes an operational definition of: (i) the system that we want to study (the whole), (ii) its parts (those relevant for the purpose of the study), and (iii) the state-space of the system for describing relevant changes, i.e. the set of attributes of the metabolic pattern considered relevant for studying (changes in) the performance of the system.
2. To provide a quantitative characterization of the *state-pressure relation* of the socio-economic system with its ecological environment. This characterization is based on the coupling of two distinct metrics in that the internal process of dissipation (the internal state) must be compatible with the existence of favorable boundary conditions in the environment (an admissible environmental pressure). The state-pressure relation is studied by linking the representation of the processes under human control inside the society (on the socioeconomic side) to the representation of processes beyond human control outside of the society (on the ecological side). A simple example would be to link the process generating the required hydroelectricity (measured in kWh) to the processes generating falling water (measured in terms of kinetic energy). This coupling allows analyzing expected impacts on the environment of the studied pressures.
3. To assess how much the state-pressure relation is altered by *trade*. The option to import allows a social-ecological system to ease:

    (a) The internal requirement of production factors by avoiding the labor and technology (funds) and secondary inputs (flows) required for the production of the imported goods and serves, thus overcoming potential *viability* constraints.
    (b) The resulting environmental pressures, by avoiding the use of local natural resources (e.g., land use) and sink capacity for the production of the imported goods, thus overcoming potential *feasibility* constraints.

    For instance, importing electricity avoids the need of building and operating power plants (end-uses inside the society) and the need for primary energy sources to fuel it (environmental pressure inside the system boundaries). The practice of externalizing required production factors to other social-ecological systems has become so common today in developed countries that there no longer is a direct relation between what is produced and what is consumed inside their borders.

An integrated quantitative representation of the metabolic pattern of social-ecological systems is obtained by combining the different 'lenses' of the analysis across non-equivalent descriptive domains:

1. Through the lens of the *macroscope* we observe and describe the metabolic characteristics of its constituent components (the parts inside the black box), as observed from the inside, using a metric useful to study the *state*. It characterizes the sizes (absolute and relative) of the individual constituent components, their interactions, and the role they play in the expression of the emergent property of the whole—i.e., the observed metabolic state.
2. Through the lens of the *mesoscope* we observe the *openness* of the system, using a metric useful to study the degree of dependence on imports. This permits to identify how much of the total internal consumption of commodities is produced domestically and how much is imported. In this way, we can assess the dependence of the system on imports for a given set of commodities and define a series of 'virtual' supply systems (operating elsewhere) required for producing what is imported. This information is then further elaborated through the lens of the virtualscope (see below).
3. Through the lens of the *microscope* we describe different views of the interaction between the socio-economic and ecological system at the local scale, using a metric useful to study the *state-pressure* relation. This description provides the set of expected profiles of: (i) funds associated with processes under human control, such as human labor, land uses, and power capacity, that define a *size* for the structural and functional elements; (ii) flows associated with processes under human control, i.e., secondary inputs (e.g., electricity, fuels, fertilizers, materials) transformed into secondary outputs (products). Specific combinations of secondary inputs map onto *end-uses* needed for the generated outputs; and (iii) primary flows associated with processes beyond human control, i.e., primary flows derived from environment and the wastes dumped into the biosphere by the *end-uses*. These primary flows map onto *environmental pressures*. A comparison of these pressures with the locally available supply and sink capacity of ecological funds (e.g., flows per unit of land use) provides an indication of the *environmental impacts*.
4. Through the lens of the *virtualscope* we describe the characteristics of a notional set of *virtual processes* that would be required to produce the imported goods and services. Indeed, by combining the information obtained through the mesoscope and the microscope, we can calculate the amount of secondary and primary production factors required by 'virtual supply systems' to produce the imported goods. This can be done in three different ways, depending on the purpose of the analysis:

    (a) Track the countries of origin of the imports and use the observed identities of the metabolic processors of the producing (exporting) countries.
    (b) Generate a notional identity for the metabolic processors of imports based on a representative (average) mix of production processes used to supply that commodity on the global market.

(c) Use the identity of the metabolic processors of the system under study (the local supply system) to calculate the amount of secondary flows (end-uses) and primary flows (environmental pressures) that would be needed to internalize the production of the imported commodities.

Thus, the analytical framework of MuSIASEM characterizes the metabolic pattern of a given social-ecological system based on four integrated sets of data:

1. Internal end-use matrix—the secondary inputs (flow and fund elements) used *inside* the socio-economic system for expressing social practices across levels (the local technical processes needed to reproduce the internal state).
2. External end-use matrix—the secondary inputs (flow and fund elements) used by the socio-economic system *outside* its borders to produce the imports (the externalized technical processes needed to reproduce the internal state).
3. Internal environmental pressure matrix—the primary flows exchanged with the biosphere *inside* the boundaries of the social-ecological system (the local environmental pressures associated with the reproduction of the internal state).
4. External environmental pressure matrix—the primary flows exchanged with the biosphere *outside* the boundaries of the social-ecological system to produce imports (the externalized environmental pressures associated with the reproduction of the internal state).

It is important to recognize that MuSIASEM is a semantically open accounting framework. While this guarantees a broad applicability, it also implies that the specific choices of accounting must be tailored to each specific case selected. A "one size fits all" protocol does not work in metabolic analysis. Further details on how to apply MuSIASEM are available in (Giampietro et al., 2021).

10.5 Selected Applications of MuSIASEM

As mentioned in Sect. 10.2, in the early years, MuSIASEM was predominantly applied to describe the feasibility, viability, and desirability of the societal metabolism of energy and food production. While this type of study remains highly relevant and has gained popularity in societies under severe biophysical pressures, such as China (Geng et al., 2011; Han et al., 2018; Lu et al., 2016) and Latin America (Silva-Macher, 2016), recent years have seen a remarkable expansion of the types of application of MuSIASEM, including:

1. Assessments aimed at improving the quality of the *analysis* of sustainability issues, for example, in relation to biofuels (Ripa et al., 2021a), desalination for irrigation (Serrano-Tovar et al., 2019), arid land crop production (Cabello et al., 2019), energy efficiency (Velasco-Fernández et al., 2020a, b), the aging of fossil energy sources (Parra et al., 2018, 2020), waste management (Chifari et al., 2017), urban metabolism (Pérez-Sánchez et al., 2019) and ecosystem health (Lomas & Giampietro, 2017). This type of study focuses on advancements in the theoretical framework in relation to specific applications.

2. Assessments aimed at improving the quality of the *conceptualization* of sustainability policies, also referred to as "quantitative story-telling," for example, in relation to biofuels (Cadillo-Benalcazar et al., 2021), alternative sources of electricity (Renner & Giampietro, 2020), aquaculture (Cadillo-Benalcazar et al., 2020a), and the circular economy (Giampietro, 2019; Giampietro & Funtowicz, 2020). These applications aim to integrate the 'politico-institutional structure' into the picture, thus responding to the criticisms on MuSIASEM raised by Gerber and Scheidel (2018) and connecting with the limits to growth discourses.
3. Assessments of the socio-economic and environmental pressure exerted by developed countries (EU and USA) on other countries, such as the labor hours (worker equivalent) embodied in EU and US imports (Pérez-Sánchez et al., 2021); the externalization of primary energy source requirements by developed countries (Ripa et al., 2021b); and the externalization of food production by the EU, showing the impossibility of achieving local food security by re-internalizing the production of imported commodities (Cadillo-Benalcazar et al., 2020b; Renner et al., 2020). This type of studies explicitly connects with the ecological distribution conflicts addressed by the Political Ecology stream of the Barcelona school.

MuSIASEM has been criticized for providing a static rather than a dynamic picture of the metabolic pattern of social-ecological systems (Gerber & Scheidel, 2018). As a matter of fact, no model of complex systems can represent real evolutionary change (see the concept of complex time in (Giampietro et al., 2006)). Conventional models can only predict futures by applying the ceteris paribus hypothesis to observed quantitative relations. To avoid this predicament, MuSIASEM uses contingent definitions of the option space of expected metabolic relations between fund and flow elements considering their metabolic characteristics, their relative size, and their combination. In this way, MuSIASEM can assess changes in structural and functional elements at different scales using historic series (comparing apples with apples and oranges with oranges across levels of analysis), as shown in several applications (Giampietro & Mayumi, 2018; Ramos-Martin & Giampietro, 2005; Velasco-Fernández et al., 2015).

10.6 Concluding Remarks

Ecological economics was established out of the belief, shared by a group of interdisciplinary scholars, that both the master narratives and the quantitative models used to describe the sustainability predicament were not only useless but outright misleading when used to inform policy. For this reason, ecological economics called for less simplistic narratives and more transdisciplinary and effective representations of the economy; representations that describe the whole system, the context, and the parts, as well as the interactions between parts/parts, parts/whole, and whole/context. Despite the significant progress made by various scholars in this regard, as eloquently shown by the contributions in this book, recent years have seen conventional economic approaches regain a foothold in the field of ecological economics (Plumecocq, 2014).

Probably the explanation for the relative lack of success of biophysical approaches to the economic process lies in the institutional setting. Given the prolonged economic stagnation of the past decade, policy-makers are thirsty for simplistic models and indicators, and win-win solutions, and many scientists have been happy to satisfy their needs. We are increasingly witnessing a change in the way decisions are made in relation to sustainability: We have moved from 'evidence-based policy' to 'policy-based evidence'. The policy legend of the circular economy is a striking example of this phenomenon (Giampietro & Funtowicz, 2020).

It is time that we recall the wisdom of great scholars. Albert Einstein observed, "Everything should be made as simple as possible, but not simpler," while Joseph Schumpeter remarked, ".. the general reader will have to make up his mind whether he wants *simple* answers to his questions or *useful* ones—in this as in other economic matters he cannot have both" (Schumpeter, 1930). It is the very aim of societal metabolism studies to address the complexity of the sustainability issue from the point of view of both analysis—avoiding the pitfalls of reductionism—and narrative—selecting adequate criteria to guarantee the quality of the process of production and use of scientific information in decision making. The recent focus on the nexus between water-energy-food-environment has clearly shown that there is a direct relation between silo-governance (a main cause of policy failures in the sustainability domain) and the lack of appropriate analytical tools capable of dealing with the complexity of the sustainability issue at hand (Giampietro, 2018). While orthodox economic research develops simplistic models and indicators aimed at the governance of complexity, the Barcelona School of Ecological Economics collectively provides a forward-looking line of research that develops decision support tools for multi-level governance in complexity.

References

Allen, T. F. H., & Starr, T. B. (1982). *Hierarchy: Perspectives for ecological complexity*. University of Chicago Press.
Cabello, V., Renner, A., & Giampietro, M. (2019). Relational analysis of the resource nexus in arid land crop production. *Advances in Water Resources, 130*(January), 258–269.
Cadillo-Benalcazar, J. J., Giampietro, M., Bukkens, S. G. F., & Strand, R. (2020a). Multi-scale integrated evaluation of the sustainability of large-scale use of alternative feeds in salmon aquaculture. *Journal of Cleaner Production, 248*, 119210.
Cadillo-Benalcazar, J. J., Renner, A., & Giampietro, M. (2020b). A multiscale integrated analysis of the factors characterizing the sustainability of food systems in Europe. *Journal of Environmental Management, 271*, 110944.
Cadillo-Benalcazar, J. J., Bukkens, S. G. F., Ripa, M., & Giampietro, M. (2021). Why does the European Union produce biofuels? Examining consistency and plausibility in prevailing narratives with quantitative storytelling. *Energy Research & Social Science, 71*, 101810.

Chifari, R., Renner, A., Lo Piano, S., Ripa, M., Bukkens, S. G. F., & Giampietro, M. (2017). Development of a municipal solid waste management decision support tool for Naples, Italy. *Journal of Cleaner Production, 161.* Accessed at https://doi.org/10.1016/j.jclepro.2017.06.074

Christensen, P. P. (1989). Historical roots for ecological economics – Biophysical versus allocative approaches. *Ecological Economics, 1*(1), 17–36.

Geng, Y., Liu, Y., Liu, D., Zhao, H., & Xue, B. (2011). Regional societal and ecosystem metabolism analysis in China: A multi-scale integrated analysis of societal metabolism(MSIASM) approach. *Energy, 36*(8), 4799–4808.

Georgescu-Roegen, N. (1971). *The entropy law and the economic process* (1971st ed.). Hardvard University Press.

Georgescu-Roegen, N. (1975). Energy and economic myths. *Southern Economic Journal, 41*(3), 347.

Georgescu-Roegen, N. (1977). Bioeconomics, a new look at the nature of economic activity. In L. Junker (Ed.), *The political economy of food and energy* (pp. 105–134). Taylor & Francis.

Gerber, J.-F., & Scheidel, A. (2018). In search of substantive economics: Comparing Today's two major socio-metabolic approaches to the economy – MEFA and MuSIASEM. *Ecological Economics, 144*, 186–194.

Giampietro, M. (1994). Using hierarchy theory to explore the concept of sustainable development. *Futures, 26*(6), 616–625.

Giampietro, M. (1997). The link between resources, technology and standard of living: A theoretical model. In L. Freese (Ed.), *Advances in human ecology* (Vol. 6, pp. 73–128). JAI Press.

Giampietro, M. (2003). *Multi-scale integrated analysis of agroecosystems.* CRC Press.

Giampietro, M. (2018). Perception and representation of the resource nexus at the interface between society and the natural environment. *Sustainability (Switzerland), 10*(7), 1–17.

Giampietro, M. (2019). On the circular bioeconomy and decoupling: Implications for sustainable growth. *Ecological Economics, 162*(May), 143–156.

Giampietro, M., & Funtowicz, S. O. (2020). From elite folk science to the policy legend of the circular economy. *Environmental Science and Policy, 109*, 64–72.

Giampietro, M., & Mayumi, K. (1997). A dynamic model of socioeconomic systems based on hierarchy theory and its application to sustainability. *Structural Change and Economic Dynamics, 8*(4), 453–469.

Giampietro, M., & Mayumi, K. (2000a). Multiple-scale integrated assesment of societal metabolism: Introducing the approach. *Population and Environment, 22*(2), 109–153.

Giampietro, M., & Mayumi, K. (2000b). Multiple-scale integrated assessments of societal metabolism: Integrating biophysical and economic representations across scales. *Population and Environment, 22*(2), 155–210.

Giampietro, M., & Mayumi, K. (2018). Unraveling the complexity of the Jevons Paradox: The link between innovation, efficiency, and sustainability. *Frontiers in Energy Research, 6*(APR). Accessed at https://doi.org/10.3389/fenrg.2018.00026

Giampietro, M., & Renner, A. (2021). The Generation of Meaning and Preservation of Identity in Complex Adaptive Systems: The LIPHE4 Criteria. In D. Braha (Ed), *Unifying themes in complex systems X: Proceedings of the tenth international conference on complex systems* (Series: Springer Proceedings in Complexity). Springer. Accessed at https://doi.org/10.1007/978-3-030-67318-5

Giampietro, M., Bukkens, S. G. F., & Pimentel, D. (1997). The link between resources, technology and standard of living: Examples and applications. In L. Freese (Ed.), *Advances in human ecology* (Vol. 6, pp. 129–199). JAI Press.

Giampietro, M., Allen, T. F. H., & Mayumi, K. (2006). The epistemological predicament associated with purposive quantitative analysis. *Ecological Complexity, 3*(4), 307–327.

Giampietro, M., Mayumi, K., & Ramos-Martin, J. (2009). Multi-scale integrated analysis of societal and ecosystem metabolism (MuSIASEM): Theoretical concepts and basic rationale. *Energy, 34*(3), 313–322.

Giampietro, M., Mayumi, K., & Sorman, A. H. (2012). *The metabolic pattern of societies : Where economists fall short*. Routledge.

Giampietro, M., Aspinall, R. J., Bukkens, S. G. F., Cadillo Benalcazar, J. J., Diaz-Maurin, F., Flammini, A., Gomiero, T., Kovacic, Z., Madrid, C., Ramos-Martin, J., & Serrano-Tovar, T. (2013a). *An innovative accounting framework for the food-energy-water nexus – Application of the MuSIASEM approach to three case studies*. FAO Environment and Natural Resources Management Working Paper, No. 56. Accessed 19 April 2020 at http://www.fao.org/3/i3468e/i3468e.pdf

Giampietro, M., Mayumi, K., & Sorman, A. H. (2013b). *Energy analysis for a sustainable future: Multi-scale integrated analysis of societal and ecosystem metabolism*. Routledge.

Giampietro, M., Aspinall, R., Ramos-Martin, J., & Bukkens, S. G. F. (2014). *Resource accounting for sustainability assessment. The nexus between energy, food, water and land use*. Routledge, cop.

Giampietro, M., J. J. Cadillo Benalcazar, L. J. Di Felice, M. Manfroni, L. Pérez Sánchez, A. Renner, M. Ripa, R. Velasco-Fernández and S. G. F. Bukkens (2021), *Report on the experience of applications of the Nexus structuring space in quantitative story-telling.*. MAGIC (H2020–GA 689669) Project Deliverable 4.4, Revision (Version 2.0). First Published 30 August 2020, Revised 25 January 2021. Accessed at http://magic-nexus.eu/documents/deliverable-44-report-nexus-structuring-space

Glansdorff, P., & Prigogine, I. (1971). *Thermodynamic theory of structure, stability and fluctuations*. Wiley-Interscience.

Hall, C. A. S., Cleveland, C. J., & Kaufmann, R. K. (1986). *Energy and resource quality: The ecology of the economic process*. Wiley.

Han, W., Geng, Y., Lu, Y., Wilson, J., Sun, L., Satoshi, O., Geldron, A., & Qian, Y. (2018). Urban metabolism of megacities: A comparative analysis of Shanghai, Tokyo, London and Paris to inform low carbon and sustainable development pathways. *Energy, 155*, 887–898.

Lomas, P. L., & Giampietro, M. (2017). Environmental accounting for ecosystem conservation: Linking societal and ecosystem metabolisms. *Ecological Modelling, 346*. Accessed at https://doi.org/10.1016/j.ecolmodel.2016.12.009

Louie, A. H. (2009). *More than life itself: A synthetic continuation in relational biology* (Vol. 1). Walter de Gruyter.

Lu, Y., Geng, Y., Qian, Y., Han, W., McDowall, W., & Bleischwitz, R. (2016). Changes of human time and land use pattern in one mega city's urban metabolism: A multi-scale integrated analysis of Shanghai. *Journal of Cleaner Production, 133*, 391–401.

Madrid, C., Cabello, V., & Giampietro, M. (2013). Water-use sustainability in socioecological systems: A multiscale integrated approach. *BioScience, 63*(1). Accessed at https://doi.org/10.1525/bio.2013.63.1.6

Madrid-López, C., & Giampietro, M. (2015). The water metabolism of socio-ecological systems: Reflections and a conceptual framework. *Journal of Industrial Ecology, 19*(5). Accessed at https://doi.org/10.1111/jiec.12340

Martínez-Alier, J., & Schlüpmann, K. (1987). *Ecological economics : Energy, environment, and society*. Blackwell Publishers.

Odum, H. T. (1971). *Environment, power, and society*. Wiley Interscience.

Parra, R., Di Felice, L. J., Giampietro, M., & Ramos-Martin, J. (2018). The metabolism of oil extraction: A bottom-up approach applied to the case of Ecuador. *Energy Policy, 122*(July), 63–74.

Parra, R., Bukkens, S. G. F., & Giampietro, M. (2020). Exploration of the environmental implications of ageing conventional oil reserves with relational analysis. *Science of the Total Environment, 749*, 142371.

Pérez-Sánchez, L., Giampietro, M., Velasco-Fernández, R., & Ripa, M. (2019). Characterizing the metabolic pattern of urban systems using MuSIASEM: The case of Barcelona. *Energy Policy, 124*(March 2018), 13–22.

Pérez-Sánchez, L., Velasco-Fernández, R., & Giampietro, M. (2021). The international division of labor and embodied working time in trade for the US, the EU and China. *Ecological Economics, 180*, 106909.

Plumecocq, G. (2014). The second generation of ecological economics: How far has the apple fallen from the tree? *Ecological Economics, 107*, 457–468.

Prigogine, I. (1980). *From being to becoming : Time and complexity in the physical sciences*. W.H. Freeman.

Ramos-Martin, J., & Giampietro, M. (2005). Multi-scale integrated analysis of societal metabolism: Learning from trajectories of development and building robust scenarios. *International Journal of Global Environmental Issues, 5*(3–4). Accessed at https://doi.org/10.1504/IJGENVI.2005.007993

Ramos-Martín, J., Giampietro, M., & Mayumi, K. (2007). On China's exosomatic energy metabolism: An application of multi-scale integrated analysis of societal metabolism (MSIASM). *Ecological Economics, 63*(1). Accessed at https://doi.org/10.1016/j.ecolecon.2006.10.020

Ramos-Martín, J., Cañellas-Boltà, S., Giampietro, M., Gamboa, G. (2009). 'Catalonia's energy metabolism: Using the MuSIASEM approach at different scales. *Energy Policy, 37*(11). Accessed at https://doi.org/10.1016/j.enpol.2009.06.028

Renner, A., & Giampietro, M. (2020). Socio-technical discourses of European electricity decarbonization: Contesting narrative credibility and legitimacy with quantitative story-telling. *Energy Research & Social Science, 59*, 101279.

Renner, A., Cadillo-Benalcazar, J. J., Benini, L., & Giampietro, M. (2020). Environmental pressure of the European agricultural system: Anticipating the biophysical consequences of internalization. *Ecosystem Services, 46*, 101195.

Renner, A., Louie, A. H., & Giampietro, M. (2021). Cyborgization of modern social-economic systems: Accounting for changes in metabolic identity. In D. Braha (Ed), *Unifying themes in complex systems X: Proceedings of the tenth international conference on complex systems* (Series: Springer Proceedings in Complexity). Springer. Accessed at https://doi.org/10.1007/978-3-030-67318-5

Ripa, M., Cadillo-Benalcazar, J. J., & Giampietro, M. (2021a). Cutting through the biofuel confusion: A conceptual framework to check the feasibility, viability and desirability of biofuels. *Energy Strategy Reviews, 35*, 100642.

Ripa, M., Di Felice, L. J., & Giampietro, M. (2021b). The energy metabolism of post-industrial economies. A framework to account for externalization across scales. *Energy, 214*, 118943.

Røpke, I. (2004). The early history of modern ecological economics. *Ecological Economics, 50*(3–4), 293–314.

Rosen, R. (1991). *Life itself : A comprehensive inquiry into the nature, origin, and fabrication of life*. Columbia University Press.

Schumpeter, J. A. (1930). Preface. In F. Zeuthen (Ed.), *Problems of monopoly and economic welfare* (pp. vii–xiii). Routledge.

Serrano-Tovar, T., Peñate Suárez, B., Musicki, A., de la Fuente Bencomo, J. A., Cabello, V., & Giampietro, M. (2019). Structuring an integrated water-energy-food nexus assessment of a local wind energy desalination system for irrigation. *Science of the Total Environment, 689*, 945–957.

Silva-Macher, J. C. (2016). A metabolic profile of Peru: An application of multi-scale integrated analysis of societal and ecosystem metabolism (MuSIASEM) to the mining sector's exosomatic energy flows. *Journal of Industrial Ecology, 20*(5), 1072–1082.

Velasco-Fernández, R., Ramos-Martín, J., & Giampietro, M. (2015). The energy metabolism of China and India between 1971 and 2010: Studying the bifurcation. *Renewable and Sustainable Energy Reviews, 41*(1). Accessed at https://doi.org/10.1016/j.rser.2014.08.065

Velasco-Fernández, R., Dunlop, T., & Giampietro, M. (2020a). Fallacies of energy efficiency indicators: Recognizing the complexity of the metabolic pattern of the economy. *Energy Policy, 137*, 111089.

Velasco-Fernández, R., Pérez-Sánchez, L., Chen, L., & Giampietro, M. (2020b). A becoming China and the assisted maturity of the EU: Assessing the factors determining their energy metabolic patterns. *Energy Strategy Reviews, 32*, 100562.

White, L. A. (1943). Energy and the evolution of culture. *American Anthropologist, 45*(3), 335–356.

Zipf, G. K. (1941). *National unity and disunity: The nation as a bio-social organism*. Principia Press.

Open Access This chapter is licensed under the terms of the Creative Commons Attribution 4.0 International License (http://creativecommons.org/licenses/by/4.0/), which permits use, sharing, adaptation, distribution and reproduction in any medium or format, as long as you give appropriate credit to the original author(s) and the source, provide a link to the Creative Commons license and indicate if changes were made.

The images or other third party material in this chapter are included in the chapter's Creative Commons license, unless indicated otherwise in a credit line to the material. If material is not included in the chapter's Creative Commons license and your intended use is not permitted by statutory regulation or exceeds the permitted use, you will need to obtain permission directly from the copyright holder.

Chapter 11
Materials Flow Analysis in Latin America

Mario Alejandro Pérez-Rincón

11.1 Introduction

Ecological economics (EE) incorporates the biophysical perspective into economic analysis through the concept of social metabolism, a term adopted from biology, extending the idea of the metabolic profile of living organisms to the functioning of economies and society (Ayres & Kneese, 1969; Ayres & Simonis, 1994; Fisher-Kowalski & Haberl, 2015; Infante-Amate et al., 2017). Material Flow Analysis (MFA) is one of the principal methods to study social metabolism by quantifying the economy in tons (Hák et al., 2012; Martínez-Alier, 2003). MFA is part of the satellite systems of natural resource accounts. It is used by industrial ecology and ecological economics to quantify the natural processes of extraction, exchange, and consumption of natural resources (Vallejo, 2015). In Europe, this method has been used academically and has also been included in official government statistics.

In Latin America and the Caribbean (LAC), material flow studies have been developed by researchers and academics without yet being incorporated into national statistical institutions. Some of the research has focused on the study of individual countries and others have made comparisons between countries in the region. However, a detailed inventory and evaluation of MFA-related academic production for LAC have not yet been carried out to identify its progress, scope, and limitations. This chapter has that purpose. Not only to learn about research

M. A. Pérez-Rincón (✉)
Universidad del Valle – Instituto CINARA, Cali, Colombia
e-mail: mario.perez@correounivalle.edu.co

developments in the region but also to see the influence of the *"Barcelona School"*[1] and that of Professor Joan Martínez-Alier in this field of work.

To achieve these objectives, a systematic and organized literature review was carried out using two search engines: *Google Scholar* and *Scopus*. Literature was searched for in Spanish, Portuguese, and English, using as search words Material Flow Analysis and Material Flow Accounting. These were filtered with the word Latin America and the Caribbean (LAC) and the names of the countries in the region. After excluding unpublished papers and articles where there was little mention of LAC or were more focused on the field of Industrial Ecology or Urban Ecology, the universe of publications analyzed was 47.

The results of the analysis of this inventory of publications are organized in four parts: first, a historical analysis of social metabolism as a pillar concept of MFA is made; then, the articles analyzed are characterized along three lines: the role of the *"Barcelona School"* in the dynamics of publications; the most studied countries and sub-regions; and the most frequent time periods analyzed and their implications for metabolic studies. Subsequently, we show the main findings on the dynamics of the flow of materials from Latin American economies, extracted from the publications evaluated. We conclude with some reflections on the important legacy of the *"Barcelona School"* and of our honored professor Joan Martínez-Alier on the work on MFA in LAC.

11.2 Social Metabolism: A Short History

After the emergence of the concept of "metabolism" associated with the biochemical behavior of cells, it was extended to organs, organisms, and even ecosystems, studying their exchanges of matter and energy (Fisher-Kowalski, 1998; Fisher-Kowalski & Hüttler, 1998). From there it is taken up by Marx, who knew the writings of the Dutch physiologist Jacob Möleschot (1822–1893) on the *"The Cycle of Life"*, extrapolating it to society (Martinez-Alier, 1987; Fischer-Kowalski, 1998).

Social Metabolism (SM) is defined as the way societies organize their exchange of matter and energy with their environment (Fischer-Kowalski, 1998, 2002). As in Marx, this concept is not a metaphor; it is an analogy, as it extends cellular and living organism behavior to society. Human beings, articulated by social relationships and institutions that are organized to guarantee our subsistence and reproduction, extract matter and energy from nature through collective structures and artifacts, and excrete a whole range of wastes or residues (González de Molina & Toledo, 2007).

Robert Ayres and Allen Kneese (1968), recover the notion of the economy as an open system and present a clear idea of the metabolism of nations. However, the author who is key to the development of the concept is Marina Fisher-Kowalski who

---

[1] It refers to the Doctorate of Environmental Sciences in its option of Ecological Economics of the Instituto de Ciencia y Tecnología Ambientales (ICTA) of the Universidad Autónoma de Barcelona (UAB) (Spain).

publishes one of the pioneering texts in SM: "*Society's Metabolism: On the Childhood and Adolescence of a Rising Conceptual Star*" (Fischer-Kowalski, 1997). From these years onwards, their work together with other colleagues (Fisher-Kowalski, 1998; Fisher-Kowalski & Huttler, 1998; Fischer-Kowalski & Haberl, 1997) are the starting point of one of the most robust tools for understanding the complex interactions between society and nature (Infante-Amate et al., 2017). Since the nineties of the last century, SM has grown in terms of followers, methodologies, and fields of action and application, with detailed evidence that allows us to better understand the biophysical functioning of societies (Toledo, 2013).

Several state-of-the-art papers on Anglo-Saxon contributions have appeared (Pauliuk & Hertwich, 2015; Gerber & Scheidel, 2018). In 2017, an article was published that inventories work on SM in several languages and since the late nineteenth century ("*Social metabolism. History, methods and main contributions*", Infante-Amate et al., 2017). This inventory shows that the concept had been used on 10,038 occasions, mostly in the period 2001–2016 (Idem, p. 132). English was until 2005 the main vehicle of dissemination, but from that year onwards, articles and publications in Spanish and Portuguese (both languages combined) were the preferred ones (Idem, p. 132).

Growing academic output has included methodological and instrumental developments to measure a society's metabolism. These try to quantify the flows of energy and materials used, transformed, and produced as pollutants or wastes to and from the environment. For added large-scale studies the methodology used is Economy-Wide MFA (EW-MFA or simply MFA). This provides indicators of extraction, consumption, and trade of materials in a territory (generally a country). There is now a standardized and recognized MFA methodology that has allowed its consolidation and dissemination (Eurostat, 2001, updated 2018). This chapter is aimed at reviewing the state of the art of the academic production on MFA for LAC (MFA-LAC).

11.3 Characterization of the Literature Analyzed[2]

11.3.1 Role of the "Barcelona School" in the Dynamics of MFA-LAC Publications

Chart 11.1 presents the academic production on MFA-LAC according to 3-year periods. It shows a trend with two growing cycles between 2004 and 2012 and with a great dynamic for the period 2016–2018. Of the 47 publications inventoried, 28 are written in English, 15 in Spanish, and 4 in Portuguese.

Academic production began in 2001 with two classic articles analyzing the environmental implications of North-South trade relations from an ecological

---

[2] Support for these results can be found in Annexures A and B.

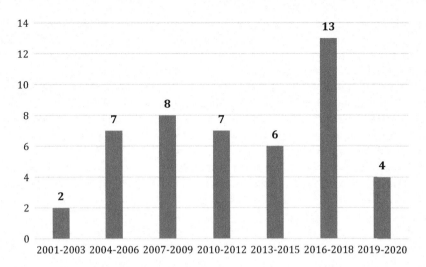

Chart 11.1 MFA-LAC: Number of publications by periods (2001–2020). (Source: MFA-LAC Inventory Database (Scopus-Google Scholar))

economics (EE) perspective.[3] These two articles are a reference for many of the following works identified in the framework of this research for LAC (Muradian & Martinez-Alier, 2001a, b).

Both are produced within the framework of the ICTA-UAB Doctorate in Environmental Sciences created in 1997, under the direction of Martinez-Alier. This fact begins to highlight the importance of the "Barcelona School" in the field of EE and specifically in the area of MFA. This influence will be much greater in the Latin American scenario, as will be shown below.

The following period of academic production on MFA-LAC marks a significant increase between 2004 and 2012. In these 8 years, 22 articles were produced (47% of the total). The influence in this period of the *"Barcelona School"*, is overwhelming. Academics or students linked to ICTA published, alone or in partnership, 77% (17) of the 22 articles. There are two additional facts that accentuate the influence of *"Barcelona School"*: (i) 11 of these articles (55%) are part of doctoral theses carried out at ICTA-UAB, many of them directed by Professor Joan Martínez-Alier; (ii) 32% (7) of the articles are published in the journal *REVIBEC* (Ibero-American Journal of Ecological Economy).

This journal is published in Spanish and Portuguese, which facilitates access to academic publications for Latin Americans. It was created in 2004 at the initiative of the first graduates of the PhD in Environmental Sciences (Fander Falconí, Roldán Muradian, and Jesús Ramos), who also promoted the creation of the *Ibero-American Network of Ecological Economics*. This network helped to promote the founding of

---

[3] Although these two articles do not strictly use MFA, they are pioneers in identifying the socio-environmental implications of North-South trade relations. They are also among the first papers to address *Ecologically Unequal Trade* by analyzing its effects in the region.

2 of the 4 regional societies of ecological economics that are part of the ISEE: the *Mesoamerican and Caribbean Society* (SMEE) founded in 2008 and the *Andean Society* (SAEE) created in 2013.

During 2013–2015, there was a small decrease in the production of articles on the subject, but the influence of the *"Barcelona School"* is maintained. Of the 6 publications made, 3 (50%) are linked to ICTA-UAB". Most important in this period is the beginning of the MFA methodology's institutionalization at the regional level through UNEP and the *Commonwealth Scientific and Industrial Research Organization* (CSIRO). The latter is strongly influenced by the *"Interdisciplinary Institute of Research and Continuing Education"* (IFF) in Vienna, Austria, where Marina Fischer-Kowalski works. Through this cooperation, a database for LAC covering the period 1970–2005 is produced. It also produced the Working Document *"Trends in Material Flows and Resource Productivity in Latin America"* (UNEP-SCIRO, 2013), which is an important reference for this topic in the region.

In the following period (2016–2018), there is a significant increase in the number of publications on the subject, reaching 13 (28% of the total analyzed). During this period, the influence of the *"Barcelona School"* is somewhat reduced, contributing with 4 articles (31% of the total for the period). At the same time, other academic origins gain weight (9), and a thematic shift towards other areas of work closer to industrial ecology can also be observed. These include specific economic sectors (construction, soya, nickel) or smaller territorial units (e.g. Galapagos Islands). The spectrum of publication sources is also diversifying.

In the last period, 2019–2020, 4 articles are published. One linked to the *"Barcelona School"* on social metabolism and environmental conflicts in the Andean countries (Pérez-Rincón et al., 2019); another, which addresses the analysis of the most extended available time series in the flow of International Trade materials for the region: *"Latin America's Open Veins in the Anthropocene Era: A Biophysical Study of Foreign Trade (1900–2016)"* (Infante-Amate et al., 2020). And the final two work conceptually on metabolism (Araújo et al., 2019) and the quantification of potassium stocks and fluxes for Brazil (Sipert et al., 2020).

11.3.2 Countries and Sub-regions Most Studied in the Studies Reviewed

Among the 47 articles evaluated on MFA-LAC, we find the following: 5 analyze LAC as an aggregate whole; 7 select the largest countries; 4 investigate the Andean countries; 1 compares the metabolic patterns of this group of countries with those of Central America (Crespo & Pérez-Rincón, 2017); 4 analyze the material dynamics of the world economy and its influence on LAC. The most studied countries are: Colombia (19), Brazil (16), and Ecuador (15) (Chart 11.2).

The inventory also finds that 17 of the 20 countries in the continental region have some study of MFA, with only Belize, Guyana, and Suriname missing. In contrast,

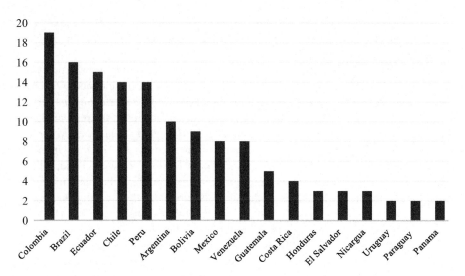

Chart 11.2 Publication of MFA-LAC by country analyzed. (Source: MFA-LAC Inventory Database (Scopus-Google Scholar))

the Caribbean region does not have any studies, although we are aware of an unpublished doctoral thesis on MFA for Cuba (Eisenhut, 2009).

11.3.3 Periodization and Analytical Implications for Social Metabolism Studies

The time period most used in this inventory is the medium term (between 20 and 40 years). Eighteen (38%) of the 47 publications deal with this period and 6 (13%) with a period between 10 and 20 years. The longest term is covered by 6 publications (13%), highlighting Infante et al. (2020) covering a period of 118 years. The sum of publications dealing with the medium and long term gives 30 articles (64% of the total). The preponderance of this time range is understood by the greater explanatory capacity to find trends, patterns, and aggregate metabolic transitions of economies.

Publications covering short periods of time are also identified: 4 that make temporal analyses between 1 and up to 10 years and 10 articles (21% of the total) that use a base year. These last 10 articles are characterized by the fact that they seek to overcome the limitations generated by the high level of aggregation of the MFA methodology. To do so, they complement MFA with other methodologies: *Life Cycle Analysis* (LCA), *Input-Output Analysis* (IOA), *Embodied Energy Analysis* (EEA), *Land-use change* (LUC), *Global Supply Chain* (GSC) and *Exergy Replacement Cost* (ERC). This allows them to better understand the black box of the material and energy processes of the economic sectors analyzed and their environmental impacts.

The analysis of the temporality of these publications shows that there is an exclusion threshold between achievable targets for the MFA. If we want to understand metabolic trends and material transition processes of economies, long-term time series analyses are essential. But these have limitations for understanding in detail the environmental impacts and metabolic processes within economies and their sectors. To go deeper into these two aspects, a combination of short-term studies with a wide range of methodologies complementing MFA is needed.

11.4 Main Findings on Material Flow Dynamics in Latin American Economies

11.4.1 On the Aggregate Dynamics of Material Resource Consumption

In all the works related to the aggregate material dynamics of the countries studied, there is evidence of a continuous and strong absolute materialization of the region's economies. This materialization is growing and permanent, but there is a significant upward break that becomes more pronounced in the 1990s. This change relates to what has been called in LAC, the extractivist model. This model corresponds to the action of massively extracting large quantities of raw materials from the soil, subsoil, or water in order to subsequently commercialize them without much transformation, mainly aimed at the global market (Gudynas, 2013). The renewed impetus that extractivism has taken on has made it a fundamental concept for understanding the contemporary reality of the region, with significant economic, social, and environmental impacts.

The evaluated publications also show that this new material impulse since the 1990s has generated a socio-metabolic transition of the region's economies towards the abiotic. In the 1970s and until the mid-1990s, extraction focused on the appropriation of biotic material (mainly biomass for export). This pattern, although maintained in some countries, has shifted in the last two decades towards economies with a greater emphasis on abiotic extraction (minerals and fossil fuels). However, this specialization bias is not homogeneous throughout the region. It is not true in Central America (Crespo & Pérez-Rincón, 2017) and is less intensive in Mexico.

11.4.2 On International Trade and Unequal Ecological Exchange

Historically, the region has been a net global exporter of natural resources and raw materials to the rest of the world. The longest-term study carried out for LAC (1900–2016) demonstrates this (Infante et al., 2020, p. 187), noting that the "great acceleration" of this biophysical deficit is due to the growth of LAC exports from the 1980s onwards. Neoliberal policies imposed by international financial institutions contributed to this. Total exports of materials increased from 7 to 115 Mt between 1900 and 1980. The jump starts in 1980 and accelerates in the 1990s. In 2016, they amounted to 1035 Mt. (Infante et al., 2020: 188).

The study of the *"Terms of Trade"* from an ecological perspective has been an important research topic of the *"Barcelona School"* and Professor Joan Martínez-Alier. This School, together with Professor Alf Hornborg of Lund University (Sweden), has put forward the hypothesis of *"Ecologically Unequal Terms of Trade"*, an environmental extension of Prebisch's thinking. ECLAC's approach pointed to a historical trend for commodity prices to lose value to imported capital goods. As southern countries specialize in exporting goods rich in natural resources, getting imported goods requires increasing exports. In doing so, they increase pressures on nature, thereby increasing environmental liabilities (Hornborg, 1998; Pérez-Rincón, 2006; Hornborg & Martinez-Alier, 2016).

The works analyzed that address this issue find that this trend cannot be generalized for all countries and that it becomes more relevant for certain periods of time. However, the growth of environmental impacts seems to be decoupled from the behavior of the *"terms of trade"*. For example, during the commodity boom (late 1990s–2009), *"terms of trade"* recovered (Muradian et al., 2012), but environmental pressures did not decrease; on the contrary, they increased. This was facilitated in two ways: (i) regulatory flexibilization of extractive concessions, environmental requirements, and capital inflows policies. This enhanced comparative advantages related to the abundance of natural resources; and (ii) high international prices incentivized the extraction of resources for export, further impacting the environment and fostering environmental conflicts (Pérez-Rincón et al., 2018). In the next phase of the post-boom cycle with declining prices (2010–2018), the terms of trade of exports declined. Despite this, the region's economies increased extraction to compensate for the price reduction, producing a double trade deficit: monetary and biophysical (Samaniego et al., 2017).

11.4.3 On Distributive Ecological Conflicts

One of the frequent themes addressed in several of the articles in this MFA-LAC inventory is that of environmental conflicts generated by increasing metabolic dynamics. This approach is clearly influenced by the *"Barcelona School"*, especially Professor Martínez-Alier. With the strengthening of extractivism in LAC since the 1990s and the increased environmental awareness of the population, environmental conflicts have become more dynamic throughout the region.

The Global Atlas of Environmental Justice (www.ejatlas.org), initiated in 2011 and led by Joan Martínez-Alier, currently (Jan/11/2023) reports 3795 cases of environmental conflicts in the world. Of these, 1041 correspond to LAC (27,4%), when the region's population only represents 8.3% of the world's inhabitants. Almost all of these conflicts in the subcontinent are related to extractive dynamics, generating an over-representation of the region. Several of the articles analyzed address this issue, pointing to a close relationship between the growing dynamics of different metabolic variables and the growth in the number of environmental conflicts (Manrique et al., 2013; Samaniego et al., 2017; Infante-Amate et al., 2017; Pérez-Rincón et al., 2018; Pérez-Rincón et al., 2019).

11.5 Final Reflections

This review of academic literature on MFA in LAC identified 47 papers. The overall balance shows the important influence of the "*Barcelona School*" on the region. Of the 47 articles published, 27 (57% of the total) have its mark; 17 (36%) are part of doctoral theses linked to the PhD in Environmental Sciences at ICTA-UAB; 30% of the papers (14) were published in journals generated within the "*Barcelona School*": *REVIBEC* and *Political Ecology*, the latter created and directed by Joan Martínez Alier since 1991.

Besides, the main topics addressed in a good part of the articles and their main reflections have the stamp of the "*Barcelona School*": absolute materialization of the region's economies, deficit biophysical trade balance, unequal exchange relationships in environmental terms, increasing ecological debts and liabilities, environmental conflicts associated with the growth of social metabolism, and several others. These characteristics adequately describe LAC's economies.

And the "*Barcelona School*", with Professor Martínez Alier at its head, has been pointing this out since the first articles produced in 2001. All this shows the relevance of this School for the region, but specifically, the influence of an academic who has promoted a joint perspective between ecological economics and the political ecology in the world. In this work, there is evidence that much of the intellectual output in these fields in the subcontinent bears the imprint of his thinking. These results are the best tribute to a master who has sown seeds that have germinated favorably in our region: Professor Joan Martínez-Alier.

Annexures

Annex (part 1): Triannual Synthesis of the Main Characteristics Found in the 47 Inventoried Articles on MFA-LAC

Period	Total number of publications	%	Influence, trajectory or institutional link						Work field				Main topic addressed					Methodologies used (Part 1)								
			ICTA-UAB	IFF	ICTA-UAB & IFF	ICTA-UAB & Others	Remaining	Is part of doctoral thesis UAB-ICTA	EE	IE	EE/IE	EE/PE	General economy	International trade	Economic sector/product	Small or regional scale	Conceptual	MFA*	MFA/LCA	MFA/IOA	MFA/MUSIASEM	MFA/SDA/IOA	MFA/WF	MFA/IOA/RME	MFA/TRM/HMF/PERF	MFA/I=PAT
2001–2003	2	4,3%	2				0	2	2					2								2				
2004–2012	22	46,8%	12	1	2	3	4	12	16	4			8	8	5		1	10	3	2	1		1	1	1	2
2013–2015	6	12,8%	3	2	0	0	1	2	6	0			4	1			1	2								
2016–2018	13	27,7%	4	0	0	0	9	1	7	0	5	1	4	3	4	1	1	5	1	2						
2019–2020	4	8,5%	1	0	0	0	3	0	2	0	1	1	2	0	1		1	2								
TOTAL	47	100%	22	3	2	3	17	16	33	4	6	2	18	14	10	1	4	19	4	4	1	2	1	1	1	2
Participation (%)			47,8%	6,4%	4,3%	6,4%	36,2%	34,0%	70,2%	8,5%	12,8%	4,3%	38,3%	29,8%	21,3%	2,1%	8,5%	40,4%	8,5%	4,3%	2,1%	4,3%	2,1%	2,1%	2,1%	4,3%

Annex (part 2): Triannual Synthesis of the Main Characteristics Found in the 47 Inventoried Articles on MFA-LAC

Period	Total number of publications	%	Methodologies used (Part 2)								Type of flow analyzed		Temporality of the analysis					Name of the journals where the article has been published and other types of publications							
			MFA/DS	MFA/EEA	MFA/LUC/IOA/GSC	MFA/DEC	MFA/LUC	MFA/ERC	MFA/Other tools	Theoretical	Inputs	Inputs and outputs	1 year	<10 years	>10 <20 years	>20 <40 years	>40 years	Ecological Economics	Industrial Ecology	Revibec	Ecología Política	Resources, Conservation and Recycling	Other journals	Book or book chapter	Institucional document
2001–2003	2	4,3%	1						2		2			1		1		1					1		
2004–2012	22	46,8%					2		11	1	18	4	5	1	5	10	1	4	5	6	1	3	2	1	
2013–2015	6	12,8%							3	1	5	0				5		1	1	2	1	0	0		1
2016–2018	13	27,7%		1	1	1	1	1	10	1	11	2	4	2	1	2	3	1	0	3	0	0	8	1	0
2019–2020	4	8,5%	0	0	0	1	0	0	1	1	2	1	1	0	0	0	2	1	0	1	0	0	2	0	0
TOTAL	**47**	**100%**	**1**	**1**	**1**	**2**	**3**	**1**	**27**	**4**	**38**	**7**	**10**	**4**	**6**	**18**	**6**	**8**	**6**	**12**	**2**	**3**	**13**	**2**	**1**
Participation (%)			**2,1%**	**2,1%**	**2,1%**	**4,3%**	**6,4%**	**2,1%**	**57,4%**	**8,5%**	**80,9%**	**14,9%**	**21,3%**	**8,5%**	**12,8%**	**38,3%**	**12,8%**	**17,0%**	**12,8%**	**25,5%**	**4,3%**	**6,4%**	**27,7%**	**4,3%**	**2,1%**

Source: MFA-LAC database (Scholar Google and Scopus)

ACRONYM: *ICTA-UAB* Instituto de Ciencias y Tecnologías Ambientales – Universidad Autónoma de Barcelona, *IFF* Interdisciplinary Institute of Research and Continuing Education, *IE* industrial ecology, *PE* political ecology, *MFA* material flow aralysis, *LCA* life cycle analysis, *IOA* input-output analysis, *MUSIASEM* multi-scale integrated analysis of societal and ecosystem metabolism, *SDA* structural decomposition analysis, *WF* water footprint, *RME* raw material equivalents, *TRM* total material requirement, *HMF* hidden material flow, *PERF* potentially environmentally relevant flows, *I* =*PAT* impact = population, affluency, technology, *DS* dynamic of systems, *EEA* embodied energy analysis, *LUC* land-use change, *GSC* global supply chain, *DEC* database of environmental conflicts, *ERC* exergy replacement cost

[a] Money flows are also regularly included

References

Araújo, A., Andrade, D., & Souza, H. (2019). Metabolismo Socioeconômico (MSE): Construção Conceitual e Convergência com a Economia Ecológica (EE). *Revista Iberoamericana de Economía Ecológica, 31*, 127–143.

Ayres, R., & Kneese, A. (1968). Environmental pollution. In *U.S. Congress, Joint Economic committee: Federal programs for the development of human resources* (Vol. 2).

Ayres, R. U., & Kneese, A. V. (1969). Production, consumption, and externalities. *The American Economic Review, 6*, 282–297.

Ayres, R. U., & Simonis, U. E. (1994). *Industrial metabolism: Restructuring for sustainable development*. United Nations University Press.

Crespo-Marín, Z., & Pérez-Rincón, M. (2017). Las economías andinas y centroamericanas vistas desde el metabolismo social: 1970–2013. *Sociedad y Economía, 36*, 53–81.

Eisenhut, S. (2009). *National material flow analysis: Cuba*. Master thesis.

EUROSTAT, Statistical Office of the European Union. (2001). *Economy-wide material flow accounts and derived indicators: A methodological guide*. Eurostat, European Commission.

EUROSTAT. (2018). Economy-wide material flow accounts Handbook 2018 Edition.

Fischer-Kowalski, M. (1997). Society's metabolism: on the childhood and adolescence of a rising conceptual star. In M. Redclift & W. Goodgate (Eds.), *The international handbook of environmental sociology*. Edward Elgar.

Fischer-Kowalski, M. (1998). Society's metabolism. The intellectual history of materials flow analysis, part I, 1860–1970. *Journal of Industrial Ecology, 2*, 61–78.

Fischer-Kowalski, M. (2002). Exploring the history of industrial metabolism. In R. U. Ayres & L. W. Ayres (Eds.), *A handbook of industrial ecology* (pp. 16–26). Edward Elgar Publishing.

Fischer-Kowalski, M., & Haberl, H. (1997). Tons, joules and money: Modes of production and their sustainability problems. *Society and Natural Resources, 10*(1), 61–85.

Fischer-Kowalski, M., & Hüttler, W. (1998). Society's metabolism. The intellectual history of materials flow analysis, part II, 1970–1998. *Journal of Industrial Ecology, 2*(4), 107–136.

Fisher-Kowalski, M., & Haberl, H. (2015). Social metabolism: A metric for biophysical growth and degrowth. In J. Martínez-Alier & R. Muradian (Eds.), *Handbook of ecological economics* (pp. 100–138). Edward Elgar.

Gerber, J. F., & Scheidel, A. (2018). In search of substantive economics: Comparing today's two major socio-metabolic approaches to the economy–MEFA and MuSIASEM. *Ecological Economics, 144*, 186–194.

González de Molina, M., & Toledo, V. (2007). El metabolismo social: las relaciones entre la sociedad y la naturaleza. In F. Garrido, M. González de Molina, J. L. Serrano, & J. L. Solana (Eds.), *El paradigma ecológico en las ciencias sociales* (pp. 85–112). Icaria-Antrazyd.

Gudynas, E. (2013). Mas allá del Nuevo extractivismo: transiciones sostenibles y alternativas al desarrollo. In *Desarrollo en cuestión* (pp. 379–410). Reflexiones desde América Latina.

Hák, T., Moldan, B., & Dahl, A. L. (Eds.). (2012). *Sustainability indicators: A scientific assessment* (Vol. 67). Island Press.

Hornborg, A. (1998). Towards an ecological theory of unequal exchange: Articulating word system theory and ecological economics. *Ecological Economics, 25*, 127–136.

Hornborg, A., & Martínez-Alier. (2016). Ecologically unequal exchange and ecological debt. *Journal of Political Ecology, 23*(1), 2016.

Infante-Amate, J., González de Molina, M., & Toledo, V. (2017). El metabolismo social. Historia, métodos y principales aportaciones. *Revista Iberoamericana de Economía Ecológica, 27*, 130–152.

Infante-Amate, J., Urrego-Mesa, A., & Tello-Aragay, E. (2020). Las venas abiertas de América Latina en la era del antropoceno: un estudio biofísico del comercio exterior (1900–2016). *Diálogos Revista Electrónica de Historia, 21*(2), 177–214.

Manrique, P., Brun, J., González-Martínez, A., Walter, M., & Martínez-Alier, J. (2013). Biophysical performance of Argentina. *Journal of Industrial Ecology, 17*, 4.

Martinez-Alier, J. (1987). *Ecological economics: Energy, environment and society*. Blackwell.

Martinez-Alier, J. (2003). *The environmentalism of the poor: A study of ecological conflicts and valuation*. Edward Elgar Publishing.

Muradian, R., & Martinez-Alier, J. (2001a). South North materials flow: History and environmental repercussions. *Innovation: The European Journal of Social Science Research, 14*(2), 171–187.

Muradian, R., & Martinez-Alier, J. (2001b). Trade and the environment: From a 'Southern' perspective. *Ecological Economics, 36*, 281–297.

Muradian, R., Walter, M., & Martinez-Alier. (2012). Hegemonic transitions and global shifts in social metabolism: Implications for resource-rich countries. Introduction to the special section. *Global Environmental Change, 22*, 559–567.

Pauliuk, S., & Hertwich, E. G. (2015). Socioeconomic metabolism as paradigm for studying the biophysical basis of human societies. *Ecological Economics, 119*, 83–93.

Pérez-Rincón, M. (2006). Colombian international trade from a physical perspective: Towards an ecological "Prebisch thesis". *Ecological Economics, 59*, 519–529.

Pérez-Rincón, Vargas-Morales, & Crespo-Marín. (2018). Trends in social metabolism and environmental conflicts in four Andean countries from 1970 to 2013. *Sustainability Science, 2018*(13), 635–648.

Pérez-Rincón, Vargas-Morales, & Martínez-Alier. (2019). Mapping and analyzing ecological distribution conflicts in Andean countries. *Ecological Economics, 157*(2019), 80–91.

Samaniego, P., Vallejo, M. C., & Martinez-Alier, J. (2017). *Commercial and biophysical deficits in South America, 1990–2013*.

Sipert, S., Cohim, E., & Alves, F. (2020). Identification and quantification of main anthropogenic stocks and flows of potassium in Brazil. *Environmental Science and Pollution Research, 27*, 32579–32593.

Toledo, V. (2013). El metabolismo social: una nueva teoría socioecológica. *Relaciones, 136*(otoño 2013), 41–71.

UNEP-SCIRO. (2013). *Tendencias del flujo de materiales y productividad de recursos en América Latina* (Número de trabajo: DEW/1578/PA).

Vallejo, M. C. (2015). *Perfiles metabólicos de tres economías andinas: Colombia, Ecuador y Perú*. FLACSO.

Open Access This chapter is licensed under the terms of the Creative Commons Attribution 4.0 International License (http://creativecommons.org/licenses/by/4.0/), which permits use, sharing, adaptation, distribution and reproduction in any medium or format, as long as you give appropriate credit to the original author(s) and the source, provide a link to the Creative Commons license and indicate if changes were made.

The images or other third party material in this chapter are included in the chapter's Creative Commons license, unless indicated otherwise in a credit line to the material. If material is not included in the chapter's Creative Commons license and your intended use is not permitted by statutory regulation or exceeds the permitted use, you will need to obtain permission directly from the copyright holder.

Chapter 12
Biophysical Approaches to Food System Analysis in Latin America

Jesus Ramos-Martin and Fander Falconí

12.1 Introduction

The Barcelona School of Ecological Economics provided the students not only with the basics of heterodox economic theory, but also insights from other disciplines, such as biology and ecology, demography, geography, politics, and international relations. Moreover, we learned about the problems of monetary valuation and were introduced to multi-criteria decision aid methods, that allow tackling the issue of incommensurability of values (Martinez-Alier et al., 1998), therefore including different valuation languages. We also learned about post-normal science (Funtowicz & Ravetz, 1994), requiring peer-evaluation and democratizing decision-making, and about the need to use non-equivalent descriptive domains in our analysis (Giampietro et al., 2001) to be able to grasp the complexity involved in describing the economic system.

Biophysical analysis complements the usual economic interpretation of the phenomena of society. Table 12.1 shows the different perspectives (or toolboxes) for examining food systems:

Generally, agricultural systems are analyzed in terms of yields (tons/ha) or economic productivity (USD/ton). However, this can lead to obscuring the interpretations of reality and to mistakes in the design, management, and implementation of public policy. A peasant agricultural system, which is carried out on a small

J. Ramos-Martin (✉)
Department of Economics and Economic History, Institute for Environmental Science and Technology (ICTA), Universitat Autònoma de Barcelona, Barcelona, Spain
e-mail: Jesus.Ramos@uab.cat

F. Falconí
Department of Development, Environment and Territory,
Facultad Latinoamericana de Ciencias Sociales – Sede Ecuador, Quito, Ecuador
e-mail: ffalconi@flacso.edu.ec

Table 12.1 A toolbox for the analysis of food systems: conventional approach vs. biophysical analysis

Conventional approach	Biophysical analysis
Yield (ton/ha)	Yield (ton/ha)
Economic productivity (USD/ton)	Economic productivity (USD/ton)
Labor productivity (ton/h)	Labor productivity (ton/h)
Market price	Energy accounting (conversion efficiency of agriculture, including inputs)
Unequal exchange (prices)	Unequal exchange (prices)
Cost-benefit analysis	Ecologically unequal exchange (tons)
	Caloric unequal exchange (calories)
	Analysis of metabolic profiles
	Land-time budget analysis
	Multi-criteria decision methods

extension of land (a smallholding) could be more efficient than a large one, because it uses fewer fossil energy inputs (machinery, fertilizers, etc.) and more energy from the sun, in addition to its diversification and rotation in various crops. Similarly, biophysical analysis broadens the perspective of international food trade. The ECLAC and Latin American dependency schools, which have been minimized in the international debate nowadays, proposed unequal exchange in prices as a result of asymmetries in international insertion and power relations. The Barcelona School has proposed ecologically unequal exchange (Muradian et al., 2002; Muradian & Giljum, 2007; Muradian & Martinez-Alier, 2001a, b; Pérez-Rincón, 2006; Samaniego et al., 2017; Vallejo, 2010) and caloric unequal exchange (Falconí et al., 2017; Ramos-Martin et al., 2017).

12.2 Food Sovereignty and Complementarity in Latin America

Latin America and the Caribbean are regions with abundant arable land and water, which accounts for 14% of global food production and 23% of the world's exports of agricultural and fisheries commodities (OECD/FAO, 2019). This fact is coupled with the prevalence of obesity and malnutrition in the region. In general terms, malnutrition affects poorer quintiles of population, while obesity is transversal to all income levels of the population (Freire et al., 2014).

The focus of the region as food exporter comes with impacts attached. Ceddia et al. (2013) analyze the intensification of agriculture in South America, concluding that the region would face a "Jevons´ Paradox" like situation; that is, improvements in efficiency would not lead to lowering the use of the resource (land), but the opposite would be true. More efficient techniques of production will use more land to export more. In the case of intensification of agriculture, at the expense of more deforestation occurring.

In sum, although the region is a net exporter, its citizens are not eating enough and properly, and export-oriented food production is having social and environmental impacts that need to be analyzed. This analysis of food systems needs to be multi-faceted, so a single discipline would never be able to cover the different angles involved.

Because of the problems and challenges mentioned above, food self-sufficiency can be considered as one of the main attainments for countries, as exemplified by the European Union, which still nowadays allocates a large fraction of the budget to the Common Agricultural Policy and has reached that goal (Guinea, 2013). South America produces all the food needed to satisfy the caloric requirements of its population and to maintain a balanced diet. However, the region experiences a systematic loss in food self-sufficiency (more imports in relation to domestic consumption, measured in volume) in the period 1961–2011 (Falconi et al., 2015). This result is because food production is ever more oriented to exports, so food exports are increasing dramatically and also changing consumption patterns in the region (see Sect. 12.3).

To see if this move towards export-oriented food production pays or not, we need to introduce the concepts of unequal exchange and ecologically unequal exchange.

The concept of unequal exchange has been critical in the modern economic history of Latin America. The concept focused on the unequal relationship found in traded goods between countries in terms of embodied labor time, which reflected in prices being higher in products coming from developed countries than in products coming from developing countries, that is, showing deteriorating terms of trade for Latin America when compared to the USA or European countries. Unequal exchange obeys asymmetric power relations between the center and the periphery (Amin, 1976; Emmanuel, 1972; Furtado, 1964, 1970; Prebisch, 1950, 1959).

Later, researchers gave a twist to the concept of unequal exchange by including the environmental variable. The exports of developing countries would be intensive in natural resources. However, their prices would not account for the value of environmental externalities involved, implying a de facto transfer of wealth from poor to rich countries. Externalities would not be seen as market failures, but rather as 'cost-shifting successes' in the words of Martínez-Alier (Muradian & Martinez-Alier, 2001a).

When analyzing food production, consumption and related trade patterns under the biophysical approach is very easy to realize that the loss in self-sufficiency driven by export-oriented food production represents a non-desired development path for the region. Population and GDP per capita continue to grow in the region, therefore domestic demand will keep increasing in the long term. One way out is encouraging agricultural complementarity, defined as the contribution each country of the region can make to achieve self-sufficiency as a block (Cango et al., 2023). In this way, not only more food is available at lower costs, but this common goal encourages economic cooperation and integration. Among the potential benefits of this increased integration, one can find: (a) improvements in transport and communications networks; (b) mutual assistance in case of emergencies; (c) improvements in economies of scale; and (d) improvements in food security (Hubbard et al., 1992). One could also add that external vulnerabilities are reduced, and outward flows of currency are also reduced, a critical aspect for economies characterized by lacking access to financial markets and having periodic liquidity problems.

In 2018, South American countries imported 24.8 million tons of food from other regions of the world. 74.1% of the imports belong to just three products: maize (39.1%), wheat (24.6%), and soybeans (10.3%). Should there be agreements in place to encourage intra-regional trade, none of those products would need to be imported, as the region exports even larger quantities of those same products to third countries. Only 19.8% of those 24.6 million tons would need to be imported, mainly wheat, beverages (no alcoholic), beer of barley, potatoes (frozen), feed and meal (gluten), etc., which is surprising, being South America the origin of potatoes. The result is indeed very humble in monetary terms, as trade diversion would only imply about four billion dollars (constant dollar 2015) per year (FAO, 2020), but the benefits of improving self-sufficiency should prevail (Falconi et al., 2015).

12.3 Caloric Unequal Exchange in Latin America

The analysis of food systems in Latin America over the years has evolved and has brought new concepts like that of caloric unequal exchange (Falconí et al., 2017; Ramos-Martin et al., 2017), which builds on the concepts of unequal exchange and ecologically unequal exchange presented in the previous section. Caloric unequal exchange means the deterioration of the terms of trade when calories of food, instead of volume, are used. That is, we calculate the cost in dollars of the calorie exported, versus the cost of the calorie imported.

We tested the hypothesis of Latin America and the Caribbean increasingly exporting food products to the rest of the world at a lower cost to the calorie, expressing a new form of unequal exchange (Falconí et al., 2017; Ramos-Martín et al., 2017). Figure 12.1 presents the cost of one million kcal exported and imported in real terms (left axis) and the ratio between the cost of the exported calorie and the imported calorie (right axis), that is, an approximation to the terms of trade measured in calories.

The hypothesis was proven right, as we found that, even though the terms of trade were still favorable to Latin America, there was a deterioration with a decrease of more than 47.7% in the period 1986–2018. Exports in volume and calories increased by 6.6 times their original size in the period, whereas its monetary value increased by 4.4 times. In the case of imports, they increased by 3.9 times in terms of volume and 3.7 in calories, while they increased by 4.7 times in monetary terms. The surplus had increased fivefold in the 28 years analyzed. Thus, the region was not only increasingly feeding the rest of the world, but it did so at a lower cost over time.

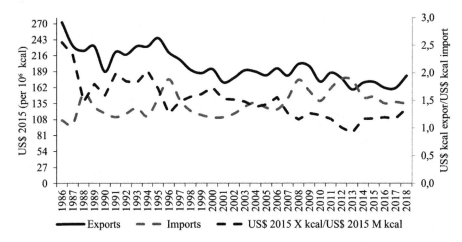

Fig. 12.1 LAC exports, imports, and terms of trade, 1986–2018. (Source: FAO, 2020)

Using calories to measure unequal trade allows us to link food trade with dietary diversity and malnutrition, food sovereignty, and environmental concerns. Volume allows us to link the monetary value of food exports with production and therefore with land use and environmental impacts, as in ecologically unequal exchange. Calories allow us to link the former with nutrition.

The analysis shows how the increase in the volume of exported food products is inducing changes in domestic consumption, towards those export-oriented products. So, not only the dietary diversity is decreasing, with 10 products accounting for 80% of the calorie intake, but also the composition is changing. Cereals, pulses, roots, and tubers are contributing less to the diet, whereas vegetable oils (soybean oil and palm oil) are increasing rapidly, impoverishing the diet (Falconí et al., 2017). More research is needed that links international trade, food production and consumption, nutrition, and health.

12.4 Bottom-Up Approaches: Using Household Types to Assess Sustainable Livelihoods

Biophysical approaches to food production can also be conducted at the level of the farmer, that is, by using bottom-up approaches, that allow to focus on the sustainability of livelihoods and allow for comparison of different production systems. This kind of approach is particularly relevant in the case of developing economies, where a large fraction of domestic demand is covered by small farmers, very often in agroecological systems (Altieri, 2009; Altieri & Nicholls, 2010, 2012; Padilla & Guzmán, 2009; Sevilla Guzmán & Soler Montiel, 2010).

This approach is found, for instance, in a livelihood sustainability assessment of coffee and cocoa producers in the northern Amazon region of Ecuador (Viteri Salazar et al., 2018). This work, built on previous work on a land-time budget analysis (Giampietro, 2003; Gomiero & Giampietro, 2001; Grünbühel et al., 2003; Grünbühel & Schandl, 2005; Pastore et al., 1999), and led by Oswaldo Viteri, another member of the Barcelona School, represents an application of societal metabolism to farming systems, following the work of other members of the School (Gomiero, 2017; Scheidel et al., 2013, 2014). This approach allowed identifying the socioeconomic and environmental restrictions implicit in different land-use patterns, the analysis of how these different land-use patterns improve livelihoods in terms of income, and the identification of how certain public policies have an impact on the income of small-scale producers as well as on the environment.

The information that is generated in this kind of analysis is very easily understood by the participants, which allows for using it with participatory methodologies for a better understanding of the complexities the communities face. It also helps policy-makers with the design of policies addressing the complex realities of rural households, not just focusing on increasing yields, but making explicit the trade-offs involved in production patterns in terms of generation of income and terms of the intensity of use of resources, and therefore the impact upon the environment.

12.5 Conclusion: The Contribution of the Barcelona School

Particularly important for the analysis of economic systems, their surrounding environment, and their evolution over time, is the application of the biophysical approach *a là* Georgescu-Roegen, that is, to describe and analyze the metabolic profiles of activities and sectors so that the trade-offs involved can be discussed in informed debates regarding public policies.

The biophysical perspective on societal metabolism is key for the Barcelona School of Ecological Economics. This means analyzing the economic process, or a particular sector, considering the physical constraints of the availability of resources, but also analyzing the consequences of the process of production and consumption (i.e., waste and pollution) also from a biophysical perspective. In this chapter, we have introduced research conducted by the authors, some with third author members as well as from the Barcelona School, focused on the study of food systems in Latin America. We have also introduced the newly coined concept of caloric unequal exchange, which follows the tradition of the ecologically unequal exchange, and we have also pointed out bottom-up research of food systems, using household categories, which fall under the land-and-time budget analysis tradition within societal metabolism studies.

We believe that the biophysical approach to societal metabolism presented in this chapter is one of the main contributions from the Barcelona School of Ecological Economics, under the leadership of and inspiration from Joan Martínez-Alier.

This work, along with that on water and energy metabolism, sustainable livelihood analysis of farming systems, and the like, are also aligned with international efforts such as the work by the International Resource Panel (IRP, 2019) or the UN System of Environmental Economic Accounting – Ecosystem Accounting, adopted officially by the 52nd United Nations Statistical Commission, on March 2021.

References

Altieri, M. A. (2009). Agroecology, smallfarms, and food sovereignty. *Monthly Review, 61*, 102–113.
Altieri, M. A., & Nicholls, C. I. (2010). Agroecología: potenciando la agricultura campesina para revertir el hambre y la inseguridad alimentaria en el mundo. *Revista de Economia Critica, 10*, 62–74.
Altieri, M. A., & Nicholls, C. I. (2012). Agroecología: Única esperanza para la soberanía alimentaria y la resiliencia socioecológica. *Agroecología, 7*, 65–83.
Amin, S. (1976). *Unequal development: An essay on the social formations of peripheral capitalism*. Monthly Review Press.
Cango, P., Ramos-Martín, J., & Falconí, F. (2023). Toward food sovereignty and self-sufficiency in Latin America and the Caribbean: Opportunities for agricultural complementarity. *Revista de Economia e Sociologia Rural, 61*. https://doi.org/10.1590/1806-9479.2021.251291
Ceddia, M. G., Sedlacek, S., Bardsley, N. O., & Gomez-y-Paloma, S. (2013). Sustainable agricultural intensification or Jevons Paradox? The role of public governance in tropical South America. *Global Environmental Change, 23*, 1052–1063.
Emmanuel, A. (1972). *Unequal exchange: A study of the imperialism of trade*. New Left Books.
Falconi, F., Cadillo Benalcazar, J., Llive, F., Ramos-Martin, J., & Liger, B. (2015). *Pérdida de autosuficiencia alimentaria y posibilidades de complementariedad agrícola en los países de UNASUR* (No. 2015_06). CEPROEC.
Falconí, F., Ramos-Martin, J., & Cango, P. (2017). Caloric unequal exchange in Latin America and the Caribbean. *Ecological Economics, 134*, 140–149. https://doi.org/10.1016/j.ecolecon.2017.01.009
FAO. (2020). *Detailed trade matrix* [WWW Document]. FAOSTAT. URL http://www.fao.org/faostat/en/#data/TM. Accessed 28 Feb 2021.
Freire, W. B., Silva-Jaramillo, K. M., Ramirez-Luzuriaga, M. J., Belmont, P., & Waters, W. F. (2014). The double burden of undernutrition and excess body weight in Ecuador. *American Journal of Clinical Nutrition, 100*, 1636S–1643S. https://doi.org/10.3945/ajcn.114.083766
Funtowicz, S. O., & Ravetz, J. R. (1994). The worth of a song-bird: Ecological economics as a post-normal science. *Ecological Economics, 10*, 197–207.
Furtado, C. (1964). *Development and underdevelopment*. University of California Press.
Furtado, C. (1970). *Obstacles to development in Latin America*. Anchor Books.
Giampietro, M. (2003). *Multi-scale integrated analysis of agro-ecosystems, crop science* (Advances in agroecology). CRC Press. https://doi.org/10.2135/cropsci2006.0003br
Giampietro, M., Mayumi, K., & Bukkens, S. G. F. (2001). Multiple-scale integrated assessment of societal metabolism: An analytical tool to study development and sustainability. *Environment, Development and Sustainability, 3*, 275–307.
Gomiero, T. (2017). Agriculture and degrowth: State of the art and assessment of organic and biotech-based agriculture from a degrowth perspective. *Journal of Cleaner Production, 197*, 1823–1839. https://doi.org/10.1016/j.jclepro.2017.03.237
Gomiero, T., & Giampietro, M. (2001). Multiple-scale integrated analysis of farming systems: The Thuong Lo Commune (Vietnamese uplands) case study. *Population and Environment, 22*, 315–352. https://doi.org/10.1023/A:1026624630569

Grünbühel, C. M., & Schandl, H. (2005). Using land-time-budgets to analyse farming systems and poverty alleviation policies in the Lao PDR. *International Journal of Global Environmental Issues, 5*, 142. https://doi.org/10.1504/IJGENVI.2005.007990

Grünbühel, C. M., Haberl, H., Schandl, H., & Winiwarter, V. (2003). Socioeconomic metabolism and colonization of natural processes in SangSaeng Village: Material and energy flows, land use, and cultural change in Northeast Thailand. *Human Ecology, 31*, 53–86. https://doi.org/10.1023/A:1022882107419

Guinea, M. (2013). El modelo de seguridad alimentaria de la Unión Europea y su dimensión exterior. *Revista UNISCI*, Num. 31, 201–223.

Hubbard, M., Merlo, N., Maxwell, S., & Caputo, E. (1992). Regional food security strategies. The case of IGADD in the Horn of Africa. *Food Policy, 17*, 7–22.

IRP. (2019). *Global resources outlook 2019: Natural resources for the future we want*. UNEP.

Martinez-Alier, J., Munda, G., & O'Neill, J. (1998). Weak comparability of values as a foundation for ecological economics. *Ecological Economics, 26*, 277–286. https://doi.org/10.1016/S0921-8009(97)00120-1

Muradian, R., & Giljum, S. (2007). Physical trade flows of pollution-intensive products: Historical trends in Europe and the world. In A. Hornborg, J. R. McNeill, & J. Martinez-Alier (Eds.), *Rethinking environmental history: World-system history and global environmental change*. AltaMira Press.

Muradian, R., & Martinez-Alier, J. (2001a). Trade and the environment: From a 'Southern' perspective. *Ecological Economics, 36*, 281–297. https://doi.org/10.1016/S0921-8009(00)00229-9

Muradian, R., & Martinez-Alier, J. (2001b). South-North materials flow: History and environmental repercussions. *Innovation: The European Journal of Social Science Research, 14*, 171–187. https://doi.org/10.1080/713670544

Muradian, R., O'Connor, M., & Martinez-Alier, J. (2002). Embodied pollution in trade: Estimating the 'environmental load displacement' of industrialised countries. *Ecological Economics, 41*, 51–67. https://doi.org/10.1016/S0921-8009(01)00281-6

OECD/FAO. (2019). *OECD-FAO agricultural outlook 2019–2028*. OECD/FAO. https://doi.org/10.1787/agr_outlook-2019-en

Padilla, M. C., & Guzmán, E. S. (2009). Aportando a la construcción de la Soberanía Alimentaria desde la Agroecología. *Ecología Política, 38*, 43–51.

Pastore, G., Giampietro, M., & Ji, L. (1999). Conventional and Land-Time budget analysis of rural villages in Hubei province, China. *Critical Reviews in Plant Sciences, 18*, 331–357.

Pérez-Rincón, M. A. (2006). Colombian international trade from a physical perspective: Towards an ecological "Prebisch thesis". *Ecological Economics, 59*, 519–529. https://doi.org/10.1016/j.ecolecon.2005.11.013

Prebisch, R. (1950). *The economic development of Latin America and its principal problems*. United Nations Department of Economic Affairs. Available online athttps://archivo.cepal.org/pdfs/cdPrebisch/002.pdf

Prebisch, R. (1959). American economic association commercial policy in the underdeveloped countries. *The American Economic Review, 49*, 251–273.

Ramos-Martín, J., Falconí, F., & Cango, P. (2017). The concept of caloric unequal exchange and its relevance for food system analysis: The Ecuador case study. *Sustainability (Switzerland), 9*, 2068. https://doi.org/10.3390/su9112068

Samaniego, P., Vallejo, M. C., & Martínez-Alier, J. (2017). Commercial and biophysical deficits in South America, 1990–2013. *Ecological Economics, 133*, 62–73. https://doi.org/10.1016/j.ecolecon.2016.11.012

Scheidel, A., Giampietro, M., & Ramos-Martin, J. (2013). Self-sufficiency or surplus: Conflicting local and national rural development goals in Cambodia. *Land Use Policy, 34*, 342–352. https://doi.org/10.1016/j.landusepol.2013.04.009

Scheidel, A., Farrell, K. N. K. N., Ramos-Martin, J., Giampietro, M., & Mayumi, K. (2014). Land poverty and emerging ruralities in Cambodia: Insights from Kampot province. *Environment, Development and Sustainability, 16*, 823–840. https://doi.org/10.1007/s10668-014-9529-6

Sevilla Guzmán, E., & Soler Montiel, M. (2010). Agroecología y soberanía alimentaria: alternativas a la globalización agroalimentaria. In M. Soler Montiel, C. Guerrero Quintero, & R. Fernández-Baca Casares (Eds.), *Patrimonio Cultural En La Nueva Ruralidad Andaluza* (pp. 190–217). Junta de Andalucía, Consejería de Cultura, Instituto Andaluz del Patrimonio Histórico.

Vallejo, M. C. (2010). Biophysical structure of the Ecuadorian economy, foreign trade, and policy implications. *Ecological Economics, 70*, 159–169. https://doi.org/10.1016/j.ecolecon.2010.03.006

Viteri Salazar, O., Ramos-Martín, J., & Lomas, P. L. (2018). Livelihood sustainability assessment of coffee and cocoa producers in the Amazon region of Ecuador using household types. *Journal of Rural Studies, 62*, 1–9. https://doi.org/10.1016/j.jrurstud.2018.06.004

Open Access This chapter is licensed under the terms of the Creative Commons Attribution 4.0 International License (http://creativecommons.org/licenses/by/4.0/), which permits use, sharing, adaptation, distribution and reproduction in any medium or format, as long as you give appropriate credit to the original author(s) and the source, provide a link to the Creative Commons license and indicate if changes were made.

The images or other third party material in this chapter are included in the chapter's Creative Commons license, unless indicated otherwise in a credit line to the material. If material is not included in the chapter's Creative Commons license and your intended use is not permitted by statutory regulation or exceeds the permitted use, you will need to obtain permission directly from the copyright holder.

Chapter 13
Ecologically Unequal Exchange: The Renewed Interpretation of Latin American Debates by the Barcelona School

Beatriz Macchione Saes

13.1 Introduction

Ecological approaches to unequal exchange were inspired by different critical perspectives that point out the negative effects of economic specialization and free trade on poor countries' development – in contrast to predictions postulated by the comparative advantage classical theory, or the neoclassical trade theory. Bunker's pioneer analysis (1985) on the exploitation of natural resources in the Brazilian Amazon, conducted in the 1970s, was built based on extensions and critical analysis of world systems, dependency, and unequal trade theories (Wallerstein, 1989, 2000; Frank, 1967; Emmanuel, 1972). The author argues that the extraction and export of natural resources from poor regions lacking political power to obtain better trade conditions or economic alternatives imply not only an unequal economic exchange but mainly a transfer of value incorporated in matter and energy to metropolitan centres.

Currently, different social theories still have influence on ecological studies of unequal trade, albeit with important variations. Jorgenson (2006) is based on the dependency theory and world-system analysis. Hornborg (1998, 2009, 2014, 2018) also investigates the asymmetric transfer of matter and energy from peripheries towards centres, but mainly by deepening the biophysical (and thermodynamic) view of the economy. Other current ecological analyses of unequal exchange (Clark & Foster, 2009; Foster & Holleman, 2014) maintain stronger roots in the Marxist labour theory of value.

This chapter investigates the innovative contribution brought by Barcelona School to the ecologically unequal exchange approach, which resulted from the interpretation and update of the Economic Commission for Latin America and the Caribbean (ECLAC) dependency theory, and Latin American environmental debates

---

B. M. Saes (✉)
Universidade Federal de São Paulo, São Paulo, Brazil

in the 1990s. Latin America is recognized for its original environmental perspectives. The colonial experience, the plundering of natural resources, and the maintenance of core-periphery inequalities constitute the basis for the emergence of a "popular environmentalism" and an autonomous environmental thought when compared to Global North perspectives (Martinez-Alier et al., 2016a; Pengue, 2017). Here, I discuss how Latin American environment and development debates were appropriated by Barcelona School and complemented with the biophysical view of unequal trade. The chapter is organized into four sections. After this introduction, I present the ECLAC's and social movements' key concepts and hypotheses around unequal exchange. Next, I discuss Barcelona School's work, aiming to explore its foundations and contributions to analyse Latin America's recent development. I end the chapter with some brief conclusions.

13.2 Unequal Exchange in Latin American Perspectives

In the late 1940s, Raúl Prebisch laid the foundations for Latin American structuralism, based on evidence that peripheral and core countries dealt with unequal conditions and opportunities when exchanging their products through international trade. In contrast to core economies, peripheries are specialized in the exploitation of natural resources based on cheap, easily imitated, and low-productivity technologies, which employ few workers and generate high unemployment and social inequality. Taking into account these structural features, Prebisch (1949) and Singer (1950) raised the hypothesis of a deteriorating trend for peripheral countries in the long run – in terms of trade – due to the decline in primary products' prices when compared to industrial goods prices. As a direct consequence of this hypothesis, peripheral economies need to export increasing product quantities to core economies over time, just to maintain the same amount of imports.

Two dynamics explain this deterioration in terms of trade. First, while core economies (with oligopolistic industries, high productivity, and unionized workers) are able to appropriate productivity gains through higher wages and profits, peripheries (with production facing high competitiveness and wide supply of cheap labour) transfer productivity gains to core regions in the form of decreasing prices (Prebisch, 1949). Second, at certain levels of per capita income, the demand for industrialized goods (or services associated to them) tends to increase faster than the demand for raw materials, likely affecting relative prices in favour of the former (Prebisch, 2000 [1952]).

Although part of Latin American countries has undergone more or less intense industrialization processes since the middle of the twentieth century, global core-periphery dynamics remained unfavourable for countries in the region. ECLAC's analyses indicate that international capital flows also played a role in reinforcing unequal exchange, by improving external accounts of the main core economies. In the early twentieth century, the United Kingdom benefited from the return on previous investments made abroad, paying for a third or more of its imports during the

1920s with rents derived from foreign investments (Prebisch, 1949). Starting in the 1960s, after the rise of the United States as a main core economy, an intense outflow of US capital substantially increased global financial liquidity. Latin America benefited from international capital inflows mainly after the rise of oil prices in 1973, which further expanded the global international liquidity, and the offer of cheap loans to peripheries. However, in face of oil-shock stagnation and inflationary pressures, the United States introduced restrictive monetary policies from 1979, which raised international interest rates. Peripheral indebted economies were no longer able to pay their debts or get new financing (Griffith-Jones & Sunkel, 1986; ECLAC, 2000 [1985]; Furtado, 1992).

Throughout the 1980s, countries in the periphery had to make huge socio-environmental sacrifices in order to honour external debt commitments, while undergoing harsh economic adjustments imposed by the International Monetary Fund and World Bank. The term "ecological debt" was coined by Latin American environmental justice organizations in the early 1990s, as a direct response to these adjustment programmes, which imposed sharp reductions in public expenditures and major exports increases. Social movements were opposed to paying the fast-growing external debt to rich countries with sacrifices to people and nature, while the Global North owned a much larger, 500-year historic ecological debt to the Global South. The political challenge was to ensure that these unquantified ecological transfers were also acknowledged and considered. Although the debate had limited success in official circles in Latin America, it gained evidence in alternative treaties prepared by NGOs and grassroots organizations during Rio-92, when it was also mentioned in Fidel Castro's speech (Martinez-Alier, 1997, 2002a; Warlenius et al., 2015a, b).

13.3 The Ecologically Unequal Exchange by the Barcelona School[1]

13.3.1 Ecological Debt

After Rio-92, ecological debt was the subject of campaigns, publications, and events led by organizations such as Acción Ecológica and Friends of the Earth. In this context, Joan Martínez-Alier was one of the main proponents of the recognition of Global North ecological debt within academic circles. Besides exports by peripheries of cheap natural resources, of which prices do not account for several local or global socio-environmental impacts, he asserts that ecological debt is also related to rich countries' historical overuse of global environmental functions, such as climate

---

[1] The analysis carried out throughout this section is the result of a Barcelona School literature review. However, some analyses from other schools were eventually included, when carried out in a dialogue with researchers from Barcelona or in line with the fundamentals presented here.

and nutrient regulation. In this sense, recognizing the ecological debt can contribute to promoting, along with environmental justice, the North "ecological adjustment" (Martinez-Alier, 1997, 2002a, b).

In 1997, a special edition of *Ecologia Política* published several contributions on the ecological debt. These contributions reflect the liveliness of this debate in that context, in which it was the subject of international meetings in Latin America, and inspired several demonstrations and statements by politicians, researchers, and social organizations, reinforcing the then-prevalent campaigns for external debt cancellation (Parlamento Latinoamericano, 1997; Acción Ecológica, 1997). The relationship between external debt and ecological debt is further explored in this special issue, with articles arguing that contemporary requirements of monetary payment to the North increased the pressure on the environment in Latin American countries, deepening the historical ecologically unequal exchange unfavourable to the Global South (Acosta, 1997; Martínez-Alier, 1997). Finally, a set of articles and an interview described specific aspects of ecological debt, such as the appropriation of knowledge related to agricultural seeds and medicinal plants by the North ("biopiracy") (Bravo, 1997; Mora, 1997).

More recently, the Environmental Justice Organisations, Liabilities and Trade (EJOLT) project made additional contributions to the ecological debt discussion (Warlenius et al., 2015a, b), with a special section published in the *Journal of Political Ecology* (2016). Here, the main debates are not anymore centred on Latin American external debt, but on the ecologically unequal exchange and its relationship to ecological debt (Hornborg & Martinez-Alier, 2016). The asymmetries between core regions and extractivist economies in the periphery are described through theoretical contributions on ecologically unequal exchange (Oulu, 2016; Jorgenson, 2016), ecological debt (Manzano et al., 2016; Warlenius, 2016), and several empirical studies based on monetary and biophysical indicators to analyse these asymmetries in South American, African, and Asian countries (Dorninger & Eisenmenger, 2016; Kill, 2016; Temper, 2016; Martinez-Alier et al., 2016b; Mayer & Haas, 2016).

13.3.2 Biophysical Studies

The Barcelona School introduces a new dimension in the analysis of core-periphery dynamics. In ECLAC's analysis, international specialization accentuates global socio-economic inequalities. In the renewed interpretation by Barcelona, inspired by the debates on ecological debt, peripheral specialization – in resource-intensive or environment-intensive products – also reinforces the unequal distribution of resources and environmental damage between countries or regions (Muradian & Martinez-Alier, 2001). This new interpretation of unequal exchange led by Joan Martínez-Alier seems to be based on two central foundations: (i) strong sustainability, which is reflected in the use of social metabolism accounting to discuss the sustainability of economic models (Fischer-Kowalski, 1998; Fischer-Kowalski &

Huttler, 1998); and (ii) core-periphery division as a key category to investigate the environmental damage associated to exports, or the displacement of environmental damage to Global South.

Monetary valuation may be appropriate for calculating environmental liabilities in certain contexts, particularly to demand compensation or prevent future damage (Rodríguez-Labajos & Martinez-Alier, 2013), and indeed it is often used in claims for ecological debt. However, in more rigorous terms, ecological debt involves a variety of social and environmental costs that cannot be reduced to a cardinal metric (Martinez-Alier et al., 1998; Falconí, 2001; Martinez-Alier, 2002b). The Barcelona School's analyses of unequal trade reflect this understanding that sustainability should be discussed through biophysical indicators, such as embodied pollution in trade (Muradian et al., 2002), soil nutrient export (Pengue, 2005), and ecological footprint (Hornborg, 2006). More frequently, as I will next present, these studies also employ material flow analysis, which describes the environmental pressures of trade and economic activities.

While biophysical indicators are used in several studies of ecologically unequal exchange (see Jorgenson, 2006; Rice, 2007), Barcelona School analyses differ in that they discuss these indicators based on centre–periphery dynamics, influenced to some extent by ECLAC theories. One important contribution of Barcelona's analyses was the extension of Prebisch's hypothesis to take into account ecological aspects of trade (Muradian et al., 2002; Pérez-Rincón, 2006). The deterioration in terms of trade in the context of an "ecological Prebisch thesis" means that there is a decrease in the volume of imports (in tonnes) that can be purchased through the sale of one tonne of export (Schatan, 1998; Pérez-Rincón, 2006). Facing up declining trend in terms of trade in peripheries, central economies can import increasing amounts of natural resources, while maintaining balanced trade relations in monetary terms.

In line with this "structuralist" foundation of ecologically unequal exchange, empirical evidence from material flow analysis shows that specialization in environmental-intensive products is on the basis of unsustainable social metabolic profiles in peripheries. In the vast majority of Latin American countries, physical trade deficits (volume of materials exported exceeding the imported volume in terms of weight) are chronic, meaning that the regional export-oriented model generates a net outflow of resources to the international market. This specialization pattern also explains why, within the last decades, most of these countries saw an increase in material intensity (the amount of material input required to produce one unit of GDP), while central economies were dematerializing – often by outsourcing resource-intensive activities to peripheries (Russi et al., 2008; Munoz et al., 2009; Vallejo, 2010; Vallejo et al., 2011; Manrique et al., 2013; West & Schandl, 2013; Dorninger & Eisenmenger, 2016; Samaniego et al., 2017; Infante-Amate et al., 2020).

Overall, this research provides a strong ecological critique of the recent Latin American development. Since the mid-1970s, neoliberal reforms and deregulation have intensified unsustainable socio-metabolic patterns in the region, by promoting environmental-intensive exports (Giljum, 2004; Russi et al., 2008; González-Martinez & Schandl, 2008). Moreover, despite the provisional improvement in raw material prices in the early 2000s, these adverse economic and ecological structures were further reinforced through an extractivist approach to economic policies (Samaniego et al., 2017; Saes, 2018; Crespo-Marín & Pérez-Rincón, 2019), probably contributing to the recent socio-economic crisis in Latin America.

13.4 Conclusion

By promoting an innovative interpretation of Latin American debates, the Barcelona School conceptualized and empirically reinforced the ecological analyses of unequal exchange. Paradoxically, as this knowledge advanced throughout the 2000s, campaigns for the acknowledgment of ecological debt lost strength, and are no longer mentioned by governments in Latin America or internationally. Even so, now there is abundant empirical evidence proving the existence of ecologically unequal exchange and ecological debt, which can be powerful tools to revive the necessary discussion on the environmental liabilities of international trade, and the need for an "ecological adjustment" of the Global North.

References

Acción Ecológica. (1997). Deuda externa-deuda ecológica:¿ quién debe a quién? *Ecología Política, 14*, 155–156.
Acosta, A. (1997). La deuda externa acrecienta la deuda ecológica. *Ecología Política, 14*, 135–137.
Bravo, E. (1997). Biotecnología: una visión andino-amazónica. *Ecología Política, 14*, 139–144.
Clark, B., & Foster, J. B. (2009). Ecological imperialism and the global metabolic rift. *International Journal of Comparative Sociology, 50*(3–4), 311–334.
Crespo-Marín, Z., & Pérez-Rincón, M. (2019). El metabolismo social en las economías andinas y centroamericanas, 1970–2013. *Sociedad y economía, 36*, 53–81.
Dorninger, C., & Eisenmenger, N. (2016). South America's biophysical involvement in international trade: The physical trade balances of Argentina, Bolivia, and Brazil in the light of ecologically unequal exchange. *Journal of Political Ecology, 23*(1), 394–409.
ECLAC. (2000 [1985]). Transformação e crise na América Latina e no Caribe. In R. Bielschowsky (Ed.), *Cinquenta anos de pensamento na CEPAL* (pp. 817–849). Record/CEPAL.
Emmanuel, A. (1972). *Unequal exchange: A study on the imperialism of trade*. Monthly Review Press.
Falconí, F. (2001). La pesada carga material de la deuda externa. In C. Johnick & P. Pazmiño (Eds.), *Otras caras de la deuda: Propuestas para la acción*. CDES/Nueva Sociedad.
Fischer-Kowalski, M. (1998). Society's metabolism: The intellectual history of material flow analysis, part I, 1860–1970. *Journal of Industrial Ecology, 2*(1), 61–78.
Fischer-Kowalski, M., & Huttler, W. (1998). Society's metabolism: The intellectual history of material flow analysis, part II, 1970–1998. *Journal of Industrial Ecology, 4*, 107–136.
Foster, J. B., & Holleman, H. (2014). The theory of unequal ecological exchange: A Marx-Odum dialectic. *Journal of Peasant Studies, 41*(2), 199–233.
Frank, A. (1967). *Capitalism and underdevelopment in Latin America*. Monthly Review Press.
Furtado, C. (1992). *Brasil: A construção interrompida*. Paz e Terra.

Giljum, S. (2004). Trade, materials flows, and economic development in the South: The example of Chile. *Journal of Industrial Ecology, 8*(1–2), 241–261.

González-Martinez, A. C., & Schandl, H. (2008). The biophysical perspective of a middle income economy: Material flows in Mexico. *Ecological Economics, 68*(1–2), 317–327.

Griffith-Jones, S., & Sunkel, O. (1986). *Debt and development crises in Latin America: The end of an illusion*. Oxford University Press.

Hornborg, A. (1998). Towards an ecological theory of unequal exchange: Articulating world system theory and ecological economics. *Ecological Economics, 25*, 127–135.

Hornborg, A. (2006). Footprints in the cotton fields: The Industrial Revolution as time–space appropriation and environmental load displacement. *Ecological Economics, 59*, 74–81.

Hornborg, A. (2009). Zero-sum world: Challenges in conceptualizing environmental load displacement and ecologically unequal exchange in the world-system. *International Journal of Comparative Sociology, 50*, 237–262.

Hornborg, A. (2014). Ecological economics, Marxism, and technological progress: Some explorations of the conceptual foundations of theories of ecologically unequal exchange. *Ecological Economics, 105*, 11–18.

Hornborg, A. (2018). The money-energy-technology complex and ecological Marxism: Rethinking the concept of "Use-value" to extend our understanding of unequal exchange, part 1. *Capitalism Nature Socialism, 30*(3), 27–39.

Hornborg, A., & Martinez-Alier, J. (2016). Ecologically unequal exchange and ecological debt. *Journal of Political Ecology, 23*(1), 328–333.

Infante-Amate, J., Urrego Mesa, A., & Tello Aragay, E. (2020). Las venas abiertas de América Latina en la era del antropoceno: un estudio biofísico del comercio exterior (1900–2016). *Diálogos Revista Electrónica de Historia, 21*(2), 177–214.

Jorgenson, A. K. (2006). Unequal ecological exchange and environmental degradation: A theoretical proposition and cross-national study of deforestation, 1990–2000. *Rural Sociology, 71*(4), 685–712.

Jorgenson, A. K. (2016). The sociology of ecologically unequal exchange, foreign investment dependence and environmental load displacement: Summary of the literature and implications for sustainability. *Journal of Political Ecology, 23*(1), 334–349.

Journal of Political Ecology. (2016). Ecologically unequal exchange and ecological debt. Special section. *Journal of Political Ecology, 23*, 328–491.

Kill, J. (2016). The role of voluntary certification in maintaining the ecologically unequal exchange of wood pulp: The Forest Stewardship Council's certification of industrial tree plantations in Brazil. *Journal of Political Ecology, 23*(1), 434–445.

Manrique, P. L. P., Brun, J., González-Martínez, A. C., Walter, M., & Martinez-Alier, J. (2013). The biophysical performance of Argentina (1970–2009). *Journal of Industrial Ecology, 17*(4), 590–604.

Manzano, J. J. I., Cardesa-Salzmann, A., Pigrau, A., & Borras, S. (2016). Measuring environmental injustice: How ecological debt defines a radical change in the international legal system. *Journal of Political Ecology, 23*(26), 381–393.

Martinez-Alier, J. (1997). Deuda ecológica y deuda externa. *Ecología Política, 14*, 157–173.

Martinez-Alier, J. (2002a). *The environmentalism of the poor: A study of ecological conflicts and valuation*. Edward Elgar Publishing Ltd.

Martinez Alier, J. (2002b). Ecological debt and property rights on carbon sinks and reservoirs. *Capitalism Nature Socialism, 13*(1), 115–119.

Martinez-Alier, J., Munda, G., & O'Neill, J. (1998). Weak comparability of values as a foundation for ecological economics. *Ecological Economics, 26*, 277–286.

Martinez-Alier, J., Baud, M., & Sejenovich, H. (2016a). Origins and perspectives of latin American environmentalism. In F. Castro, B. Hogenboom, & M. Baud (Eds.), *Environmental governance in Latin America* (pp. 29–57). Palgrave Macmillan.

Martinez-Alier, J., Demaria, F., Temper, L., & Walter, M. (2016b). Changing social metabolism and environmental conflicts in India and South America. *Journal of Political Ecology, 23*(1), 467–491.

Mayer, A., & Haas, W. (2016). Cumulative material flows provide indicators to quantify the ecological debt. *Journal of Political Ecology, 23*(1), 350–363.

Mora, E. (1997). Medicina alternativa, indígenas y nómadas sabios. *Ecología Política, 14*, 145–151.

Muñoz, P., Giljum, S., & Roca, J. (2009). The raw material equivalents of international trade. Empirical evidence for Latin America. *Journal of Industrial Ecology, 13*(6), 881–897.

Muradian, R., & Martinez-Alier, J. (2001). Trade and the environment: From a 'Southern' perspective. *Ecological Economics, 36*(2), 281–297.

Muradian, R., O'Connor, M., & Martinez-Alier, J. (2002). Embodied pollution in trade: Estimating the 'environmental load displacement' of industrialised countries. *Ecological Economics, 41*(1), 51–67.

Oulu, M. (2016). Core tenets of the theory of ecologically unequal exchange. *Journal of Political Ecology, 23*(1), 446–466.

Parlamento Latinoamericano. (1997). La deuda externa y el fin del milenio Encuentro Internacional por una estrategia común. *Ecología Política, 14*, 131–133.

Pengue, W. A. (2005). Deuda ecológica con la agricultura: Sustentabilidad débil y futuro incierto en la Pampa argentina. *Ecología Política, 29*, 55–74.

Pengue, W. A. (Ed.). (2017). *El pensamiento ambiental del Sur: Complejidad, recursos y ecología política latinoamericana*. Universidad Nacional de General, Sarmiento.

Pérez-Rincón, M. A. (2006). Colombian international trade from a physical perspective: Towards an ecological "Prebisch thesis". *Ecological Economics, 59*(4), 519–529.

Prebisch, R. (1949). El desarrollo económico de la América Latina y algunos de sus principales problemas. *El Trimestre Económico, 35*(1), 347–374.

Prebisch, R. (2000 [1952]). Problemas teóricos e práticos do crescimento econômico. In R. Bielschowsky (org.), *Cinquenta anos de pensamento da CEPAL*. Editora Record.

Rice, J. (2007). Ecological unequal exchange: International trade and uneven utilization of environmental space in the world system. *Social Forces, 85*(3), 1369.

Rodríguez Labajos, B., & Martinez-Alier, J. (2013). The economics of ecosystem and biodiversity. When is money valuation appropriate. In H. Healy, J. Martinez-Alier, L. Temper, M. Walter, & J. F. Gerber (Eds.), *Ecological economics from the ground up* (pp. 488–512). Routledge.

Russi, D., Gonzalez-Martinez, A. C., Silva-Macher, J. C., Giljum, S., Martinez-Alier, J., & Vallejo, M. C. (2008). Material flows in Latin America: A comparative analysis of Chile, Ecuador, Mexico, and Peru, 1980–2000. *Journal of Industrial Ecology, 12*(5–6), 704–720.

Saes, B. M. (2018). *Comércio ecologicamente desigual no século XXI: evidências a partir da inserção brasileira no mercado internacional de minério de ferro*. Editora Garamond.

Samaniego, P., Vallejo, M. C., & Martinez-Alier, J. (2017). Commercial and biophysical deficits in South America, 1990–2013. *Ecological Economics, 133*, 62–73.

Schatan, J. (1998). El balance material de la deuda externa. *Ecología Política, 16*, 133–139.

Singer, H. W. (1950). The distribution of the gains between investing and borrowing countries. *American Economic Review, 40*, 473–485.

Temper, L. (2016). Who gets the HANPP (Human Appropriation of Net Primary Production)? Biomass distribution and the bio-economy in the Tana Delta, Kenya. *Journal of Political Ecology, 23*(1), 410–433.

Vallejo, M. C. (2010). Biophysical structure of the Ecuadorian economy, foreign trade, and policy implications. *Ecological Economics, 70*(2), 159–169.

Vallejo, M. C., Pérez Rincón, M. A., & Martinez-Alier, J. (2011). Metabolic profile of the Colombian economy from 1970 to 2007. *Journal of Industrial Ecology, 15*(2), 245–267.

Wallerstein, I. (1989). *The modern world-system, vol. III: The second great expansion of the capitalist world economy, 1730–1840s*. Academic Press.

Wallerstein, I. (2000). Globalisation or the age of transition: A long-term view of the trajectory of the world system. *International Sociology, 15*(2), 249–265.

Warlenius, R. (2016). Linking ecological debt and ecologically unequal exchange: Stocks, flows, and unequal sink appropriation. *Journal of Political Ecology, 23*(1), 364–380.

Warlenius, R., Pierce, G., Ramasar, V., Quistorp, E., Martinez-Alier, J., Rijnhout, L., & Yanez, I. (2015a). *Ecological debt*. History, meaning and relevance for environmental justice. EJOLT Report No. 18, 48 p.

Warlenius, R., Pierce, G., & Ramasar, V. (2015b). Reversing the arrow of arrears: The concept of "ecological debt" and its value for environmental justice. *Global Environmental Change, 30*, 21–30.

West, J., & Schandl, H. (2013). Material use and material efficiency in Latin America and the Caribbean. *Ecological Economics, 94*, 19–27.

Open Access This chapter is licensed under the terms of the Creative Commons Attribution 4.0 International License (http://creativecommons.org/licenses/by/4.0/), which permits use, sharing, adaptation, distribution and reproduction in any medium or format, as long as you give appropriate credit to the original author(s) and the source, provide a link to the Creative Commons license and indicate if changes were made.

The images or other third party material in this chapter are included in the chapter's Creative Commons license, unless indicated otherwise in a credit line to the material. If material is not included in the chapter's Creative Commons license and your intended use is not permitted by statutory regulation or exceeds the permitted use, you will need to obtain permission directly from the copyright holder.

Chapter 14
Flow/Fund Theory and Rural Livelihoods

Jose Carlos Silva-Macher

14.1 Introduction

Studies on the relationships between social metabolism and socio-environmental conflicts, or, more accurately, ecological distribution conflicts, are a central theme in the work of Professor Joan Martínez-Alier and the Barcelona School, where the fields of ecological economics and political ecology are united with the other environmental social sciences (Martínez-Alier, 2019; Martínez-Alier & Silva-Macher, 2021): namely, environmental history, industrial ecology, urban ecology, agroecology, and ethnoecology. Of special interest is the study of the metabolism of industrial society at the global level, as well as ecological distribution conflicts (Martínez-Alier et al., 2010; Muradian et al., 2012; Martínez-Alier, 2020) at the frontiers of raw material extraction (Moore, 2000) where different rural societies live – among which *campesinos* and indigenous peoples stand out for their historical, cultural, and political importance. As a frame of reference, these studies can utilize, first, a more biophysical/ecological dimension expressed through the concept of the Anthropocene (Steffen et al., 2011), the age in which changes in human activity created transformations on planet Earth, which hastened greatly after World War II. And, second, they can draw on a more institutional/cultural dimension expressed through the evolution of a modern expression of capitalist colonial power that began with the creation of the Americas (Quijano, 2014; Neyra-Soupplet, 2019), whose components include racial discrimination as a basic expression of colonial domination, an economy articulated around capital and the world market, and a colonization of imaginaries and subjectivities regarding a hegemonic perspective of western knowledge.

J. C. Silva-Macher (✉)
Department of Economics, Pontificia Universidad Católica del Perú, Lima, Peru

© The Author(s) 2023
S. Villamayor-Tomas, R. Muradian (eds.), *The Barcelona School of Ecological Economics and Political Ecology*, Studies in Ecological Economics 8, https://doi.org/10.1007/978-3-031-22566-6_14

Based on this frame of reference, in the present chapter, I seek to describe this relationship between social metabolism and conflicts by drawing on Georgescu-Roegen's (1971) flow/fund theory, one of the fundaments of ecological economics. The core idea in this chapter was the focus of two published scientific articles that I co-wrote with Katharine N. Farrell as part of my doctoral thesis (Silva-Macher & Farrell, 2014; Farrell & Silva-Macher, 2017), applied to case studies about extraction frontiers in the Peruvian Andes and Amazonia. This idea stems from flow/fund theory and the subsequent proposal of Farrell and Mayumi (2009) and posits that to better understand an economic process one must first comprehend the economic *Anschauung*,[1] the underlying intuition or perspective of intentionality (Farrell, 2021), which means exploring the economic purpose of the *social actor* associated with this specified economic process. According to this conceptualization, in the case of ecological distribution conflicts – such as those that unfold between multinational mining companies and *campesinos* at the metal extraction frontiers of the Peruvian Andes – what emerges are opposite visualizations and perspectives about the territory in dispute.

The rest of the chapter is structured as follows: First, I describe Georgescu-Roegen's flow/fund theory, placing an emphasis on the relationship between economic process and purpose; second, using the language of flows and funds, I present the case of the Conga conflict, an emblematic example of the environmentalism of the poor (Martínez-Alier, 2002) in which campesinos resist and defend the life of *Pachamama* against a mining company that seeks to extract gold and copper with the support of the Peruvian government. I take into account the experience of a recent field visit to the home of Máxima Acuña, a *campesina* leader who lives in the place where Yanacocha is looking to establish an open pit. Finally, I conclude the chapter with some brief reflections, pondering on the realities of raw-material exporting countries since colonial times and, in particular, on campesinos and indigenous peoples in extraction frontiers.

14.2 The Flow/Fund Theory and the Tree Metaphor

A central theme in the debate between the ecological vision of the economy and the neoclassical vision of the economy (Cavalcanti, 2018) is the operational meaning of substitution in the production function (Mayumi et al.1998). The neoclassical production functions assume that any factor can be substituted with another, with some exceptions, and therefore that an increase in the input of any factor always means a rise in production. However, as Mayumi et al. (1998) observe, "those neoclassical economists adopting the substitution assumption have not paid due attention to the essential distinction between flows (=quantities of materials qualitatively transformed in the process) and funds (=agents transforming a given set of inflows into a

---

[1] Georgescu-Roegen (1971: 362).

given set of outflows) in the material production process (Georgescu-Roegen, 1971: 115)". For example, to increase the production of fresh milk (*outflow*), having more fodder (*inflow*) counts for nothing if the number of cattle (*fund*) is the limiting factor or, conversely, if the bottleneck lies in the supply of fodder. This is to pay due attention to the biophysical foundation of economic activities – that is, the *real–real economy* of ecological economists (Martinez-Alier, 2008).

However, before describing a material production process using the flow/fund theory, it is necessary to first define the parameters of the process, which have two analytical components. The first is the *frontier* – which establishes the process vis-a-vis its environment at any moment (the process/non-process distinction). This is not the same as a geographical frontier, in that it entails the mental construction of a partial process that can hardly be said to be enclosed in a defined space. The second component is the *duration* – the points at which the process in question begins and ends, and which show the passage of time through our consciousness.[2] This is the component that defines the identity of an element, as *flow* or *fund*, in the analytical representation of a material economic process. Finally, the frontier and duration of the process depend on the *Anschauung*, or the perspective of intentionality, which shapes how the specific purpose of a given economic process is defined. That is, they depend on a primary notion that can be clarified by way of discussion and examples, but cannot be reduced by way of a formal definition (Georgescu-Roegen, 1971). These points are stressed by Farrell and Mayumi (2009) in their argument that the general theory of flows and funds is designed to be used based on a consideration of how the parameters of a given process are determined, and of the fact that these parameters are strongly related to time and tradition.

To illustrate this idea, it is worth thinking about the metaphor of a tree and how it is observed by two social actors: a carpenter and a farmer (Silva-Macher & Farrell, 2014). For the carpenter, the tree is a source of wood since its economic purpose is the production of furniture, and to this end it is necessary to cut down the tree, transforming it in a process that can take some hours. On the other hand, for the farmer, the tree is a source of food, since its economic purpose is the gathering of fruit, and to this end it is necessary to sustain and protect the life of the tree, conserving it in a process that can take many years. Therefore, the tree is a *flow* in the carpenter's process but a *fund* in the farmer's process. This situation also relates to a typical socio-environmental conflict in which a single natural resource has two mutually exclusive purposes that, in the language of flows and funds, would represent a dispute over the identity of the tree (Silva-Macher & Farrell, 2014). In the case of an ecological distribution conflict between a mining company and a farming community in the Andes, the focus of the dispute would be the identity of a *mountain*, which depends on the *Anschauung*, or perspective of intentionality (Farrell, 2021;

---

[2] "Let E (T1) and E (T2) be the entropies of the universe at two different moments in Time, T1 and T2 respectively; if E (T1) < E (T2), then T2 is later in Time than T1 – and conversely. But, clearly, if we did not know already what later means, the statement would be vacuous. The full meaning of the law is that the entropy of the universe increases as Time flows through the observer's consciousness" (Georgescu-Roegen, 1971: 133).

Farrell & Mayumi, 2009), that characterizes each social actor, while also taking into account a long history of power relations in which certain perspectives are imposed on other ones. It is at this level of analysis that ecological economics and political ecology are united in pursuit of environmental justice. The fundamental question that Martínez-Alier (2002: 271) poses is: "Who has the power to simplify complexity, ruling some languages of valuation out of order?"

14.3 A Dispute over the Identity of the *Mountain*

The Andes has historically been a mineral extraction frontier for the world economy, dating back to colonial times in the case of gold and silver, to which copper, zinc, lead, lithium, and others have more recently been added. Peru witnessed a pronounced expansion of this extraction frontier from the 1990s as a result of the country's Washington Consensus-inspired neoliberal reforms, to which the indicators of domestic extraction clearly attest in the material flow accounts (Russi et al., 2008). Moreover, the vast majority of the metals extracted and concentrated are exported to the world's industrial centres, such as China, while the tailings and other hazardous waste materials remain in the Andes as environmental liabilities, with severe repercussions for the health and livelihoods of local communities. This results in an ecologically unequal trade that characterizes the pattern of metabolism in industrial society and is reflected in the profound inequalities within raw-material exporting countries (Pérez-Rincón et al., 2019; Infante-Amate et al., 2020).

In this context, the Conga gold and copper mining project in Peru is an emblematic conflict between a multinational mining company and campesinos in the Andes (Silva-Macher & Farrell, 2014). The mining company is Yanacocha, a joint venture between Newmont Corporation (51.35%), based in the United States; Compañía de Minas Buenaventura (43.65%), based in Peru; and Sumitomo Corporation (5%), based in Japan. Campesinos are made up of small-scale farmers who live around the headwaters of the Cajamarca region, which flow into the Marañón River and, from there, into the Amazon. The landscape is typically Andean and includes lakes, natural pasture, potatoes and other crops, livestock such as horses, cattle, sheep, and guinea pigs, adobe houses, and improved unpaved roads, juxtaposed with trucks, metal fencing, gates restricting access along rural tracks, CCTV cameras, and other installations put in place by Yanacocha.

For an analytical representation with which to describe the economic purposes and processes in dispute, and thus to better understand the anatomy of the Conga conflict, the flow/fund theory is useful (Silva-Macher & Farrell, 2014). This entails an exploration of the gold and copper extraction process carried out by the Yanacocha mining company, and the fresh milk production process of the Cajamarca campesinos. On the one hand, there is the economic *Anschauung* of a multinational corporation oriented toward the maximization of private gain, which regards the *mountain* as a potential source of income through the sale of minerals to the world market. The problem is that, to realize this economic purpose, the company must extract and concentrate the gold and copper, for which it inevitably transforms the Andean landscape – over a period of some 20 years – through blasts, excavations, and mineral crushing and concentration processes that are intensive in the use of water, energy, and chemical products, leaving large open pits in the land and generating an equivalent proportion of waste. This alteration of the mountain ecosystem integrity implies the disappearance of lakes, pastures, crops, animals, and campesinos' homes. Therefore, the *mountain* would be an *inflow* from the perspective of the Yanacocha mining company, transformed into two types of *outflow*: on the one hand, the product, represented by gold and copper for export; and on the other, the waste and the devastated Andean landscape, whose social and environmental costs are transferred to the campesinos of Cajamarca, bolstering the profits of the mining corporation all the more.

In this context, there is particular interest in the land and house of Máxima Acuña, an activist and recipient of the Goldman Environmental Prize who lives in the very place earmarked for one of the open pits of the Conga project – in Tragadero Grande and the Laguna Azul[3] – whose case undoubtedly expresses the mutually exclusive purposes of agriculture and mining. During a field visit in March 2020 to Máxima Acuña's home – which, given the overlap, is also a visit to the Conga project concession area – I had the opportunity to accompany Ms. Acuña's son Daniel Chaupe, a young campesino who lives in the same house, on his daily work routine. This brought me closer to the economic *Anschauung,* or perspective, of a campesino family in relation to the *mountain* landscape[4] whose identity is in dispute. Through this lens, the *mountain* is observed as a source of life given that its purpose is associated with *buen vivir* in the Andes, which implies growing crops, rearing animals, spending time with one's family, and maintaining strong social relations, among other aspects that I was unable to measure entirely through observation. It should be noted that there is no mathematical model with which to describe this immaterial flow for the enjoyment of life, or *buen vivir*, associated with economic activity (Georgescu-Roegen, 1971; Cavalcanti, 2018). However, what can be affirmed is that, to carry out agricultural economic processes, it is necessary to conserve the integrity of the *mountain* for several generations. In this regard, the fresh milk production process, which is a simple sign of campesino life in the vicinity of the Conga project, requires that the cattle have access to sufficient flows of water and pasture, which are natural products of the mountain. Therefore, the mountain has the identity of a *fund* element from the perspective of the Acuñas and other campesino families. It is notable that this same *fund* can provide multiple services for different economic processes and for human activities in a broader sense, as part of the campesino life; studies on the multifunctionality of agriculture and rural livelihoods are of relevance here.

This perspective could be validated by other, the local inhabitants, most of them also campesinos, in the area of influence of the Conga project – perhaps by way of

---

[3] https://photos.app.goo.gl/s5aqYoY9esN5DC499
[4] Tragadero Grande: Cerro del Águila, Cerro Colorado, and Laguna Azul.

a popular referendum (Conde, 2017; Urkidi & Walter, 2011; Bebbington & Williams, 2008; Muradian et al., 2003). However, a similar viewpoint can be discerned from the testimonies of other female campesino leaders in Peru, who face similar conflicts. Two of them, Yanet Caruajulca and Zulma Zamora, told their stories during the "Territorial Struggles for Women's Socio-Environmental Justice in the face of Extractivism" panel, as part of the Second Congress of the Andean Region Society for Ecological Economics,[5] held at the Pontificia Universidad Católica del Perú in April 2019. This campesino perspective was also glimpsed through Yanet Caruajulca's presentation during the "Indigenous and Black Communities and the Impact of Covid-19" session, as part of the *Global On-line Symposium of the International Degrowth Network and the International Society for Ecological Economics,*[6] University of Manchester, in early September 2020. To borrow one of the slogans from the protests against mining in the Andes, one might recapitulate that, for life, *el agua vale más que el oro* (water is worth more than gold).

14.4 Final Reflections

The ecological vision of economics means recognizing the biophysical reality of economic processes, but this is not the only element to be observed. The flow/fund theory of Georgescu-Roegen shows us that to better understand a given economic process, we must first understand the underlying purpose. This, in turn, means exploring the *Anschauung*, or perspective of intentionality, of the social actor linked to a certain partial economic process. In this regard, the flow/fund theory can serve as a useful tool for studying the relationship between social metabolism and ecological distribution conflicts.

To prevent socio-environmental conflicts between mining and agriculture, adequate land-use planning is recommendable. Any such public policy would have to pay due attention to the different perspectives and perceptions present in a territory, which could be in conflict or, in the best-case scenario, in harmony. In the case of the Conga conflict, in which the extraction of gold and copper and the production of fresh milk dispute the identity of a mountain, in terms of *flow* and *fund*, the conflict could have been predicted in advance. For reasons of ethics and the long-term sustainability of the socio-ecological system, those perspectives that assert respect for life, fair distribution, and ecosystem integrity have to be prioritized.

However, history since colonial times has shown us that the languages of valuation respond to power relations and are often imposed through the exercise of violence. In the case of Conga, despite the campesino resistance, which has kept the project on hold to this day, five local inhabitants died during protests against the

---

[5] https://educast.pucp.edu.pe/video/11244/ii_congreso_de_la_sociedad_andina_de_economia_ecologica__parte_12

[6] http://www.confercare.manchester.ac.uk/events/degrowth2020/

mine in 2012. Mining companies, the Peruvian government, and economic elites use the arguments of economic growth and development to justify this type of extractive project, which gives us pause to reflect on the relevance of discourses of transition and post-development, such as those of degrowth in Europe and *buen vivir* in South America, to offer new narratives that place care and respect for life at the heart of the debate.

References

Bebbington, A., & Williams, M. (2008). Water and mining conflicts in Peru. *Mountain Research and Development, 28*(3/4), 190–195.
Cavalcanti, C. (2018). De la economía convencional a la economía ecológica: el significado de Nicholas Georgescu-Roegen y la encíclica Laudato Si' del Papa Francisco. *Gestión y Ambiente, 21*(supl. 1), 49–56.
Conde, M. (2017). Resistance to mining: A review. *Ecological Economics, 132*, 80–90.
Farrell, K. N. (2021). Mejorar la vida ecológica, mejora la vida económica. In A. Rincón-Ruiz, P. Arias-Arévalo, & M. Clavijo-Romero (Eds.), *Hacia una valoración incluyente y plural de la biodiversidad y los servicios ecosistémicos: visiones, avances y retos en América Latina*. Centro Editorial – Facultad de Ciencias Económicas, Universidad Nacional de Colombia.
Farrell, K. N., & Mayumi, K. (2009). Time horizons and electricity futures: An application of Nicholas Georgescu-Roegen's general theory of economic production. *Energy, 34*, 301–307.
Farrell, K. N., & Silva-Macher, J. C. (2017). Exploring futures for Amazonia's Sierra del Divisor: An environmental valuation triadics approach to analyzing ecological economic decision choices in the context of major shifts in boundary conditions. *Ecological Economics, 141*, 166–179.
Georgescu-Roegen, N. (1971). *The entropy law and the economic process*. Harvard University Press.
Infante-Amate, J., Urrego, A., & Tello, E. (2020). Las venas abiertas de América Latina en la era del Antropoceno: un estudio biofísico del comercio exterior (1900–2016). *Diálogos Revista Electrónica de Historia, 21*(2), 177–214.
Martinez-Alier, J. (2002). *The environmentalism of the poor: A study of ecological conflicts and valuation*. Edward Edgar Publishing.
Martinez-Alier, J. (2008). La crisis económica vista desde la economía ecológica. *Ecología Política, 36*, 23–32.
Martinez-Alier, J. (2019). La enseñanza de las ciencias socioambientales. *Observatorio del Desarrollo: investigación, reflexión y análisis, 8*(22), 29–36.
Martinez-Alier, J. (2020). A global environmental justice movement: Mapping ecological distribution conflicts. *Disjuntiva, 1*(2), 83–128.
Martinez-Alier, J., & Silva-Macher, J. C. (2021). Las ciencias socio-ambientales. In Azamar, Silva-Macher, & Zuberman (Eds.), *Una mirada desde la economía ecológica Latinoamericana frente a la crisis socioecológica*. CLACSO y Siglo XXI.
Martinez-Alier, J., Kallis, G., Veuthey, S., Walter, M., & Temper, L. (2010). Introduction: Social metabolism, ecological distribution conflicts, and valuation languages. *Ecological Economics, 70*, 153–158.
Mayumi, K., Giampietro, M., & Gowdy, J. (1998). Georgescu-Roegen/Daly versus Solow/Stiglitz revisited. *Ecological Economics, 27*, 115–117.
Moore, J. W. (2000). Sugar and the expansion of the early modern world-economy: Commodity frontiers, ecological transformation, and industrialization. *Review, 23*(3), 409–433.
Muradian, R., Martínez-Alier, J., & Correa, H. (2003). International capital versus local population: The environmental conflict of the Tambogrande mining project, Peru. *Society & Natural Resources: An International Journal, 16*(9), 775–792.

Muradian, R., Walter, M., & Martinez-Alier, J. (2012). Hegemonic transitions and global shifts in social metabolism: Implications for resource-rich countries, Introduction to the special section, global environmental change part A. *Human and Policy Dimensions, 22*(3), 559–567.

Neyra-Soupplet, R. (2019). Violencia y Extractivismo en el Perú contemporáneo. *Historia Ambiental Latinoamericana y Caribeña (HALAC) Revista de la Solcha, 9*(2), 210–236.

Pérez-Rincón, M., Vargas-Morales, J., & Martinez-Alier, J. (2019). Mapping and analyzing ecological distribution conflicts in Andean countries. *Ecological Economics, 157*, 80–91.

Quijano, A. (2014). Colonialidad del poder, eurocentrismo y América Latina. In *Cuestiones y horizontes: de la dependencia histórico-estructural a la colonialidad/descolonialidad del poder*. CLACSO.

Russi, D., González-Martinez, A. C., Silva-Macher, J. C., Giljum, S., Martinez-Alier, J., & Vallejo, M. C. (2008). Material flows in Latin America: A comparative analysis of Chile, Ecuador, Mexico, and Peru, 1980–2000. *Journal of Industrial Ecology, 12*(5/6), 704–720.

Silva-Macher, J. C., & Farrell, K. N. (2014). The flow/fund model of Conga: Exploring the anatomy of environmental conflicts at the Andes-Amazon commodity frontier. *Environment, Development and Sustainability, 16*, 747–768.

Steffen, W., Grinevald, J., Crutzen, P., & McNeill, J. (2011). The Anthropocene: Conceptual and historical perspectives. *Philosophical Transactions of the Royal Society A, 369*, 842–867.

Urkidi, L., & Walter, M. (2011). Dimensions of environmental justice in anti-gold mining movements in Latin America. *Geoforum, 42*(6), 683–695.

Open Access This chapter is licensed under the terms of the Creative Commons Attribution 4.0 International License (http://creativecommons.org/licenses/by/4.0/), which permits use, sharing, adaptation, distribution and reproduction in any medium or format, as long as you give appropriate credit to the original author(s) and the source, provide a link to the Creative Commons license and indicate if changes were made.

The images or other third party material in this chapter are included in the chapter's Creative Commons license, unless indicated otherwise in a credit line to the material. If material is not included in the chapter's Creative Commons license and your intended use is not permitted by statutory regulation or exceeds the permitted use, you will need to obtain permission directly from the copyright holder.

Chapter 15
Deceitful Decoupling: Misconceptions of a Persistent Myth

Alevgul H. Sorman

15.1 Introduction

The long-standing tension revolving around the "possibility of decoupling" among the green growth versus post/de-growth narratives has muddied the waters of tackling the climate crisis we are currently facing. These "decoupling wars" (Jackson & Victor, 2019) have, on the one hand, been seeking to answer whether reductions in resource and energy use and respective emissions are possible without modifying our accustomed growth trajectories. On the other hand, post/de-growth narratives argue for doing things differently: urging for the reconsideration of a reduction of resource and energy dependency as objective in itself, valuing collective well-being (People's Conference on Climate Change and the Rights of Mother Earth, 2010) instead of indicators such as GDP as a metric to be maximised (Costanza et al., 2014; van den Bergh, 2009) on a resource-constrained planet.

The recent COVID-19 pandemic has resurfaced the debate yet once again, creating yet another bifurcation in the road. The question remains on whether we can achieve a post-COVID-19 green economic recovery (UNEP, 2020) based on assumptions of efficiency and decoupling while re-growing; confronted with movements mirrored by globalised Fridays for Future or Extinction Rebellion[1] urging for

---

[1] Youth movement since August 2018, stemming from the actions of Gretha Thunberg and other youth activists, drawing attention to the climate crisis and reclaiming their future https://fridaysforfuture.org/ and Extinction Rebellion https://rebellion.earth/

A. H. Sorman (✉)
Basque Centre for Climate Change (BC3), Leioa, Spain

IKERBASQUE, Basque Foundation for Science, Bilbao, Spain
e-mail: alevgul.sorman@bc3research.org

© The Author(s) 2023
S. Villamayor-Tomas, R. Muradian (eds.), *The Barcelona School of Ecological Economics and Political Ecology*, Studies in Ecological Economics 8,
https://doi.org/10.1007/978-3-031-22566-6_15

profound transformation in the way things are done to tackle the ecological emergency and breakdown (Monbiot, 2018).

The debate, albeit has occupied the agenda over decades in the form of the infamous Environmental Kuznets Curve (EKC) (Kuznets, 1955; Stern, 2004)[2] and the IPAT formula[3] advocating for the possibility of a green economy. Such claims, attempting to overcome "limits to growth" (Meadows et al., 1972) or the "spaceship earth" (Boulding, 1966) claustrophobia, have been contested by many scholars and the ecological economics community (Martinez-Alier, 1995, 2012). Affluence, capital accumulation (Hornborg, 1998, 2009), and accumulation by dispossession (Harvey, 2004) indeed occupy a significant role in shaping our interactions between socio-economic systems and the environment (Scheidel et al., 2018). Materials and energy flows (Fischer-Kowalski & Amann, 2001) help self-organise, maintain and develop internal functions and structures of societies forming the backbone of our societal metabolisms (Giampietro et al., 2011; Şorman, 2014). Nevertheless, societal metabolisms have associated socio-ecological interdependencies, mostly stemming from the unequal distribution of ecological goods and services (Martínez-Alier, 2002). Studies indicate that even if the metabolisms of industrial countries were kept stable at 2000 levels, for the rest of the world to catch up would result in a quadrupling of global emissions by 2050 (Fischer-Kowalski et al., 2011). Moreover, the evolution of the global North-South (McGregor & Hill, 2009) divide has reinforced commodity chains and extraction frontiers (Martinez-Alier et al., 2010), manifestations of exploitative labour and trade relationships (Hornborg, 2020) surfacing via contentious political processes and powerful multilateral institutions. Therefore, as previously argued, the win-win promise of a "sustainability" scenario of letting humankind "*have our cake and eat it*" (Rees, 1990, p. 435) has not been achieved in the last 30 years, despite all goodwill and green growth promises with a decoupling intent.

The decoupling debate reappeared over the years in the form of eco-efficiency, eco-innovation, and the circular economy, fostering sustainable consumption, especially relevant in the European Union's Action Plans (EU Circular Economy Action Plan, 2020). The European Green Deal (EGD),[4] for example, aims to radically transform economic activities to make substantial progress towards creating a circular economy. However, full circularity is unattainable since there is entropic decay in materials, as discussed in further detail in Sect. 15.4. Moreover, the same notion is engraved within the Sustainable Development Goals (SDGs), where "Decent Work and Economic Growth" (Goal 8) (United Nations, 2020), although inclusive

---

[2] A hypothesis suggesting that countries follow an inverted U-shaped pathway, suggesting that environmental degradation occurs in the early stages of development; yet as income rises and countries become more affluent, environmental conditions improve. For a critique of the rise and fall of the EKC, see (Stern, 2004).

[3] Where Environmental impact (I) is expressed and directly proportional to population (P), affluence (A), and technology (T).

[4] https://ec.europa.eu/info/strategy/priorities-2019-2024/european-green-deal_en

and sustainable in principle, base their premise on an ever-increasing pie of economic growth. However, the Barcelona School of ecological economics has long contested that developing growth-oriented policies around the expectation that decoupling is or will be possible has been a misleading policy while also criticising the use of GDP as a proxy for well-being (Ward et al., 2016). Recently, Hickel and Kallis (2020) have once again claimed that green growth is a misguided objective; that absolute decoupling from carbon emissions is highly unlikely to be achieved at a rate rapid enough to prevent global warming over 1.5 °C or 2 °C, even under favourable policy conditions.

In essence, decoupling has a foundational role within the Barcelona School of Ecological Economics and Political Ecology. First, due to the lack of evidence in absolute decoupling between resource and energy dependency and growth (Parrique et al., 2019), there is a crucial need to look at "*how the world operates*" and rethink alternative pathways of living within planetary boundaries. This calls for new ways of doing economics which interrogates economic processes limited by biophysical constraints both on the supply side in terms of resources and the sink side recognising environmental limits from the local to the global. Second, decoupling (and its lack thereof) calls for scrutinising embedded societal relations that interrogate "*why we do what we do*" that cover socio-political factors such as power dynamics, institutional arrangements, cultural variables, and economic and financial drivers. These questions also explore the unequal access to and distribution of goods and services, benefits, and burdens (Robbins, 2011) while inquiring into participation and decision-making mechanisms over how the world's resources are (un)used. Third, the absence of decoupling also calls for deliberation over individual and collective action on "*how we envision alternative imaginaries*" in post normal times (Funtowicz & Ravetz, 1993). This means creating spaces and opening up the discussion for new actors and different futures departing from the business as usual growth-based scenarios.

Along my academic journey – deeply rooted in the Barcelona school Ecological Economics and Political Ecology – I try to scrutinise these notions of decoupling, having closely worked on *energy metabolism*, the study of energy flows that are required to sustain societies; *energy justice* calling for a re-evaluation of ethical and gender concerns in energy decision making and my research on *energy cultures*, delving deeper on role individual and collective behaviour toward energy in transformational research and action.

In the remainder of this chapter, Sect. 15.2 focuses on the different concepts of decoupling and system boundaries; Sect. 15.3 reviews and synthesises further empirical evidence that analyses trends of decoupling both in terms of resources and emissions; Sect. 15.4 dissects claims for a circular economy and rebound effects, a phenomenon closely observed during the COVID-19 global pandemic that goes hand in hand with the decoupling narrative; Sect. 15.5 wraps up the discussion of decoupling as a deceitful narrative that is prolonged as a persistent myth hindering genuine systemic and transformative change.

15.2 Different Decoupling Concepts and Accounting Mechanisms

Decoupling is typically categorised based on environmental pressures stemming from the production side, referring to *resource decoupling* including materials, energy, and the less obvious water; and the impact side of our actions framed around *impact decoupling* including greenhouse gasses, land, water pollutants, and biodiversity loss (Parrique et al., 2019).

Relative (or weak) decoupling indicates that the rate at which materials, energy use, or emissions increase is lower than the rate at which GDP increases (Burton, 2015), or in other terms that the economic growth outpaces environmental impact. Although this may seem like a favourable veneer, it still maps out as extractivism or greenhouse gases accumulating in the atmosphere, beating our overall target to tackle climate change and live within our ecological boundaries in the long run. Already, the UNEP emissions gap report indicates (UNEP, 2019) that now in 2020, we need to reduce emissions by 7.6% per annum globally every year until 2030; otherwise, limiting global warming to 1.5 °C will be a missed chance (see also (Patterson et al., 2018)). Similarly, the Production Gap Report (SEI et al., 2019), assessing the world's current pace of fossil fuel extraction to align with Paris Agreement goals, indicates that the world is to produce far more coal, oil, and gas than is consistent with limiting warming to 1.5 °C or 2 °C, creating a "production gap" that makes climate goals much harder to reach.

Absolute (or strong) decoupling, on the other hand, claims that economic performance behaves independently from material or energy extraction or emissions. Newer terms extending the potential boundaries and glossary definitions of decoupling have also been defined such as *"virtual decoupling"* (Moreau & Vuille, 2018) referring to developed countries outsourcing intensive industrial production chains to lesser developed countries, also known as the carbon leakage phenomena. The role of increased "tertiarisation" (or dematerialisation through services) (Heiskanen & Jalas, 2000) as a complementary angle also shifts attention given to emissions, with the know-how (immaterial) sectors occupying a greater weight in the composition of the more "developed" economies. Vadén et al. (2020) also argue that relationships between resources and emissions decoupling might not be so straightforward and linear; such that there may be instances of material efficiency (Schandl et al., 2016) or a boost in "financialisation" (the role and weight of the financial sphere within the economy) (Kovacic et al., 2018), which may all lead to somewhat decoupling with very different implications.

In terms of accounting for impact, discussions center around shifting current accounting mechanisms[5] from one based on *territorial emissions* (production-based accounting with GHG emissions assigned based on the source localisation) to one based on a *consumption-based accounting* (CBA) (Lininger, 2015; Davis & Caldeira, 2010; Peters, 2008; Munksgaard & Pedersen, 2001). CBA, initially used

---

[5] As used by the International Energy Agency (IEA) via the national measurements methodology.

for Carbon Footprint measurements (Ireland, 2018), takes into account the outsourcing effect (Bastianoni et al., 2014); somewhat[6] internalising responsibilities based on re-integrating externalities. It is often argued that CBA should be mainstreamed in climate policy for disclosing "real" corresponding emissions per country as it will serve for constructing policies with a holistic perspective for crucial innovations (Wiedenhofer et al., 2020) for tackling the global climate emergency and ecological crisis. Approximate numbers indicate that production-based emissions in the Global South are 10–15% higher than consumption-based emissions, and vice versa for the Global North (Fuhr, 2019). Similarly, research illustrates (Wood et al., 2018) that approximately one-quarter of the global land use (Weinzettel et al., 2013), 40% of materials (Wiedmann et al., 2015), 20–30% of global water use (Lenzen et al., 2013), and over 20% of greenhouse gas (GHG) emissions (Peters & Hertwich, 2008) reside embodied in trade.

In terms of "truth-ful" accounting and defining adequate policy mechanisms based on political realities (Afionis et al., 2017) for tackling issues of equity and justice appropriately and for proposing alternative exit strategies, such realities must urgently be confronted.

15.3 Results from Empirical Evidence and Reviews

Recent literature argues that there is little or no evidence in terms of absolute decoupling looking into embodied energy in trade, material consumption, resource use, and emissions.

Akizu-Gardoki et al. (2018) devise an alternative "Decoupling Index" that uses 126 countries' *total primary energy footprints* (rather than total primary energy supplies), taking into account embodied energy imported with goods and services. Within a 14-year period of analysis (2000–2014), the authors empirically show that 93 countries[7] disprove decoupling; while 27 show absolute decoupling for the analysis period, with only 6 countries (ESP, ITA, HUN, GBR, JPN, and FRA) with a Human Development Index of 0.8 above maintain absolute decoupling over time (*ibid.*).

Regarding *global material flows* covering over half a century of analysis (1950–2010) Schaffartzik et al. (2014) reveal that although industrial metabolic profiles stabilise over time with equal shares of biomass, fossil energy, and construction minerals; they are surpassed by other regions like Asia, replicating patterns of industrial growth engines. However, this does not translate into "per capita" affluence or material consumption and instead adds to the growing bubble of global

---

[6] There are different methodologies and discussions on how trade-related GHG emissions should be accounted for. For an overview of alternative approaches to allocating GHGs or proposed shared allocation schemes, see (Peters, 2008).

[7] Decoupled countries reduce to 27 (from 40); relatively decoupled countries reduce to 17 (from 29); and conversely, recoupling rises in 80 countries from 55.

resource use and extraction (*ibid.*). On a similar note, after an analysis of 40 years of resource productivity analysis Krausmann et al. (2017) summarise that although some countries may be decoupling in absolute terms, this value is cancelled out when trade is taken into account. Wood et al. (2018) confirm that impacts embodied in trade, especially regarding material goods, led by clothing and footwear, are growing tremendously, with energy and GHG emissions following in a less pronounced manner. Wu et al. (2018) highlight the differences in a decoupling index, where "developed countries" primarily continue to develop high-tech and high-efficiency GDP growth drivers, whereas "developing countries" have not undergone a transformation to absolute decoupling due to a lack in energy efficiency measures, disorganised industrial structures, and absence of Information Technologies (IT) in capital investment. These studies highlight a threefold causality: (1) we have lengthened commodity supply chains due to the abundance of readily available transport fuel; (2) this has been made possible by notably shifting and outsourcing more significant impacts to developing regions that take on primary extractive and secondary industries; (3) meanwhile, we have bought into the promise of circularity regarding the reuse of materials. Hass et al. (2015) illustrate that circularity is low mainly because most materials are not available for recycling in the first place and that the growth of materials injection into our systems outpaces any potential recycling attempt (See Sect. 15.4 for more details).

Similarly, recently conducted reviews of empirical research conclude no robust evidence regarding decoupling. (For an exhaustive literature review of decoupling literature, see Parrique et al., 2019; Koirala et al., 2011; Mardani et al., 2019 among others). The Decoupling Debunked document (Parrique et al., 2019) concludes that there is no indication of decoupling that is *absolute, global, permanent,* and *sufficiently fast and large enough both in terms of resources or impacts.* Likewise, a review of 179 decoupling articles (Vadén et al., 2020) that appeared between 1990 and 2019 concluded with no evidence of economy-wide, national/international absolute resource decoupling and reckons that "the goal of decoupling rests partly on faith." A recent exhaustive review, composed of a bibliometric mapping of 835 peer-reviewed articles (published in two parts, part I (Wiedenhofer et al., 2020) and part II (Haberl et al., 2020)) also highlight the need for substantial advances in both theoretical and empirical research required while also being complemented by alternative goals and ambitions.

With unsuccessful robust and systemic evidence, reaffirming the findings of Hickel and Kallis (2020), *absolute decoupling from resource use* cannot be achieved on a global scale against a global scale backdrop of continued economic growth.

In terms of accounting for *emissions decoupling*, a recent study by the Breakthrough Institute (Hausfather, 2021) highlights that 32 countries have managed to demonstrate economic growth while CO_2 emissions declined since 2005, even when accounting for emissions embodied in the goods consumed in a country. However, they note with caution that these economies are already wealthy-service driven economies and that very few examples are present from low- or middle-income countries based on extractive industries and energy-intensive manufacturing to date. While this study can present a departure narrative of consumption-based emissions and absolute decoupling as we switch to clearer technologies, we must not forget that extraction and resource dependence remain intact on the supply side – supplies and sinks being the two sides of the same coin of the decoupling phenomena.

15.4 The "Not-So-Circular" Economy and Rebound Effects

The industrial economy is not as circular as claims make it be; but rather is entropic (Georgescu-Roegen, 1971) with the depletion of low entropy materials from the environment, resulting in an accumulation of high entropy wastes and exotic materials in the environment (Daly, 1992; Kerschner, 2010). This, also highlighted in a review of decoupling literature by Wiedenhofer et al. (2020), is frequently disregarded in decoupling studies, where the phenomena itself is usually approached from a statistical/econometric viewpoint, often overlooking thermodynamic principles of energy and materials and their core role and function in defining societal metabolisms.

As Haas et al. (2015) show, on a global metabolism, only 6% of all processed materials are recycled, against the backdrop of global material consumption increasing by 3.6% in a decade (Schaffartzik et al., 2014). These numbers only slightly improve for the European Union amidst convincing circular economy narratives (Haas et al., 2015). Over half of the total solid material throughput in economies (52% or 3.5 GT/year) (Giampietro & Funtowicz, 2020) is composed of either food or energy inputs that are subject to entropic decay (especially in terms of energy quality). Only when biomass is included among the recycled solid flows, total circularity only increases to 37% (Haas et al., 2015). The remaining 45% of materials that are used form the backbones of social systems, becoming infrastructure and construction and "immovables" of societies; while only a minute percent of material flow, 3% or 0.7 GT/year, is associated with consumable and durable products (Giampietro & Funtowicz, 2020). Moreover, all of these social structures and metabolisms come at the expense of ecosystem services at our disposal (Costanza et al., 2017), with water throughput often ignored in stabilising socio-ecological systems (Giampietro & Funtowicz, 2020).

The COVID-19 pandemic and confinement period has also been an occasion to observe decoupling and rebounds (Alcott, 2005; Polimeni, 2012) within societies adjusting to a dramatic, involuntary downscaling of socio-economic activities due to global lockdowns. In April 2020, in comparison to 2019 values, daily global CO_2 emissions had decreased by around 17% (Le Quéré et al., 2020). Values exhibited in the peak low of emissions during COVID-19 were equal to those corresponding to 2006 values – 14 years ago (Canadell et al., 2020). However, several months after, studies detected a rapid rebound (Harvey, 2020), and in 2021 global energy-related CO_2 emissions are projected to grow once again by 4.8% (IEA, 2021). It is often argued that decoupling is bound by *temporal features* such that decoupling is usually followed by periods of no decoupling or even recoupling, making it a challenge for becoming a permanent and continuous matter (Vadén et al., 2020; Williamson, 2021). The notion of rebound is also captured where in some instances, the very act of solving environmental problems in itself spends resources and may even create additional problems that were previously unthinkable (Allen et al., 2003).

15.5 Discussion and Conclusions

Most empirical literature and reviews illustrate little to no evidence of resource or impact decoupling in absolute terms that is extensive and effective enough to be accepted as given. Moreover, when externalities are to some degree accounted for, decoupling attempts are cancelled out by the impacts and consequences of trade or the exertion of built-in power asymmetries and extractive frontiers prolonging the North-South divide and socio-environmental inequalities in terms of benefits and burdens – primary research loci of the Ecological Economics and Political Ecology disciplines.

The study of decoupling needs to be scrutinised holistically, such that both demand-side and supply-side analyses go hand in hand. For embarking on a transformative path, we require a holistic vision regarding tracing the origins of resources, internalising externalities, and, better yet, downscaling dependencies. This implies moving away from "certain" narratives (Lazarevic & Valve, 2017), like decoupling or the circular economy. Instead, the intersectionality of resource use and socio-ecological well-being needs scrutiny for transformative policy and change. Recent directions for example taken by the European Environmental Agency (2021) evaluate alternative narratives other than growth accepting a long-lasting, absolute decoupling of economic growth and environmental pressures.

Decoupling is not only a biophysical constraint or a matter of efficiency limited to the technosphere but rather is one of distribution, one where far-reaching lifestyle changes (Wiedmann et al., 2020) complement technological advancements and one which embraces principles of environmental and social justice (Parrique et al., 2019). Ivan Illich had alluded to principles of sufficiency and justice in terms of putting limits to energy use back in the 1970s, where he recalled that a ceiling on energy use could indeed bring upon social relations characterised with high levels of equity (Illich, 1974). Some studies argue that a 2–6 times increase in sustainable resources at the global level would universally attain more qualitative goals (O'Neill et al., 2018); others (Millward-Hopkins et al., 2020) indicate that in the year 2050, final energy consumption globally could be reduced to 1960 levels, despite a tripling in the population.

Pathways that entail less dependency on resources need to be formulated based on alternative policies that comply with sufficiency-oriented strategies with strict enforcement of absolute reduction targets (Haberl et al., 2020). These, as such, may promote alternative forms of existence like that of degrowth (Kallis et al., 2020) or other prosperous ways down (Odum & Odum, 2008) that link resource use and emissions to collective well-being rather than provoking ecological destruction.

What is clear, however, as per COVID-19 pandemic times faced with unprecedented conditions, we are indeed able to adapt genuinely and responsively to change. Similarly, when tackling the climate crises and our dependence on resources, sufficiency for defining alternative futures is an alternative worth considering. While technological advancements proceed in all sectors, behavioural and narrative changes are crucial, if not more, in guiding this transformation.

References

Afionis, S., Sakai, M., Scott, K., Barrett, J., & Gouldson, A. (2017). Consumption-based carbon accounting: Does it have a future? *Wiley Interdisciplinary Reviews: Climate Change, 8*(1), e438.

Akizu-Gardoki, O., Bueno, G., Wiedmann, T., Lopez-Guede, J. M., Arto, I., Hernandez, P., & Moran, D. (2018). Decoupling between human development and energy consumption within footprint accounts. *Journal of Cleaner Production, 202*, 1145–1157.

Alcott, B. (2005). Jevons' paradox. *Ecological Economics, 54*(1), 9–21.

Allen, T. F. H., Tainter, J. A., & Hoekstra, T. W. (2003). *Supply-side sustainability series: Complexity in ecological systems* (p. 440). Columbia University Press. https://doi.org/10.7312/alle10586

Bastianoni, S., Caro, D., Borghesi, S., & Pulselli, F. M. (2014). The effect of a consumption-based accounting method in national GHG inventories: A trilateral trade system application. *Frontiers in Energy Research, 2*, 4.

Boulding, K. E. (1966). *The economics of the coming spaceship earth* (pp. 1–17). New York.

Burton, M. (2015). *The decoupling debate: Can economic growth really continue without emission increases?* Available Online: https://www.degrowth.info/en/2015/10/the-decoupling-debate-can-economic-growth-really-continue-without-emission-increases/. Accessed on: 26 Oct 2020.

Canadell, P., Le Quéré, C., Creutzig, F., Peters, G., William Jones, M., Friedlingstein, P., Jackson, R., & Shan, Y. (2020, May 19). Coronavirus is a 'sliding doors' moment. What we do now could change Earth's trajectory. *The Conversation*. Available Online: https://theconversation.com/coronavirus-is-a-sliding-doors-moment-what-we-do-now-could-change-earths-trajectory-137838. Accessed on: 01 July 2020.

Costanza, R., Kubiszewski, I., Giovannini, E., Lovins, H., McGlade, J., Pickett, K. E., et al. (2014). Development: Time to leave GDP behind. *Nature News, 505*(7483), 283.

Costanza, R., De Groot, R., Braat, L., Kubiszewski, I., Fioramonti, L., Sutton, P., et al. (2017). Twenty years of ecosystem services: How far have we come and how far do we still need to go? *Ecosystem Services, 28*, 1–16.

Daly, H. E. (1992). *Steady-state economics*. Earthscan Publications Ltd.

Davis, S., & Caldeira, K. (2010). Consumption-based accounting of CO2 emissions. *Sustainability Science, 107*, 5687–5692. https://doi.org/10.1073/pnas.0906974107

EU Circular Economy Action Plan. (2020). *A new circular economy action plan for a cleaner and more competitive Europe*. Available Online: https://ec.europa.eu/environment/circular-economy/. Accessed on: 01 July 2020.

European Environmental Agency. (2021, June, 21). Growth without economic growth Briefing no. 28/2020 Title: Title: HTML – TH-AM-20-028-EN-Q – ISBN 978-92-9480-321-4 – ISSN 2467-3196. https://doi.org/10.2800/781165. Available on: https://www.eea.europa.eu/publications/growth-without-economic-growth. Retrieved on: 30 June 2021.

Fischer-Kowalski, M., & Amann, C. (2001). Beyond IPAT and Kuznets curves: Globalization as a vital factor in analysing the environmental impact of socio-economic metabolism. *Population and Environment, 23*(1), 7–47.

Fischer-Kowalski, M., Krausmann, F., Giljum, S., Lutter, S., Mayer, A., Bringezu, S., Moriguchi, Y., Schütz, H., Schandl, H., & Weisz, H. (2011). Methodology and indicators of economy-wide material flow accounting. *Journal of Industrial Ecology, 15*, 855–876.

Fuhr, H. (2019, June 20). *The Global South's contribution to the climate crisis – And its potential solutions*. OECD Development Matters. Available Online: https://oecd-development-matters.org/2019/06/20/the-global-souths-contribution-to-the-climate-crisis-and-its-potential-solutions/. Accessed on: 26 Oct 2020.

Funtowicz, S. O., & Ravetz, J. R. (1993). Science for the post-normal age. *Futures, 25*(7), 739–755.

Georgescu-Roegen, G. (1971). *The entropy law and the economic process*. Harvard University Press.

Giampietro, M., & Funtowicz, S. O. (2020). From elite folk science to the policy legend of the circular economy. *Environmental Science & Policy, 109*, 64–72.

Giampietro, M., Mayumi, K., & Sorman, A. H. (2011). *The metabolic pattern of societies: Where economists fall short* (Vol. 15).

Haas, W., Krausmann, F., Wiedenhofer, D., & Heinz, M. (2015). How circular is the global economy? An assessment of material flows, waste production, and recycling in the European Union and the world in 2005. *Journal of Industrial Ecology, 19*(5), 765–777.

Haberl, H., Wiedenhofer, D., Virág, D., Kalt, G., Plank, B., Brockway, P., et al. (2020). A systematic review of the evidence on decoupling of GDP, resource use and GHG emissions, part II: Synthesizing the insights. *Environmental Research Letters, 15*(6), 065003.

Harvey, D. (2004). The 'new' imperialism: Accumulation by dispossession. *Socialist Register, 40*, 63–87.

Harvey, F. (2020, June 11). 'Surprisingly rapid' rebound in carbon emissions post-lockdown. *The Guardian*. Available Online: https://www.theguardian.com/environment/2020/jun/11/carbon-emissions-in-surprisingly-rapid-surge-post-lockdown. Accessed on: 01 July 2020.

Hausfather, Z. (2021, April 6). *Absolute decoupling of economic growth and emissions in 32 countries*. Available online: https://thebreakthrough.org/issues/energy/absolute-decoupling-of-economic-growth-and-emissions-in-32-countries. Retrieved on: 25 June 2021.

Heiskanen, E., & Jalas, M. (2000). *Dematerialization through services-A review and evaluation of the debate*. Available Online: https://helda.helsinki.fi/bitstream/handle/10138/40558/FE_436.pdf?sequence=1. Accessed on: 26 Oct 2020.

Hickel, J., & Kallis, G. (2020). Is green growth possible? *New Political Economy, 25*(4), 469–486.

Hornborg, A. (1998). Toward an ecological theory of unequal exchange: Articulating world system theory and ecological economics. *Ecological Economics, 25*(1), 127–136.

Hornborg, A. (2009). Zero-sum world: Challenges in conceptualizing environmental load displacement and ecologically unequal exchange in the world system. *International Journal of Comparative Sociology, 50*(3–4), 237–262.

Hornborg, A. (2020). *The commodification of human life: Labor, energy, and money in a deteriorating biosphere* In The Palgrave Handbook of Environmental Labour Studies (pp. 677–697). Palgrave Macmillan, Cham.

IEA. (2021). *Global energy review 2021*. IEA. Available online: https://www.iea.org/reports/global-energy-review-2021. Accessed on: 30 June 2021.

Illich, I. (1974). *Energy and equity* (p. 84). Harper & Row.

Ireland, R. (2018, November 14). *The carbon dioxide embodied in imports: A different way to measure CO2 emissions*. Regulating for Globalization, Trade, Labor and EU Law Perspectives. Available online: http://regulatingforglobalization.com/2018/11/14/the-carbon-dioxide-embodied-in-imports-a-different-way-to-measure-co2-emissions/?doing_wp_cron=1593175035.8672249317169189453125. Accessed on: 26 Oct 2020.

Jackson, T., & Victor, P. A. (2019). Unraveling the claims for (and against) green growth. *Science, 366*(6468), 950–951.

Kallis, G., Paulson, S., D'Alisa, G., & Demaria, F. (2020). *The case for degrowth*. Polity Press.

Kerschner, C. (2010). Economic de-growth vs. steady-state economy. *Journal of Cleaner Production, 18*(6), 544–551.

Koirala, B. S., Li, H., & Berrens, R. P. (2011). Further investigation of environmental Kuznets curve studies using meta-analysis. *Journal of Ecological Economics and Statistics, 22*, 13–32.

Kovacic, Z., Spanò, M., Piano, S. L., & Sorman, A. H. (2018). Finance, energy and the decoupling: An empirical study. *Journal of Evolutionary Economics, 28*(3), 565–590.

Krausmann, F., Wiedenhofer, D., Lauk, C., Haas, W., Tanikawa, H., Fishman, T., et al. (2017). Global socioeconomic material stocks rise 23-fold over the 20th century and require half of annual resource use. *Proceedings of the National Academy of Sciences, 114*(8), 1880–1885.

Kuznets, S. (1955). Economic growth and income inequality. *American Economic Review, 49*(1955), 1–28.

Lazarevic, D., & Valve, H. (2017). Narrating expectations for the circular economy: Towards a common and contested European transition. *Energy Research & Social Science, 31*, 60–69.

Le Quéré, C., Jackson, R., Jones, M., Smith, A., Abernethy, S., Andrew, R., De-Gol, A., Shan, Y., Canadell, J., Friedlingstein, P., Creutzig, F., & Peters, G. (2020). *Supplementary data to: Le Quéré et al (2020), temporary reduction in daily global CO2 emissions during the COVID-19 forced confinement* (Version 1.0). Global Carbon Project. https://doi.org/10.18160/RQDW-BTJU

Lenzen, M., Moran, D., Bhaduri, A., Kanemoto, K., Bekchanov, M., Geschke, A., & Foran, B. (2013). International trade of scarce water. *Ecological Economics, 94*, 78–85.

Lininger, C. (2015). *Consumption-based approaches in international climate policy*. Springer.

Mardani, A., Streimikiene, D., Cavallaro, F., Loganathan, N., & Khoshnoudi, M. (2019). Carbon dioxide (CO2) emissions and economic growth: A systematic review of two decades of research from 1995 to 2017. *Science of the Total Environment, 649*, 31–49. https://doi.org/10.1016/j.scitotenv.2018.08.229

Martinez-Alier, J. (1995). The environment as a luxury good or 'too poor to be green'. *Ecological Economics, 13*, 1–10.

Martínez-Alier, J. (2002). *The environmentalism of the poor: A study of ecological conflicts and valuation*. Edward Elgar.

Martinez-Alier, J. (2012). Environmental justice and economic degrowth: An alliance between two movements. *Capitalism Nature Socialism, 23*(1), 51–73.

Martinez-Alier, J., Kallis, G., Veuthey, S., Walter, M., & Temper, L. (2010). Social metabolism, ecological distribution conflicts, and valuation languages. *Ecological Economics, 70*(2), 153–158.

McGregor, A., & Hill, D. (2009). North–South. In *International encyclopaedia of human geography* (pp. 473–480). Elsevier.

Meadows, D. H., Meadows, D. L., Randers, J., & Behrens, W. (1972). *The limits to growth*. Universe Books.

Millward-Hopkins, J., Steinberger, J. K., Rao, N. D., & Oswald, Y. (2020). Providing decent living with minimum energy: A global scenario. *Global Environmental Change, 65*, 102168.

Monbiot, G. (2018, November 14). The Earth is in a death spiral. It will take radical action to save us. *The Guardian*. https://www.theguardian.com/commentisfree/2018/nov/14/earth-death-spiral-radical-action-climate-breakdown. Accessed on: 26 Oct 2020.

Moreau, V., & Vuille, F. (2018). Decoupling energy use and economic growth: Counter evidence from structural effects and embodied energy in trade. *Applied Energy, 215*, 54–62. https://doi.org/10.1016/j.apenergy.2018.01.044

Munksgaard, J., & Pedersen, K. A. (2001). CO2 accounts for open economies: Producer or consumer responsibility? *Energy Policy, 29*(4), 327–334.

O'Neill, D. W., Fanning, A. L., Lamb, W. F., & Steinberger, J. K. (2018). A good life for all within planetary boundaries. *Nature Sustainability, 1*, 88.

Odum, H. T., & Odum, E. C. (2008). *A prosperous way down: Principles and policies*. University Press of Colorado.

Parrique, T., Barth, J., Briens, F., Kerschner, C., Kraus-Polk, A., Kuokkanen, A., & Spangenberg, J. H. (2019). *Decoupling debunked: Evidence and arguments against green growth as a sole strategy for sustainability*. European Environmental Bureau.

Patterson, J. J., Thaler, T., Hoffmann, M., Hughes, S., Oels, A., Chu, E., et al. (2018). Political feasibility of 1.5° C societal transformations: The role of social justice. *Current Opinion in Environmental Sustainability, 31,* 1–9.

People's Conference on Climate Change and the Rights of Mother Earth. (2010). https://pwccc.wordpress.com/

Peters, G. P. (2008). From production-based to consumption-based national emission inventories. *Ecological Economics, 65*(1), 13–23.

Peters, G., & Hertwich, E. (2008). CO2 embodied in international trade with implications for global climate policy. *Environmental Science & Technology, 42*(5), 1401–1407.

Polimeni, J. M. (2012). Empirical evidence for the Jevons Paradox. The Jevons paradox and the myth of resource efficiency improvements. In B. Alcott, M. Giampietro, K. Mayumi, & J. Polimeni (Eds.), *The Jevons paradox and the myth of resource efficiency improvements* (pp. 141–171). Routledge.

Rees, J. A. (1990). *Natural resources: Allocation, economics and policy* (2nd ed.). Routledge and Kegan Paul.

Robbins, P. (2011). *Political ecology: A critical introduction* (Vol. 16). Wiley.

Schaffartzik, A., Mayer, A., Gingrich, S., Eisenmenger, N., Loy, C., & Krausmann, F. (2014). The global metabolic transition: Regional patterns and trends of global material flows, 1950–2010. *Global Environmental Change, 26,* 87–97.

Schandl, H., Hatfield-Dodds, S., Wiedmann, T., Geschke, A., Cai, Y., West, J., et al. (2016). Decoupling global environmental pressure and economic growth: Scenarios for energy use, materials use and carbon emissions. *Journal of Cleaner Production, 132,* 45–56.

Scheidel, A., Temper, L., Demaria, F., & Martínez-Alier, J. (2018). Ecological distribution conflicts as forces for sustainability: An overview and conceptual framework. *Sustainability Science, 13*(3), 585–598.

SEI, IISD, ODI, Climate Analytics, CICERO, & UNEP. (2019). *The production gap: The discrepancy between countries' planned fossil fuel production and global production levels consistent with limiting warming to 1.5°C or 2°C.* Available Online: http://productiongap.org/. Accessed on: 26 Oct 2020.

Şorman, A. H. (2014). Metabolism, societal. In G. D'Alisa, F. Demaria, & G. Kallis (Eds.), *Degrowth: A vocabulary for a new era.* Routledge.

Stern, D. I. (2004). The rise and fall of the environmental Kuznets curve. *World Development, 32*(8), 1419–1439.

UNEP. (2019). *Emissions gap report 2019. Executive summary.* United Nations Environment Programme. Available Online: http://www.unenvironment.org/emissionsgap. Accessed on: 26 Oct 2020.

UNEP. (2020, July 16). Green economy & COVID-19 recovery. *Sustainable Development Goals Stories.* https://www.unenvironment.org/news-and-stories/story/green-economy-covid-19-recovery. Accessed on: 26 Oct 2020.

United Nations. (2020). *Department of Economic and Social Affairs, Disability #Envision2030 Goal 8: Decent work and economic growth.* Available Online: https://www.un.org/development/desa/disabilities/envision2030-goal8.html. Accessed on: 01 July 2020.

Vadén, T., Lähde, V., Majava, A., Järvensivu, P., Toivanen, T., Hakala, E., & Eronen, J. T. (2020). Decoupling for ecological sustainability: A categorisation and review of research literature. *Environmental Science & Policy, 112,* 236–244.

van den Bergh, J. (2009). The GDP paradox. *Journal of Economic Psychology, 30*(2), 117–135. https://doi.org/10.1016/j.joep.2008.12.001

Ward, J. D., Sutton, P. C., Werner, A. D., Costanza, R., Mohr, S. H., & Simmons, C. T. (2016). Is decoupling GDP growth from environmental impact possible? *PLoS One, 11*(10), e0164733.

Weinzettel, J., Hertwich, E. G., Peters, G. P., Steen-Olsen, K., & Galli, A. (2013). Affluence drives the global displacement of land use. *Global Environmental Change, 23*(2), 433–438.

Wiedenhofer, D., Virág, D., Kalt, G., Plank, B., Streeck, J., Pichler, M., et al. (2020). A systematic review of the evidence on decoupling of GDP, resource use and GHG emissions, part I: Bibliometric and conceptual mapping. *Environmental Research Letters, 15*(6), 063002.

Wiedmann, T. O., Schandl, H., Lenzen, M., Moran, D., Suh, S., West, J., & Kanemoto, K. (2015). The material footprint of nations. *Proceedings of the National Academy of Sciences of the United States of America, 112*(20), 6271–6276.

Wiedmann, T., Lenzen, M., Keyßer, L. T., & Steinberger, J. K. (2020). Scientists' warning on affluence. *Nature Communications, 11*(1), 1–10.

Williamson, P. (2021). De-globalisation and decoupling: Post-COVID-19 myths versus realities. *Management and Organization Review, 17*(1), 29–34.

Wood, R., Stadler, K., Simas, M., Bulavskaya, T., Giljum, S., Lutter, S., & Tukker, A. (2018). Growth in environmental footprints and environmental impacts embodied in trade: Resource efficiency indicators from EXIOBASE3. *Journal of Industrial Ecology, 22*(3), 553–564.

Wu, Y., Zhu, Q., & Zhu, B. (2018). Comparisons of decoupling trends of global economic growth and energy consumption between developed and developing countries. *Energy Policy, 116*, 30–38.

Open Access This chapter is licensed under the terms of the Creative Commons Attribution 4.0 International License (http://creativecommons.org/licenses/by/4.0/), which permits use, sharing, adaptation, distribution and reproduction in any medium or format, as long as you give appropriate credit to the original author(s) and the source, provide a link to the Creative Commons license and indicate if changes were made.

The images or other third party material in this chapter are included in the chapter's Creative Commons license, unless indicated otherwise in a credit line to the material. If material is not included in the chapter's Creative Commons license and your intended use is not permitted by statutory regulation or exceeds the permitted use, you will need to obtain permission directly from the copyright holder.

Part IV
Environmental Justice Conflicts and Alternatives

Chapter 16
Does the Social Metabolism Drive Environmental Conflicts?

Arnim Scheidel

16.1 Introduction

The proposition that changes in social metabolism drive environmental conflicts is frequently found in studies of ecological distribution conflicts. Martinez-Alier (2009) identified a *"three-tier relation between the increasing social metabolism of human economies pushed by population and economic growth, the resulting ecological distribution conflicts among human groups, and the different languages of valuation deployed historically and currently by such groups when they reaffirm their rights to use the environmental services and products in dispute."* Diverse studies and attempts to shed light on these relations have stimulated much interdisciplinary research at the interface of political ecology and ecological economics (see special issues by Martinez-Alier et al., 2010; Muradian et al., 2012; Temper et al., 2018a). Yet, questions remain regarding why, how, and when the proposition *'more metabolism, more conflicts'* is useful and valid to understand the emergence of environmental conflicts. At first sight, the countries with higher consumption of energy and materials per capita and year seem not to be the countries with more environmental conflicts.

This chapter provides a brief overview of some of the theoretical foundations underlying this proposition and the main pathways through which increases in socio-metabolic processes may trigger environmental conflicts, often at distant locations where energy and materials are consumed. Environmental conflicts can emerge over socio-environmental impacts and injustices that arise at the input, throughout, and output stages of the global social metabolism, hence, at the stages of resource extraction, transport and processing, and waste disposal. The chapter points also to several other properties of socio-metabolic processes beyond the

---

A. Scheidel (✉)
Institute of Environmental Science and Technology,
Universitat Autónoma de Barcelona (ICTA-UAB), Barcelona, Spain

increase in resource flows in the economy, which are expected to co-shape conflict dynamics together with other social and biophysical dynamics. Three additional propositions are outlined: the potential of socio-metabolic processes to trigger conflicts increases with (i) the degree of toxicity and ecological harmfulness of the materials extracted, processed, and consumed; (ii) the temporal immediacy of the perceived risks of adverse impacts from societal resource uses; and (iii) the spatial proximity of social groups to adverse impacts resulting from resource uses. Finally, the chapter places the socio-metabolic perspective into context with other 'grand explanations' of environmental conflicts (i.e. the expansion of capitalism under a neo-Marxist perspective), and points to the important role of other social, political and cultural variables in shaping conflict dynamics.

Overall, the chapter argues that a socio-metabolic perspective has much to offer to explain some of the drivers of environmental conflicts because it allows linking local conflicts to the resource use profiles of economies as well as to global production and consumption systems and their 'commodity extraction frontiers' (Moore, 2000). However, these processes must be placed into context with the political economy governing them, as well the specific social, economic, historic, and cultural contexts in which conflicts unfold and which shape how contestations manifest and develop. Thus, the nuanced study of environmental conflicts requires the consideration of the interaction of both biophysical and social aspects in a dynamic manner to understand how human, non-human, and more-than-human natures dynamically co-produce and co-constitute the socio-material worlds in which we live (Kolinjivadi, 2019).

16.2 More Metabolism, More Conflicts? Theoretical Foundations

The proposition that changes, specifically increases, in social metabolism drive environmental conflicts has originated with the parallel development of ecological economics and political ecology. The social metabolism, a central field of study in ecological economics (Gerber & Scheidel, 2018), refers to the processes of appropriation, transformation, and disposal of energy and materials by an economy or social system, in short, the material exchange relations between society and nature (Haberl et al., 2019). Environmental conflicts, which are at the core of political ecology research, refer to social conflicts over the use of the environment and natural resources (Robbins, 2012). They manifest through mobilizations by individuals or groups in response to perceived environmental injustices and detrimental socio-environmental impacts of resource use (Scheidel et al., 2020). Environmental injustices can include issues of unjust distributions of environmental benefits and burdens, procedural injustices in how decisions affecting the environment were made, or a lack of recognition of the worldviews and values of different social groups, including their material and cultural relations with the environment

(Schlosberg, 2004). Joan Martinez-Alier was the first one who put forward the proposition that higher rates of social metabolism would lead to more environmental conflicts (Martinez-Alier, 2007, 2009).

This proposition rests on theoretical insights from ecological economics, particularly from the observation that the economy is entropic, and not circular (Georgescu-Roegen, 1971). Materials are recycled only to a small extent (Haas et al., 2015). This implies that even a non-growing economy would constantly need new resources on the input side, while creating unrecycled waste, pollution, and emissions on the output side. The resulting pressures on material sources and sinks arguably amplify with an increase in the social metabolism, globally, as well as at the country level. Because of (pre-existing) power relations and inequalities, this leads frequently to unequal distributions of environmental benefits (such as access to resources) and burdens (such as exposure to pollution) across different social groups, triggering social conflicts (Martinez-Alier, 2007). The resulting 'ecological distribution conflicts' (Martínez-Alier & O'Connor, 1996) can be observed, and analysed at the input, throughput, and output side of the economy (i.e. at the stages of resource extraction, transport and processing, and waste disposal).

At the input side of the economy, the expansion and deepening of resource extraction frontiers to satisfy the global resource demand (Banoub et al., 2020) frequently triggers environmental conflicts over dispossession and displacement of local social groups from their territories and resources. While, it could be argued that growing resource demand could potentially produce higher revenues for local resource producers, in practice, the increasing competition over access to resources driven by relative resource scarcity has commonly provoked the dispossession of customary users to make way for extractivist projects because of unequal power relations (Muradian et al., 2012). Global and national resource demand may also exceed the capacity of local customary resource use systems to provide surplus flows at the speed demanded by a growing economy (Scheidel et al., 2013). To obtain larger surplus flows per area of land use, and to appropriate the economic and material benefits resulting from these flows, states and corporations have pursued through the process of 'development' and industrialization a fundamental restructuring of what Nicholas Georgescu-Roegen (1969) termed the 'funds' of the economic production process (i.e. a reorganization of labour, land, and technology). Karl Marx referred to these processes as a 'metabolic rift', pointing to the rupture between humanity and nature caused by industrial production methods.

This restructuring of funds caused by the metabolic rift – for example, from customary farming to intensive agribusiness (Dell'Angelo et al., 2021), from artisanal miners to large-scale excavations (Geenen, 2014), and from decentral energy uses based on local biomass to large centralized energy provision infrastructures such as dams (Del Bene et al., 2018) or fossil fuel explorations (Orta-Martínez & Finer, 2010) – has created vast distributional conflicts over who is able to benefit from the environmental benefits and burdens resulting from these transformations. The specific sectors in which these ecological distribution conflicts emerge tend to coincide with the changes in the metabolic profile of the economies undergoing industrialization (Pérez-Rincón et al., 2019; Spiric, 2018). Such processes of so-called development have also provoked many conflicts over the violent transformations of worldviews, values, and livelihood systems towards industrial modes of societies that have left many people with hunger and in poverty (Escobar, 2012).

Conflicts at the extraction sites can also arise from contamination and environmental degradation of adjunct ecosystems, even if there is no direct dispossession or displacement of local groups. The extraction at high rates of minerals, fossil fuels, and biomass through industrial agriculture and other resources produces significant levels of pollution that are often not contained within the formal concession boundaries of an extractivist project, but expand through ecological processes such as water and air flows into adjunct ecosystems. For example, the massive mining spill at the *Padcal mine* in the Philippines, causing the release of 20.6 million tons of toxic tailings into water bodies, created a vast environmental disaster in the surrounding areas (EJAtlas, 2015). This caused conflicts over the severe impacts on customary groups whose lives and livelihoods depended on the larger ecosystems, such as through health impacts from exposure to environmental pollution, livelihood impacts through loss of key resources (fish, wildlife, etc.), as well as cultural impacts through the degradation of sacred landscapes, decline in traditional knowledge or loss of sense of place. Such environmental burdens tend to be unequally distributed across different groups because of inequalities in power, locally, as well as internationally (Martinez-Alier, 2007)

Neo-classical economists call these adverse impacts 'externalities' that arise from 'market failures' (Martinez-Alier, 1995). William Kapp (1950), one of the intellectual fathers of ecological economics, described them more appropriately as cost-shifting processes through which powerful groups are able to make large benefits, because they shift some of the social and environmental costs and impacts resulting from resource uses to more vulnerable groups (e.g. Demaria, 2010). Local communities and social movements contest the resulting distributional injustices locally, nationally, and globally through protests and mobilizations that shed light on the causes of unsustainable and unjust resource uses (Martinez-Alier et al., 2016; Walter & Urkidi, 2017). This 'environmentalism of the poor' (Martinez-Alier, 2002) has become a powerful social force for more sustainable and just resource uses (Scheidel et al., 2018; Temper et al., 2018b)

Environmental conflicts are not limited to the extraction sites, but occur along the entire resource use chain, from the cradle to grave of commodities (Martinez-Alier et al., 2010). Conflicts over pollution, environmental destruction, dispossession of livelihood resources, or disrespect of local customary uses occur also along transport routes and infrastructures. A well-known example is the conflict around the Dakota Access Pipeline (DAPL) that created the #NoDAPL movement. Thousands of protesters led by members of the Standing Rock Sioux Tribe have mobilized against the pipeline construction since 2016. The concerns voiced by different groups and people include not only issues of lacking consultation, or safety concerns, but also fundamental issues of recognition of different values and worldviews, and repeated injustices committed against the Tribe within the history of US colonialism (Whyte, 2017).

At the output side of the economy, increases in social metabolism translate into growing amounts of solid, liquid, and gaseous wastes. Only a few actors are able to take advantage of the growing waste production as an emerging commodity frontier (Demaria & Schindler, 2016). For most urban and rural dwellers, garbage and waste pollution represents a threat to the environment and their health. This can cause environmental conflicts over growing and disproportionate exposure to waste across different social groups, whereas vulnerable and marginalized actors along the lines of race and class are frequently most affected (Bullard, 1990; Mohai & Saha, 2015). A well-known historic example of a waste conflict is the rise of the US environmental justice movement in the early 1980s, which contested the burdens of waste and pollution disproportionally imposed on poor black neighbourhoods (Bullard, 1990, 1994). Nowadays, one of the biggest global waste problems is the massive release of CO_2 and other climate gases into the atmosphere, caused by fossil resource consumption. Also, here strong and diverse social movements have emerged that contest these unsustainable resource uses, thus contributing to climate change mitigation (Temper et al., 2020; Thiri et al., 2022; Tramel, 2016).

16.3 Further Propositions on the Links Between Social Metabolism and Environmental Conflicts

As illustrated so far, increases in the *quantity* of social metabolism can act as key drivers of environmental conflicts across all stages of social metabolism. However, also the *qualities* of resources used by societies, as well as the characteristics of how they are extracted, transported, and processed further shape whether environmental conflicts emerge, or not. Consider for example the degree of toxicity of extracted materials, the temporal immediacy (or latency) of adverse health and environmental impacts resulting from waste disposal and pollution, or the spatial proximity (or distance) of social groups exposed to adverse impacts from resource processing plants. These characteristics also have an important role in shaping the dynamics of environmental conflicts. In addition to the general proposition of *more metabolism, more conflicts*, the following three propositions may be further useful to conceptualize and understand the interactions between socio-metabolic processes and environmental conflicts.

The first proposition is that *the more ecologically harmful the extracted, processed, and disposed materials are, the higher their potential to provoke social conflict*. Not only the quantity but also the types and qualities of materials metabolized by an economy shape the social, environmental and health risks posed to different social groups, and thus the potential to produce social conflicts. This proposition applies to the input, throughput, and output stages of the social metabolism. For example, compared to the large amounts of construction materials such as sand and gravel that are constantly processed by economies for infrastructure development, the extraction, transport, and disposal of much smaller amounts of highly toxic substances such as uranium or nuclear waste may provoke the perception of high risks for social groups exposed to these substances, thus triggering conflicts (e.g. EJAtlas, 2021; Litmanen, 1996). In short, we may expect that the more ecologically harmful the social metabolism, the more socially conflictive. Future research on the links between social metabolism and environmental conflicts may further consider such qualitative aspects.

The second proposition is that *the more immediate the risk perception of adverse impacts resulting from resource uses is, the higher their potential to provoke social conflict*. Not only the quantity and type of materials shape the conflict potentials of socio-metabolic processes, but also the temporal immediacy (or latency) of adverse impacts resulting from them across all stages of the social metabolism. For example, agrochemicals linked to the extraction of biomass may accumulate in ecosystems over time and provoke health impacts only after continuous exposure to them. Related conflicts may thus emerge only after many years, when impacts are being felt (Navas et al., 2018). In addition to the immediacy of impacts, also their risk perception influences whether social conflicts arise or not. For example, the most severe impacts of climate change are not immediate but will happen in the future. However, the perceived risks of these impacts are high, leading already now to conflicts over fossil fuel extraction and climate change concerns (Temper et al., 2020).

The third proposition is that *the greater the proximity of social groups to adverse impacts from resource uses, the higher their potential to provoke social conflict*. The spatial proximity of human settlements to pollution and environmental degradation occurring at the input, throughput, and output sides of the social metabolism may play an important role in shaping conflict dynamics. Questions such as how distance to conflictive events shapes social conflict dynamics, and the capacity of groups to mobilize, are discussed with much detail in social movement studies within the branch of spatial ecology studies (Tilly, 2000; Zhang & Zhao, 2018) Generally, exposure to adverse impacts from resource extraction, processing, and waste disposal can be expected to be higher when these processes are located closer to human settlements, which in turn is related, among other factors, to population densities (Muradian et al., 2012). Greater spatial proximity to the adverse impacts of socio-metabolic processes thus translates into a higher potential for conflict and social mobilizations.

Finally, also the scale of analysis of the social metabolism must be considered, as socio-metabolic changes at the national level translate in distinct ways to the local level where conflict occurs. For example, from a local perspective, conflicts over conservation areas occur not because of an increase in local social metabolism, but a radical decrease (i.e. the prohibition of customary resource uses), leading often to restrictions and evictions (Brockington & Igoe, 2006). Yet from a national perspective, conservation areas are sometimes developed in response to growing resource extraction elsewhere, to spare some land for recovery while intensifying other land uses to increase resource extraction. An example is the conflict about the Tanintharyi Nature Reserve in Myanmar. The establishment of the conservation area is closely linked to the establishment of three gas pipelines running through it and was funded by the gas companies who aimed to secure the pipelines and compensate for the environmental damages produced elsewhere (EJAtlas, 2018). Hence, countrywide increases in social metabolism can lead to spatial segregation processes with distinct implications for territories and resource uses at the local level, including both local increases and decreases in social metabolism. The hypothesis *more metabolism, more conflicts* thus applies best to the national and global level, while at the local level also decreases can provoke environmental conflicts.

16.4 Other 'Grand Explanations': Social Metabolism and Neo-Marxist Perspectives

The socio-metabolic changes and the transformations in local production systems leading to conflicts resemble many of the processes described by neo-Marxist scholars when explaining the development and expansion of capitalism. Karl Marx used the term *Ursprüngliche Akkumulation* (primitive accumulation) to refer to a *"process which takes away from the labourer the possession of his means of production; a process that transforms, on the one hand, the social means of subsistence and of production into capital, on the other, the immediate producers into wage laborers."* (Marx, 1887, p. 508). Marx suggested that primitive accumulation was a historical and transitory phase of societies moving to capitalist systems, which then would be replaced by accumulation based on expanded reproduction (i.e. growth). Neo-Marxists suggest however that 'primitive' accumulation is a persistent process, central to capitalist accumulation in general and not only in its origin. Harvey (2004) refers to it as *accumulation by dispossession,* a concept commonly used to discuss environmental conflicts at the nodes of material extraction (e.g. Holden et al., 2011; Veuthey & Gerber, 2012). Jason Moore's idea of *commodity frontiers* (Moore, 2000) draws further attention to how the expansion of global capitalism restructures not only the social relations of production but also the transformation of distant environments connected to the cores of hegemonic centres through commodity chains, thus, through flows of materials and energy.

In this context, some may wonder whether a neo-Marxist perspective on the expansion of capitalism may serve as a more profound explanation for the rise of environmental change and conflicts. It is important to recognize that these are two sides of the same coin. Increases in social metabolism come together with capitalist economic growth and represent thus the material connectors of political-economic processes at global and national levels and local environmental conflicts. (cf. Muradian et al., 2012). Resources mediated by the specific socio-metabolic configurations are a means to power and accumulation. However, this applies not only to the expansion of capitalism as the dominant socio-economic system. For instance, an industrialized or resource-intensive planning economy or autocratic monarchy would also require large amounts of materials and energy. For the same reasons described above, they would likely trigger environmental conflicts, although the ways how these conflicts manifest could be quite different and may range, for example, from open confrontations and mobilizations to repressed and indirect expressions of discontent (Martinez-Alier, personal communication). The same applies for organized illicit forms of resource extraction, territories under control by armed groups and struggles, or whatever other forms of parallel economies and insurgencies that extract resources for their maintenance and expansion.

The social metabolism resembles therefore a material proxy to track the means for the maintenance and expansion of social systems, power, and profit, no matter in which institutional, political, and economic context they operate and are organized. In that sense, a socio-metabolic perspective is powerful in explaining the emergence of environmental conflicts, because it is not limited to a certain type of social organization such as capitalism. At the same time, this is also its weakness for explaining the ultimate drivers of environmental conflicts. Looking only at the social metabolism does not provide insights on the reasons behind the increases in terms of the social, political, cultural, and economic reconfigurations of the systems of production and consumption that drive the social metabolism.

Analyses of the social metabolism as a driver of environmental conflicts could therefore be more strongly combined with Neo-Marxist approaches, and more generally, with analyses of the political economy in which the social metabolism unfolds (Gerber & Scheidel, 2018). Such combinations can be found, for example, in empirical studies on international trade that combine political economy perspectives with quantitative biophysical studies (e.g. Dorninger et al., 2021; Pérez-Rincón, 2006). These studies illustrate how the international division between centre and periphery countries is based on both an unequal economic and unequal ecological change, through which countries of the global North simultaneously generate monetary surplus and appropriate resources from countries in the global South, where environmental degradation and environmental conflicts arise consequently (Dorninger et al., 2021). Further integration of quantitative socio-metabolic studies with political economy perspectives represents a promising and necessary path to unveil the combined socio-metabolic and political-economic processes at play that co-produce environmental inequalities and environmental conflicts.

16.5 Towards a Balanced View in Environmental Conflict Research

This chapter has summarized some of the main arguments for why, how, and when the social metabolism may provoke environmental conflicts. A socio-metabolic perspective on environmental conflicts is useful because it allows to identify structural causes of conflicts, such as the broader changes in the resource use patterns of economies, and their relation to conflicts over the use of the environment. It also enables linking local environmental conflicts to global production and consumption systems and commodity chains through the material flows and commodities that connect them. Given that material flows continue to rise globally (Schaffartzik et al., 2014), it is also a timely perspective that sheds light on the impacts of growing global resource use on local socio-ecological systems, as well as the role that diverse local communities and social movements play in contesting them at the input, throughput and output side of social metabolism (Scheidel & Schaffartzik, 2019).

Integration of insights from multiple disciplines and theoretical perspectives will further benefit the nuanced understanding of the drivers and dynamics of environmental conflicts. As Martinez-Alier et al. (2010) have argued, *"Ecological economics explain why environmental conflicts arise shedding light on the material origins of conflicts, whereas 'post-structuralist political ecology' (Escobar, 1996) complements this with insights by looking at cultural discourses shaping material outcomes."* Furthermore, as discussed by Muradian et al. (2012, p. 565) *"between the material and energy flows in the economy and the actual occurrence of socio-environmental conflicts there is a large variety of "mediating" variables involved."* This chapter has addressed a few of these variables directly related to socio-metabolic processes, yet there are many more. These are, for example, geographical, ecological, technological, and socio-cultural contexts, the exposure and distribution of impacts across and within diverse social groups, vulnerability of local ecosystems to environmental change, people's risk perception, benefits distribution, how corporations behave and operate, whether there is Free Prior Informed Consent (FPIC) or not, histories of land and resource use and related claims from different actors, local perceptions of justice and injustice, and so on. Close attention must be paid to the political economy and the institutions of societies that shape, and are shaped by, socio-metabolic processes, and which govern the modes of appropriation, distribution, and disposal of materials and energy (Gerber & Scheidel, 2018). Furthermore, cultural aspects centrally shape how conflict and protest manifest and express themselves (della Porta et al., 1999; Hanna et al., 2016).

While it has been beyond the scope of this brief chapter to review the many social, cultural, political, and economic variables that shape the intersections of social metabolism and environmental conflicts, the chapter closes by recalling the need to seek a balanced integration of the social and biophysical processes that co-produce environmental conflicts (Scheidel et al., 2022). The combination of perspectives and the integration of knowledge from different fields as diverse as ecological economics, neo-Marxist approaches or post-structuralist political ecology might create epistemological tensions, but also 'fruitful frictions' (Zimmerer, 2015) that provoke deep learning processes and the careful discussion and consideration of the manifold factors involved in the dynamics of environmental conflicts.

References

Banoub, D., Bridge, G., Bustos, B., Ertör, I., González-Hidalgo, M., & de los Reyes, J. A. (2020). Industrial dynamics on the commodity frontier: Managing time, space and form in mining, tree plantations and intensive aquaculture. *Environment and Planning*. https://doi.org/10.1177/2514848620963362

Brockington, D., & Igoe, J. (2006). Eviction for conservation: A global overview. *Conservation and Society, 4*, 424–470. https://doi.org/10.1126/science.1098410

Bullard, R. D. (1990). *Dumping in Dixie: Race, class, and environmental quality*. Westview Press.

Bullard, R. D. (1994). *Unequal protection: Environmental justice and communities of color*. Random House.

Del Bene, D., Scheidel, A., & Temper, L. (2018). More dams, more violence? A global analysis on resistances and repression around conflictive dams through co-produced knowledge. *Sustainability Science, 13*, 617–633. https://doi.org/10.1007/s11625-018-0558-1

Dell'Angelo, J., Navas, G., Witteman, M., D'Alisa, G., Scheidel, A., & Temper, L. (2021). Commons grabbing and agribusiness: Violence, resistance and social mobilization. *Ecological Economics, 184*, 107004. https://doi.org/10.1016/j.ecolecon.2021.107004

della Porta, D., Kriesi, H., & Rucht, D. (1999). *Social movements in a globalizing world*. Macmillan.

Demaria, F. (2010). Shipbreaking at Alang-Sosiya (India): An ecological distribution conflict. *Ecological Economics, 70*, 250–260. https://doi.org/10.1016/j.ecolecon.2010.09.006

Demaria, F., & Schindler, S. (2016). Contesting urban metabolism: Struggles over waste-to-energy in Delhi, India. *Antipode, 48*, 293–313. https://doi.org/10.1111/anti.12191

Dorninger, C., Hornborg, A., Abson, D. J., von Wehrden, H., Schaffartzik, A., Giljum, S., Engler, J. O., Feller, R. L., Hubacek, K., & Wieland, H. (2021). Global patterns of ecologically unequal exchange: Implications for sustainability in the 21st century. *Ecological Economics, 179*, 106824. https://doi.org/10.1016/j.ecolecon.2020.106824

EJAtlas. (2015). *Philex's Padcal mine, the biggest mining disaster of the Philippines* [WWW Document]. https://ejatlas.org/conflict/philex-padcal-mining-disaster-benguet-philippines

EJAtlas. (2018). *Tanintharyi Nature Reserve conservation area funded by gas pipeline developers, Myanmar* [WWW Document]. https://ejatlas.org/print/tanintharyi-nature-reserve-tanintharyi-region-myanmar

EJAtlas. (2021). *Uranium conflicts* [WWW Document]. https://ejatlas.org/commodity/uranium

Escobar, A. (2012). *Encountering development: The making and unmaking of the Third World* (2012 Ed.). Princeton University Press.

Geenen, S. (2014). Dispossession, displacement and resistance: Artisanal miners in a gold concession in South-Kivu, Democratic Republic of Congo. *Resources Policy, 41*, 90–99. https://doi.org/10.1016/j.resourpol.2013.03.004

Georgescu-Roegen, N. (1969). Process in farming versus process in manufacturing: A problem of balanced development. In U. Papi & C. Nunn (Eds.), *Economic problems of agriculture in industrial societies* (pp. 497–533). Palgrave Macmillan UK. https://doi.org/10.1007/978-1-349-08476-0_24

Georgescu-Roegen, N. (1971). *The entropy law and the economic process*. Harvard University Press.

Gerber, J.-F., & Scheidel, A. (2018). In search of substantive economics: Comparing today's two major socio-metabolic approaches to the economy – MEFA and MuSIASEM. *Ecological Economics, 144*, 186–194. https://doi.org/10.1016/j.ecolecon.2017.08.012

Haas, W., Krausmann, F., Wiedenhofer, D., & Heinz, M. (2015). How circular is the global economy? An assessment of material flows, waste production, and recycling in the European union and the world in 2005. *Journal of Industrial Ecology, 19*, 765–777. https://doi.org/10.1111/jiec.12244

Haberl, H., Wiedenhofer, D., Pauliuk, S., Krausmann, F., Müller, D. B., & Fischer-Kowalski, M. (2019). Contributions of sociometabolic research to sustainability science. *Nature Sustainability*. https://doi.org/10.1038/s41893-019-0225-2

Hanna, P., Vanclay, F., Jean, E., & Arts, J. (2016). Conceptualizing social protest and the significance of protest actions to large projects. *The Extractive Industries and Society, 3*, 217–239. https://doi.org/10.1016/j.exis.2015.10.006

Harvey, D. (2004). The 'new' imperialism: accumulation by dispossession. *Socialist Register, 40*, 63–87. https://doi.org/10.1215/01642472-18-1_62-1

Holden, W., Nadeau, K., & Jacobson, R. D. (2011). Exemplifying accumulation by dispossession: Mining and indigenous peoples in the philippines. *Geografiska Annaler: Series B, Human Geography, 93*, 141–161. https://doi.org/10.1111/j.1468-0467.2011.00366.x

Kapp, K. W. (1950). *The Social Costs of Private Enterprise*. Harvard University Press.

Kolinjivadi, V. (2019). Avoiding dualisms in ecological economics: Towards a dialectically-informed understanding of co-produced socionatures. *Ecological Economics, 163*, 32–41. https://doi.org/10.1016/j.ecolecon.2019.05.004

Litmanen, T. (1996). Environmental conflict as a social construction: Nuclear waste conflicts in Finland. *Society and Natural Resources, 9*, 523–535. https://doi.org/10.1080/08941929609380991

Martinez-Alier, J. (1995). Distributional issues in ecological economics. *Review of Social Economy, 53*, 511–528. https://doi.org/10.1080/00346769500000016

Martinez-Alier, J. (2002). *The environmentalism of the poor: A study of ecological conflicts and valuation*. Edwar Elgar Publishing.

Martinez-Alier, J. (2007). Social metabolism and environmental conflicts. *Socialist Register, 43*.

Martinez-Alier, J. (2009). Social metabolism, ecological distribution conflicts, and languages of valuation. *Capitalism Nature Socialism, 20*, 58–87.

Martínez-Alier, J., & O'Connor, M. (1996). Ecological and economic distribution conflicts. In *Getting down to earth: Practical applications of ecological economics* (pp. 153–183). Island Press.

Martinez-Alier, J., Kallis, G., Veuthey, S., Walter, M., & Temper, L. (2010). Social metabolism, ecological distribution conflicts, and valuation languages. *Ecological Economics, 70*, 153–158. https://doi.org/10.1016/j.ecolecon.2010.09.024

Martinez-Alier, J., Temper, L., Del Bene, D., & Scheidel, A. (2016). Is there a global environmental justice movement? *Journal of Peasant Studies, 43*, 731–755. https://doi.org/10.1080/03066150.2016.1141198

Mohai, P., & Saha, R. (2015). Which came first, people or pollution? Assessing the disparate siting and post-siting demographic change hypotheses of environmental injustice. *Environmental Research Letters, 10*. https://doi.org/10.1088/1748-9326/10/11/115008

Moore, J. W. (2000). Sugar and the expansion of the early modern world-economy. Commodity frontiers, ecological transformation, and industrialization. *Review, 23*, 409–433.

Muradian, R., Walter, M., & Martinez-Alier, J. (2012). Hegemonic transitions and global shifts in social metabolism: Implications for resource-rich countries. Introduction to the special section. *Global Environmental Change, 22*, 559–567. https://doi.org/10.1016/j.gloenvcha.2012.03.004

Navas, G., Mingorría, S., & Aguilar, B. (2018). Violence in environmental conflicts: The need for a multidimensional approach. *Sustainability Science, 13*, 649–660. https://doi.org/10.1007/s1162

Orta-Martínez, M., & Finer, M. (2010). Oil frontiers and indigenous resistance in the Peruvian Amazon. *Ecological Economics, 70*, 207–218. https://doi.org/10.1016/j.ecolecon.2010.04.022

Pérez-Rincón, M. A. (2006). Colombian international trade from a physical perspective: Towards an ecological "Prebisch thesis". *Ecological Economics, 59*, 519–529. https://doi.org/10.1016/j.ecolecon.2005.11.013

Pérez-Rincón, M., Vargas-Morales, J., & Martinez-Alier, J. (2019). Mapping and analyzing ecological distribution conflicts in Andean countries. *Ecological Economics, 157*, 80–91. https://doi.org/10.1016/j.ecolecon.2018.11.004

Robbins, P. (2012). *Political ecology: A critical introduction* (2nd ed.). Wiley.

Schaffartzik, A., Mayer, A., Gingrich, S., Eisenmenger, N., Loy, C., & Krausmann, F. (2014). The global metabolic transition: Regional patterns and trends of global material flows, 1950–2010. *Global Environmental Change, 26*, 87–97. https://doi.org/10.1016/j.gloenvcha.2014.03.013

Scheidel, A., & Schaffartzik, A. (2019). A socio-metabolic perspective on environmental justice and degrowth movements. *Ecological Economics, 161*, 330–333. https://doi.org/10.1016/j.ecolecon.2019.02.023

Scheidel, A., Giampietro, M., & Ramos-Martin, J. (2013). Self-sufficiency or surplus: Conflicting local and national rural development goals in Cambodia. *Land Use Policy, 34*, 342–352. https://doi.org/10.1016/j.landusepol.2013.04.009

Scheidel, A., Temper, L., Demaria, F., & Martínez-Alier, J. (2018). Ecological distribution conflicts as forces for sustainability: An overview and conceptual framework. *Sustainability Science, 13*, 585–598. https://doi.org/10.1007/s11625-017-0526-1

Scheidel, A., Del Bene, D., Liu, J., Navas, G., Mingorría, S., Demaria, F., Avila, S., Roy, B., Ertör, I., Temper, L., & Martínez-Alier, J. (2020). Environmental conflicts and defenders: A global overview. *Global Environmental Change, 63*, 102104. https://doi.org/10.1016/j.gloenvcha.2020.102104

Scheidel, A., Liu, J., Del Bene, D., Mingorria, S., Villamayor-Tomas, S., (2022) Ecologies of contention: how more-than-human natures shape contentious actions and politics, The Journal of Peasant Studies, https://doi.org/10.1080/03066150.2022.2142567

Schlosberg, D. (2004). Reconceiving environmental justice: Global movements and political theories. *Environmental Politics, 13*, 517–540. https://doi.org/10.1080/0964401042000229025

Spiric, J. (2018). Socio-environmental conflicts and sustainability: Lessons from the post-socialist European semi-periphery. *Sustainability Science, 13*, 661–676. https://doi.org/10.1007/s11625-017-0505-6

Temper, L., Demaria, F., Scheidel, A., Del Bene, D., & Martinez-Alier, J. (2018a). The Global Environmental Justice Atlas (EJAtlas): Ecological distribution conflicts as forces for sustainability. *Sustainability Science, 13*, 573–584. https://doi.org/10.1007/s11625-018-0563-4

Temper, L., Walter, M., Rodriguez, I., Kothari, A., & Turhan, E. (2018b). A perspective on radical transformations to sustainability: Resistances, movements and alternatives. *Sustainability Science, 13*, 747–764. https://doi.org/10.1007/s11625-018-0543-8

Temper, L., Avila, S., Del Bene, D., Gobby, J., Kosoy, N., Le Billon, P., Martinez-Alier, J., Perkins, P., Roy, B., Scheidel, A., & Walter, M. (2020). Movements shaping climate futures: A systematic mapping of protests against fossil fuel and low-carbon energy projects. *Environmental Research Letters, 15*, 123004. https://doi.org/10.1088/1748-9326/abc197

Thiri, M. A., Villamayor-Tomás, S., Scheidel, A., & Demaria, F. (2022). How social movements contribute to staying within the global carbon budget: Evidence from a qualitative meta-analysis of case studies. *Ecological Economics, 195*, 107356. https://doi.org/10.1016/j.ecolecon.2022.107356

Tilly, C. (2000). Spaces of contention. *Mobilization. An International Quarterly, 5*, 135–159. https://doi.org/10.17813/maiq.5.2.j6321h02n200h764

Tramel, S. (2016). The road through Paris: Climate change, carbon, and the political dynamics of convergence. *Globalizations, 13*, 960–969. https://doi.org/10.1080/14747731.2016.1173376

Veuthey, S., & Gerber, J. F. (2012). Accumulation by dispossession in coastal Ecuador: Shrimp farming, local resistance and the gender structure of mobilizations. *Global Environmental Change, 22*, 611–622. https://doi.org/10.1016/j.gloenvcha.2011.10.010

Walter, M., & Urkidi, L. (2017). Community mining consultations in Latin America (2002–2012): The contested emergence of a hybrid institution for participation. *Geoforum, 84*, 265–279. https://doi.org/10.1016/j.geoforum.2015.09.007

Whyte, K. P. (2017). The Dakota access pipeline, environmental injustice, and U.S. colonialism. *Red Ink: International Journal of Indigenous Literature, Arts, & Humanities, 19*, 1–6. https://doi.org/10.2307/j.ctv26d8h0.1

Zhang, Y., & Zhao, D. (2018). The ecological and spatial contexts of social movements. In D. A. Snow, S. A. Soule, H. Kriesi, & H. J. McCammon (Eds.), *The Wiley Blackwell companion to social movements* (pp. 98–114). Wiley. https://doi.org/10.1002/9781119168577.ch5

Zimmerer, K. (2015). Methods and environmental science in political ecology. In T. Perreault, G. Bridge, & J. McCarth (Eds.), *The Routledge handbook of political ecology* (pp. 172–190). Routledge.

Open Access This chapter is licensed under the terms of the Creative Commons Attribution 4.0 International License (http://creativecommons.org/licenses/by/4.0/), which permits use, sharing, adaptation, distribution and reproduction in any medium or format, as long as you give appropriate credit to the original author(s) and the source, provide a link to the Creative Commons license and indicate if changes were made.

The images or other third party material in this chapter are included in the chapter's Creative Commons license, unless indicated otherwise in a credit line to the material. If material is not included in the chapter's Creative Commons license and your intended use is not permitted by statutory regulation or exceeds the permitted use, you will need to obtain permission directly from the copyright holder.

Chapter 17
Critical Mapping for Researching and Acting Upon Environmental Conflicts – The Case of the EJAtlas

Daniela Del Bene and Sofia Ávila

17.1 Introduction

Cartography and mapping practices hold a close connection to Environmental Justice (hereafter, EJ). Since the inception and mobilization of the concept of EJ in the 1980s, EJ scholars have developed maps to expose the unequal distribution of environmental burdens and benefits, showing how toxic dumpsites and polluting industries are largely located adjacent to vulnerable and disadvantaged communities (Bullard, 1990, 1999; Pellow et al., 2002; Mohai et al., 2009). Environmental Justice Organizations (EJOs) and activist scholars have also engaged in mapping work as both a research and mobilizing tool (Walker, 2009; Drozdz, 2020).

The Global Atlas of Environmental Justice (hereafter EJAtlas) is a reflection of an evolving approach taking place in the field of EJ studies and activism. The EJAtlas was created in 2011 to collect and systematize data around socio-environmental conflicts arising in response to multiple forms of environmental degradation across the globe. Initially made possible by the EJOLT project,[1] the work of the EJAtlas has continued through two milestone projects led by directors Joan Martinez-Alier (ERC Advanced Grant ENVJUSTICE 2016–2021) and Leah Temper (ISSC-sponsored ACKnowl-EJ project, codirected with Ashish Kothari 2016–2018). These projects have expanded the scope of the conflicts database and

---

[1] EJOLT (Environmental Justice Organizations, Liabilities, and Trade) was a European FP7 project, coordinated by Prof. Joan Martinez Alier at ICTA-UAB (2011–14).

D. Del Bene (✉)
Institute of Environmental Science and Technology (ICTA),
Autonomous University of Barcelona (UAB), Catalonia, Spain

S. Ávila
Institute of Social Studies in the National Autonomous University of Mexico (UNAM),
Mexico City, Mexico

© The Author(s) 2023
S. Villamayor-Tomas, R. Muradian (eds.), *The Barcelona School of Ecological Economics and Political Ecology*, Studies in Ecological Economics 8, https://doi.org/10.1007/978-3-031-22566-6_17

its engagement in both academic research and activist advocacy, providing the most comprehensive database on EJ conflicts globally (Scheidel et al., 2020).

Based on the work of Martínez-Alier and O'Connor (1996), the EJAtlas conceptualizes "socio-environmental conflicts" or "ecological distribution conflicts" as the result of the unfair distribution of environmental 'goods' -i.e. clean water and air, as well as access to fertile land – and 'bads' – i.e. exposure to pollution, as well as risks and threats to health, livelihoods, and social and cultural identities (Scheidel et al., 2020; Temper et al., 2015). However, it also expands this definition to encompass those situations where conflicting worldviews or planes de vida clash and reveal antagonizing political ontologies and epistemologies around the environment, the territory, and all its life forms (Álvarez & Coolsaet, 2020). At its core, the EJAtlas seeks to bridge ecological economics and political ecology to shed light on conflicts that emerge from a changing global social metabolism (Martínez-Alier et al., 2010). By covering a large set of activities and commodities, the EJAtlas dataset offers a socio-metabolic approach to environmental justice, identifying how and why energy and materials are increasingly consumed, but also how they are unevenly allocated, consumed, and disposed of within and across geographical scales and social groups (Scheidel et al., 2018).

As an academic and activist platform, the main aim of the EJAtlas is twofold. First, it offers empirical materials coupled with a methodology that fosters systematic, comparative, and statistical research on socio-environmental conflicts (Temper et al., 2018). Second, it provides an action-oriented tool for organizing, campaigning, and teaching. The ongoing process of data collection is extensive and collaborative, involving several hundreds of individuals and organizations worldwide. The EJAtlas has now become the largest existing global database of socio-environmental conflicts, with over 3800 documented cases by December 2022, and with plans to continue to expand its thematic and geographical coverage over the coming years.

The codified information for each case (including both qualitative and quantitative data) as well as the diversity of cases covered by the database has provided empirical data for several academic publications, and has supported specific campaigns around the world (Avila, 2018; Del Bene et al., 2018; Navas et al., 2018; Scheidel et al., 2020; Špirić, 2018; Temper et al., 2020; Le Tran et al., 2020; Walter & Wagner, 2021). This has enabled conversations around the scale and scope of the EJ movement (Martínez-Alier et al., 2016), the growing relevance of popular environmental concerns spanning different geographical scales, and the connections of EJ with other emerging debates around environmental sustainability (Akbulut et al., 2019; Scheidel et al., 2020; Temper et al., 2018).

This chapter offers an overview of the EJAtlas trajectory and reflections over the theoretical, methodological, and political backgrounds, implications, and challenges of this tool for Environmental Justice studies. We write this chapter using plural pronouns as we reflect and acknowledge the work, reflections, and contributions of the large team of colleagues and collaborators that builds and sustains the EJAtlas.

17.2 Critical Cartography and Environmental Justice

Environmental Justice is both a movement and a community-led science emphasizing how environmental injustices are unequally distributed across social groups and space (Walker, 2009; Temper et al., 2015). As such, the connections between critical cartography and environmental justice are long standing and have evolved over time as the movement has expanded.

The concept of EJ originated in struggles during the 1960's–1980's against the disproportionate effects of toxic chemical use and dumping on people of color in the US. This led to increased academic attention on studying unequal exposures to risks and hazards resulting from policies and practices that discriminated against individuals and communities based on class and race (Bullard, 1990, 1994; Pellow et al., 2002). Key to this "first generation of EJ studies" was the use of spatial analysis – providing critical evidence on the "proximity" of industrial and waste-disposal facilities to particular individuals and communities (Hurley, 1995; Dobson, 1998; Bullard et al., 2007; Mohai et al., 2009).

Over the last two decades, the EJ movement has flourished globally through multiple place-based struggles and the creation of large alliances between communities, organizations, and networks at different scales. In the process, material and sociological concerns have extended beyond the local distribution of pollution, risk, and race to include many other environmental concerns and forms of social difference (Chaix et al., 2006; Charles & Thomas, 2007; Stein, 2004; Wolch et a., 2014; Whyte, 2016). What emerged is the "second generation" of EJ studies, commonly characterized by different features. Firstly, it embraces a wider understanding of justice, in which the distribution of environmental "goods" and "bads" goes hand in hand with the recognition and participation of communities affected by environmental change (Schlosberg, 2013). Secondly, it demands deeper understandings on how specific investments and forms of environmental change are not isolated objects, but are rather connected spaces where value flows, accumulation occurs, and injustices expand (Robbins, 2014). Thirdly, "second generation" EJ studies demand more granular spatial analysis of these processes, through critical mapping interventions at different scales (Pellow, 2017). An increasing literature is today pushing critical EJ research to further engage with decolonial and, ecofeminist studies, and emotional political ecology from research of all continents, languages, and cultural traditions (González-Hidalgo et al., 2022; McGregor et al., 2020; Rodríguez & Inturias, 2018; Pulido & De Lara, 2018; Álvarez & Coolsaet, 2020).

This evolution of EJ studies happened alongside the development of multiple activist-oriented spatial research and cartographic platforms. The production of cartographic knowledge for Environmental Justice has been part of a wider movement to contest, challenge and revert dominant representations of space produced by state and corporate actors (Crampton, 2009). Critical mapping for EJ follows the assumption that, as powerful representations of the world, maps are in fact not neutral depictions of space, but rather are filled with political content (Wainwright & Bryan, 2009). If maps have been a unique source of power for the powerful, then they can also be effective tools of protest and support for social movements reclaiming space (Drozdz, 2020; Lee Peluso, 2011).

In recent decades, EJ research and activism has taken advantage of the spread and accessibility of spatial information and software (such as GIS), providing new epistemological tools for questioning, confronting, and reestablishing the legitimacy of claims for EJ (Elwood & Leszczynski, 2013). Examples of spatial analysis in activist academia include the map of conflicts related to environmental injustices and health by FIOCRUZ in Brazil (Porto et al., 2013; Da Rocha et al., 2018), the spatial analysis on green gentrification in cities by the Barcelona Lab for Urban Environmental Justice and Sustainability – BCNUEJ (Anguelovski et al., 2018), and the mapping work by the CICADA project around extractivism in indigenous territories (cicada.world), among others.

The EJAtlas is positioned in this context as a global-scale project on critical cartography and Environmental Justice, with the aim of mapping and systematizing information on mobilizations against contentious projects and activities happening along the commodity chain, from extraction to disposal.

As we will discuss in the following sections, the EJAtlas has evolved since its creation, becoming a large collaborative dataset with multiple possibilities for academic and activist outlets, some of which are presented in detail hereafter. We take the next sections as a space to reflect on the ways in which the EJAtlas is becoming a "repository of environmental justice stories" (Martinez-Alier dixit), that raises new opportunities for EJ, critical cartography, and activist research development. As a large team of collaborators, we are particularly interested in exploring the challenges and possibilities of spatial, comparative, and statistical political ecology (Scheidel et al., 2020), as well as grassroots cartographic approaches to big data (Mah, 2017; Robinson et al., 2017).

17.3 The EJAtlas. Origins, Goals, and Methods

17.3.1 *Origins, Motivations, and Scope*

The EJAtlas draws on initiatives led by social movements and organizations in mapping conflicts. These include the aforementioned *Fundacao Oswaldo Cruz* (FIOCRUZ) with its work on health and environment with the *Rede Brasileira de Justiça Ambiental* in Brazil, the *Centro di Documentazione sui Conflitti Ambientali* (CDCA-A Sud) in Italy which has been documenting emblematic ecological conflicts globally since 2007, the *Observatorio Latino-Americano de Conflictos Ambientales* (OCMAL), GRAIN, the World Rainforest Movement (WRM), and Oilwatch.

These databases represent impressive sets of cases but are often limited to specific geographic regions or themes and activities (oil extraction, mining, etc). They provide excellent overviews and understanding of specific sectors, cover a country or a region in great detail, and are important references in their field. Thus, they constituted the main source of inspiration for embarking on the ambitious effort of creating a global and transversal database able to draw trends of trade flows, corporate investments, the internationalization of EJ struggles or features of mobilization that increasingly transcend national borders and single continents (Pellow, 2007; Walker, 2009; Sikor & Newell, 2014).

While in-depth case study analysis is a common approach for academic EJ and political ecology studies (Urkidi, 2010; Veuthey & Gerber, 2012), a spatial, comparative, and statistical political ecology can develop tools for identifying patterns and relationships between cases as well as how these cases are shaped by the larger political economy. The EJAtlas aims therefore to provide such a platform to complement previous work with a global and cross-sectorial dataset.

Comparative or statistical approaches have recently been proposed in political ecology (Gerber, 2011; Haslam et al., 2018). Without dismissing the importance and richness of in-depth case studies, these new approaches represent an effort towards a greater plurality in EJ and political ecology research methodology. Moreover, they call for the integration of (critical) cartographic tools in the spatial analysis of environmental conflicts, as well as in data collection.

Based on the EJAtlas dataset, academic articles have provided global overviews on EJ conflicts (Martinez-Alier et al., 2016), as well as regional and thematic analyses on topics such as resistance and violence in Central America (Navas et al., 2018), struggles against mega wind power projects (Ávila, 2018), violence around hydropower dams (Del Bene et al., 2018), resistance to conventional and non-conventional energy projects (Temper et al., 2020), and global patterns of violence against environmental defenders and their role in building sustainability (Scheidel et al., 2020).

Beyond the academic scope, the EJAtlas also serves as a tool for organizing and mobilizing around EJ struggles, including creating alliances and analyzing threats and common patterns of impacts by specific activities or companies. Cartographic information can also be integrated into Featured Maps for visualizing and analyzing spatial implications of such alliances, threats, and impacts (see https://ejatlas.org/featured) (see also Sect. 17.4.1).

Lastly, the pedagogical use of the EJAtlas has also been consolidated in the last few years. Teachers and professors increasingly use the online platform in their courses for illustrating patterns of injustices and resistance, for encouraging their students to engage with EJ issues, and eventually to write about struggles they might be directly involved in (Walter et al., 2020).

17.3.2 Methodology and Co-production of Knowledge

> We collect the contemporary history of the global environmental justice movement. We are like recycling scavengers. Environmental activists might not spend hours with a microphone in hand or on a computer, they have to be active in the demonstrations, they travel to remote places, etc., but they publish blogs, statements, petitions, pictures with banners, sometimes they write songs and poems. We are making an archive of this material to be studied now and in the future. – Joan Martínez Alier

The role of scavengers in nature is to search for and to digest unwanted scraps on the ground, and release them back into the ecological cycle after death. According to Martínez Alier, the EJAtlas takes up a similar role, especially with old cases, or those conflicts that remain unreported, or forgotten, either by scientists, the media, or even by concerned groups.

Like the work of historians, data collection at the EJAtlas includes relevant pieces of evidence and information from online sources, including scientific articles, books, literature produced by EJOs, governments, and international agencies. Both general and specialized media is collected, along with offline archive material such as interviews from fieldwork or from the personal experience and knowledge of academic researchers, activists on the ground, journalists, etc. For cases that are ongoing, or relatively recent, reaching out to local people and activists offers a unique chance to include first-hand information; when this is not feasible, the entry draws on the material produced by local groups and EJOs.

The data is organized into an online database form and sent to one or more rounds of moderation by the EJAtlas editors. This process is intended to ensure that the entry meets a high standard for quality, clarity, and adequate referencing to related sources. The entry can also be translated into different languages and updated whenever necessary.

In 2016, a new feature was introduced to allow co-authorship and co-moderation of entries. Far from being a mere technical add to the platform, it encouraged further collaboration and exchanges among users and fostered a plurality of views and perspectives. An increasing number of entries are today being reviewed, enriched with information, or commented upon by local authors and social movements that become key collaborator and creators of grounded and situated knowledge (Ashwood et al., 2014).

This 'co-production of knowledge' allows for more relational-symmetrical approaches between researchers and local activists, facilitates the integration of diverse experiences and knowledges around environmental conflicts, and contributes to the blurring of established scientific boundaries between academic and activist research (Temper & Del Bene, 2016).

17.4 Towards Statistical and Spatial Political Ecologies: Political and Methodological Challenges

17.4.1 Digging into the Dots: Featured Maps and Multi-layered Analysis

> We have been trying for a long time to mobilize rural communities over the imminent sell out of their lands to the agribusiness sector. However, it was hard for people to grasp the magnitude of this land grabbing process. We then showed people a map showing the time lapse of land concessions given away to corporations. That was stunning to them. We all realized how serious the situation was and the mobilization became then stronger. – Thailand-based researcher and activist[2]

With these words, in 2015, a Thailand-based researcher shared insights and experiences about the relevance of maps in organizing and training communities around socio-environmental injustices with the EJAtlas team. At that time, we had already been collecting and systematizing data around ecological conflicts over a span of four years. Filling in information about the dynamics of conflicts and georeferencing contentious projects was certainly a valuable task. However, to grasp their complexity, the dots sometimes need to be interconnected and their relations explained. How can such a large georeferenced database generate atlas maps that properly explore the complexity around EJ issues while mobilizing EJ research and action?

To respond to this need, the EJAtlas Featured Maps focus on specific topics or questions, trying to 'make sense of the dots' and explain the interlinkages between conflicts, as well as their geospatial features. Explanatory texts are published along with the maps to provide an analysis of the data, research methodology, and references.

Examples include maps of conflicts involving specific corporations, such as Chevron-Texaco, Vale (Saes et al., 2021), Pan American Silver, and Salini-Impregilo/WeBuild (Bontempi et al., 2021), which highlight the systematic violations perpetrated in different localities across the world. These maps were created with the active participation of local communities and transnational organizations, thus strengthening mutual knowledge and networking. This body of work also contributes to the field of Business Economics and Management, by providing examples of environmental liability and corporate social irresponsibility (CSIR) (Riera & Iborra, 2017).

Other featured maps focus on specific features and dimensions of environmental injustices, such as discrimination and environmental racism against the Roma people, or the Blockadia map, depicting politics of mobilization and resistance (Roy & Martínez Alier, 2017). The featured maps on hydropower-related mobilization in

---

[2] For reasons of safety and privacy, the identity of this person is not disclosed.

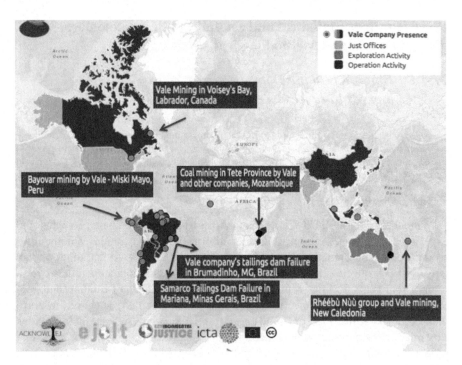

Fig. 17.1 Vale featured map global. (Retrieved from https://ejatlas.org/featured/envconflictsvale. Authors' own elaboration)

the Himalayas, or on conflicts in protected areas in India (Fanari, 2021), also include geospatial data, and allow a better understanding of the geographical and physical context of mobilizations (Fig. 17.1).

17.4.2 Dealing with Spatial Reductionism: Combining Conflict Mapping and GIS Data

Mapping ecological conflicts presents many political and methodological challenges. As critical cartographers pointed out, the cartographic gaze *produces* reality through its representation (Crampton & Krygier, 2006). How can sensitive, delicate, and complex local dynamics be spatially represented by points on a map? How can the extent of the impacted area be represented or even calculated? Spatial representation of complex processes like conflicts can indeed be difficult, and many methodological questions remain unresolved.

For instance, should dam or mining-related conflicts be geolocated on the site of the project infrastructure/concession area, on the impacted land, or on the land where impacted communities live? What if the conflict happens far away from where impacts are felt and perhaps not in the same spot as the actual damaging activity? What if the damaging activity evicts populations from a very large biodiversity-protected area? What if the socio-environmental conflict actually takes shape around the pollution of a river and around the bad management of contaminated water, but the social claims do not question the original causes of the pollution, like industrial activities or mining sites? What if the conflict happens around public policies or laws that affect the whole national territory? What cartographic representations can provide justice to these complex dynamics?

The ways conflicts are narrated and located in space have important political implications and can give rise to different interpretations. In order to ensure consistency across types of activities and levels of accuracy in data sources, the EJAtlas has to opt as much as possible for a common and transversal methodology across the database. By adopting the socio-metabolic perspective described in previous sections, the EJAtlas geolocates the cases either on the contentious infrastructure, on the place where the impacting activity takes place, or on the location where the main demonstrations occurred (such as in the case of opposition to laws, or national referendum campaigns).

Another way to tackle spatial complexity is by incorporating additional GIS data. For example, in the case of the featured map on Vale, GIS vector data was included on a static map to give a sense of the extension of mining concessions the company has in countries like Brazil (by 2020, Vale had 1630 mining concessions in the country, covering a total area of 53,977 km^2, equivalent to the surface of Croatia). Combining data on resistance, conflictive activity, and GIS information can better capture the spatial dimension of conflicts and claims (Fig. 17.2).

17.4.3 Are there Acceptable Limits in Representation and Coverage?

The issues of representation and thoroughness are another key challenges for conflict mapping. How many conflicts are there in the world? What is a statistically reliable sample – a minimum number per national population, per tons of material extracted, or per other indexes? How is selection bias to be avoided? Is having one global transversal criterion enough for establishing a minimum number?

As was previously pointed out, "it is essential to approach maps of protest with an informed critical understanding and view them in light of the sources involved in their making" (Drozdz, 2020: 368). Nobody actually knows how many conflicts over the environment have occurred and are currently unfolding in the world. Additionally, not all conflicts are being covered or reported by locals or in the news. In statistical terms, the dataset can be therefore defined as a "convenience sample of recent and previously documented conflicts from an unknown total number of environmental conflicts worldwide" (Scheidel et al., 2020: 4).

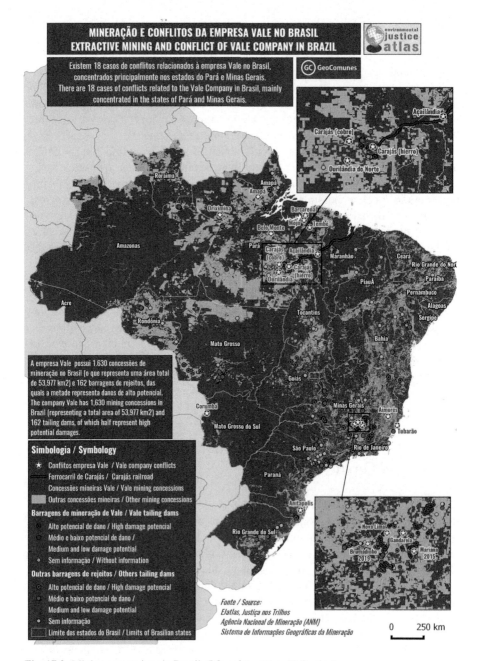

Fig. 17.2 Mining concessions in Brazil. (Map elaboration: Y. Deniau)

Moreover, the EJAtlas, through its ongoing process of documentation and its openness to expanding coverage to more areas and topics, represents a 'litmus test' of current publicly available information on conflicts, of the evolution of conflict types and conflictive sectors, and of the outreach communication capacity of local groups and EJOs. For example, several collaborators have recently expressed concern over emerging urgent issues such as the expansion of large-scale renewable energy technology and its reliance on metal mining, deep-sea mining, geoengineering, and the expansion of the commodity extraction frontier in the Amazon or the Arctic (Hanaček et al., 2022). In terms of geographical coverage, the database shows important gaps such as in the Democratic Republic of the Congo and many other African countries, Pakistan, and former Soviet countries, among others. However, missing data on a map is as interesting as the map itself, as it speaks for the gaps in our knowledge. This is often the case for contexts with weaker local civil society organizations or with difficult political circumstances for resistance, besides the limitations among the EJAtlas team to outreach with these organizations.

17.4.4 Conflicts as Complex Processes and Their Temporal Dimension

As the testimony on the time-lapse map in Cambodia shows, mapping environmental conflicts requires a certain time perspective. Conflicts are not discrete units of analysis; rather they evolve over time. This comes with multiple geographical implications and represents one of the largest challenges for the EJAtlas. One of the ways in which the EJAtlas has partially tackled this issue is by creating a sense of continuity with collaborators when reporting cases in the database. This approach has enabled contributors to update cases when conflicts continue evolving over time. However, updating cases is sometimes difficult to manage considering the size of the database and the resources available to support such a process. From a spatial perspective, this temporal dimension also requires creative ways of representation and specific tools to create such images. A set of geolocated dots of conflicts cannot show, for example, the *process* of the expansion of a commodity extraction frontier (Moore, 2000), i.e. how and where it started and along what routes it evolved.

This is one of the reasons for which critical cartography of environmental conflicts goes beyond geolocating places of resistance towards a process of additional data collection. In the EJAtlas, such temporal dimension has to be searched for inside the database form and often needs the support of other external visualization tools.

A recent collaboration between EJAtlas and Geocomunes, however, sought to overcome part of these spatio-temporal and visual challenges (see also Bracco & Genay, 2021). In aiming to understand the expansion of large-scale wind and solar power and its implications for a just transition, this mapping project included different datasets and cartographic layers. By focusing on the case of Mexico, several

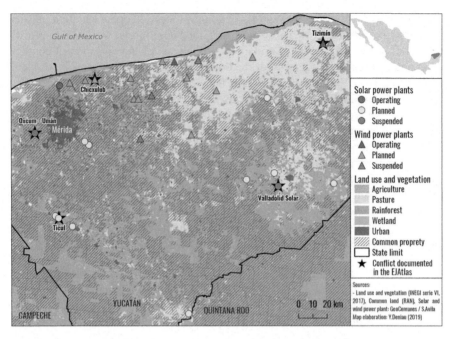

Fig. 17.3 Solar and wind plants in Yucatan, Mexico. (Map elaboration: S. Ávila and Y. Deniau)

maps were produced to juxtapose the polygonal location of mega wind and solar projects with layers on land tenure, land cover, and areas of ecological conservation. This data was produced in parallel with a temporal analysis on the evolution of policies and investments on renewable energies in the country, as well as with the systematization of data on conflicts arising against specific facilities. Together, these elements enabled more granular approaches to the expansion of new energy frontiers and the factors that contribute to the production of environmental injustices on the ground (Avila et al., 2022) (Fig. 17.3).

17.5 Conclusions

In this chapter, we reflect on the origins and evolution of the Global Atlas of Environmental Justice as one of the core components of the Barcelona School of Ecological Economics. The EJAtlas represents a unique global project that seeks to bridge disciplines such as ecological economics, political ecology, environmental justice, and critical cartography for activism and research with the stories, voices, and struggles of local communities, grassroots movements, and transnational networks mobilizing for EJ. As a documentation process, the EJAtlas shows the many challenges associated with the mapping of conflicts and the knots of statistical, comparative, and spatial political ecology, but also its potential.

As the EJAtlas evolves and expands, we find new routes to explore multiple forms of cartographic representations, data analysis, and conceptual approaches. Yet, at the same time, we face renewed challenges and epistemic tensions that, we believe, are intrinsic to activist scholarship and EJ. Do researchers always have the legitimacy to narrate underreported cases or do they sometimes risk exposing people and sensitive processes? Is the EJAtlas an activist tool corroborated by academic research, or is it a platform for scientific research based on activists' knowledge (Escobar, 2008; Martínez-Alier et al., 2014)? How do we overcome the scientific/activist dichotomy, avoid academic data extractivism, and establish methods based on academic and 'political rigor' (Temper et al., 2019)? Should aggregated data access be fully open for all kinds of uses or do EJAtlas editors have the responsibility of ensuring data are correctly interpreted and used under acceptable principles?

These questions do not have straightforward answers and we believe that constant dialogues and bridges between scholars and activists is key to provideing collaborative answers to complex and situated challenges. These questions are themselves part of disputes in the current and future EJ research and action agenda, as well as part of researching and conducting science in an engaged academia. As such, we believe that the EJAtlas not only represents one of the key pillars under which The Barcelona School of Ecological Economics rests. As a theoretical and methodological contribution, the EjAtlas is both a unique space for activists, engaged journalists, teachers and researchers to construct sound campaigns and knowledge around EJ, but also to continue exploring creative and collaborative ways to visually represent and systematically provide knowledge for socioecological justice.

References

Akbulut, B., Demaria, F., Gerber, J. F., & Martínez-Alier, J. (2019). Who promotes sustainability? Five theses on the relationships between the degrowth and the environmental justice movements. *Ecological Economics, 165*, 106418.

Álvarez, L., & Coolsaet, B. (2020). Decolonizing environmental justice studies: A Latin American perspective. *Capitalism, Nature, Socialism, 31*(2), 50–69. https://doi.org/10.1080/10455752.2018.1558272

Anguelovski, I., Connolly, J. T., Masip, L., & Pearsall, H. (2018). Assessing green gentrification in historically disenfranchised neighborhoods: A longitudinal and spatial analysis of Barcelona. *Urban Geography, 39*(3), 458–491. https://doi.org/10.1080/02723638.2017.1349987

Ashwood, L., Harden, N., Bell, M. M., & Bland, W. (2014). Linked and situated: Grounded knowledge. *Rural Sociology, 79*, 427–452.

Avila, S. (2018). Environmental justice and the expanding geography of wind power conflicts. *Sustainability Science, 13*, 1–18.

Avila, S., Deniau, Y., Sorman, A. H., & McCarthy, J. (2022). (Counter)mapping renewables: Space, justice, and politics of wind and solar power in Mexico. *Environment and Planning E: Nature and Space, 5*(3), 1056–1085. https://doi.org/10.1177/25148486211060657

Bontempi, A., Del Bene, D., & Di Felice, L. J. (2021). Counter-reporting sustainability from the bottom up: the case of the construction company WeBuild and dam-related conflicts. *Journal of Business Ethics*, 1–26. https://doi.org/10.1007/s10551-021-04946-6

Bracco, D., & Genay, L. (2021). Entretien avec Yannick Deniau: les projets GeoComunes et EJAtlas. In *Contre-cartographier le monde. Collection Espaces Humain*. Université de Limoges.

Bullard, R. D. (1990). *Dumping in Dixie: Race, class, and environmental quality*. Westview Press.

Bullard, R. D. (1994). *Unequal protection: Environmental justice and communities of color*. Random House.

Bullard, R. D. (1999). Dismantling environmental justice in the USA. *Local Environment, 4*(1), 5–20.

Bullard, R. D., Mohai, P., Saha, R., & Wright, B. (2007). *Toxic wastes and race at twenty: 1987–2007*. United Church of Christ Justice and Witness Ministries.

Chaix, B., Gustafsson, S., Jerret, M., Kristerson, H., Lithman, T., Boalt, A., & Merlo, J. (2006). Childrens exposure to nitrogen dioxide in Sweden: Investigating environmental justice in an egalitarian country. *Journal of Epidemiology and Community Health, 60*, 234–241.

Charles, A., & Thomas, H. (2007). Deafness and disability—Forgotten components of environmental justice: Illustrated by the case of Local Agenda 21 in South Wales. *Local Environment, 12*(3), 209–221.

Crampton, J. W. (2009). Cartography: Performative, participatory, political. *Progress in Human Geography, 33*, 840–848.

Crampton, J. W., & Krygier, J. (2006). *An introduction to critical cartography*. ACME.

da Rocha, D. F., Porto, M. F., Pacheco, T., & Leroy, J. P. (2018). The map of conflicts related to environmental injustice and health in Brazil. *Sustainability Science, 13*(3), 709–719.

Del Bene, D., Scheidel, A., & Temper, L. (2018). More dams, more violence? A global analysis on resistances and repression around conflictive dams through co-produced knowledge. *Sustainability Science, 13*, 617–633.

Dobson, A. (1998). *Justice and the environment: Conceptions of environmental sustainability and theories of distributive justice*. Clarendon Press.

Drozdz, M. (2020). Maps and protest. *International Encyclopedia of Human Geography, 367*–378. https://doi.org/10.1016/B978-0-08-102295-5.10575-X

Elwood, S., & Leszczynski, A. (2013). New spatial media, new knowledge politics. *Transactions of the Institute of British Geographers, 38*, 544–559.

Escobar, A. (2008). *Territories of difference: Place, movements, life, redes*. Duke University Press.

Fanari, E. (2021). Struggle for just conservation: An analysis of biodiversity conservation conflicts, India. *Political Ecology, 28*(1).

Gerber, J. F. (2011). Conflicts over industrial tree plantations in the South: Who, how and why? *Global Environmental Change, 21*, 165–176.

González-Hidalgo, M., Del Bene, D., Iniesta-Arandia, I., & Piñeiro, C. (2022). Emotional healing as part of environmental and climate justice processes: Frameworks and community-based experiences in times of environmental suffering. *Political Geography, 98*, 102721.

Hanaček, K., Kröger, M., Scheidel, A., Rojas, F., & Martínez-Alier, J. (2022). On thin ice–The Arctic commodity extraction frontier and environmental conflicts. *Ecological Economics, 191*, 107247. https://doi.org/10.1016/j.ecolecon.2021.107247

Haslam, P. A., Ary Tanimoune, N., & Razeq, Z. M. (2018). Do Canadian mining firms behave worse than other companies? Quantitative evidence from Latin America. *Canadian Journal of Political Science, 51*, 521–551.

Hurley, A. (1995). *Environmental inequalities: Class, race, and industrial pollution in Gary, Indiana, 1945-1980*. Univ of North Carolina Press.

Le Tran, D., Martinez-Alier, J., Navas, G., & Mingorria, S. (2020). Gendered geographies of violence: A multiple case study analysis of murdered women environmental defenders. *Journal of Political Ecology, 27*, 1189–1212.

Lee Peluso, N. (2011). Whose woods are these? Counter-mapping Forest territories in Kalimantan, Indonesia. In M. Dodge, R. Kitchin, & C. Perkins (Eds.), *The map reader: Theories of mapping practice and cartographic representation* (pp. 422–429). John Wiley & Sons. https://doi.org/10.1002/9780470979587.ch55

Mah, A. (2017). Environmental justice in the age of big data: Challenging toxic blind spots of voice, speed, and expertise. *Environmental Sociology, 3*, 122–133.

Martínez-Alier, J., & O'Connor, M. (1996). Economic and ecological distribution conflicts. In R. Costanza, O. Segura, & J. Martinez-Alier (Eds.), *Getting down to earth: Practical applications of ecological economics*. Island Press.

Martinez-Alier, J., Kallis, G., Veuthey, S., Walter, M., & Temper, L. (2010). Social metabolism, ecological distribution conflicts, and valuation languages. *Ecological Economics, 70*, 153–158.

Martinez-Alier, J., Anguelovski, I., Bond, P., Del Bene, D., Demaria, F., Gerber, J. F., Greyl, L., Haas, W., Healy, H., Marín-Burgos, V., Ojo, G., Porto, M., Rijnhout, L., Rodríguez-Labajos, B., Spangenberg, J., Temper, L., Warlenius, R., & Yánez, I. (2014). Between activism and science: Grassroots concepts for sustainability coined by environmental justice organizations. *Journal of Political Ecology, 21*, 19–60.

Martinez-Alier, J., Temper, L., Del Bene, D., & Scheidel, A. (2016). Is there a global environmental justice movement? *Journal of Peasant Studies, 43*, 731–755.

McGregor, D., Whitaker, S., & Sritharan, M. (2020). Indigenous environmental justice and sustainability. *Current Opinion in Environment Sustainability, 43*, 35–40.

Mohai, P., Pellow, D., & Roberts, J. T. (2009). Environmental justice. *Annual Review of Environment and Resources, 34*, 405–430.

Moore, J. W. (2000). Sugar and the expansion of the early modern world-economy. Commodity frontiers, ecological transformation, and industrialization. *Review: A Journal of the Fernand Braudel Center, 23*, 409–433.

Navas, G., Mingorria, S., & Aguilar-González, B. (2018). Violence in environmental conflicts: The need for a multidimensional approach. *Sustainability Science, 13*, 649–660.

Pellow, D. N. (2007). *Resisting global toxics: Transnational movements for environmental justice*. MIT Press.

Pellow, D. N. (2017). *What is critical environmental justice?* Wiley.

Pellow, D. N., Weinberg, A., & Schnaiberg, A. (2002). The environmental justice movement: Equitable allocation of the costs and benefits of environmental management outcomes. *Social Justice Research, 14*, 423–439.

Porto, M. F., Pacheco, T., & Leroy, J. (2013). *Injustica Ambiental e Saúde no Brasil: O Mapa de Conflitos*. Fiocruz. https://doi.org/10.7476/9788575415764

Pulido, L., & De Lara, J. (2018). Reimagining 'justice' in environmental justice: Radical ecologies, decolonial thought, and the black radical tradition. *Environment and Planning E: Nature and Space, 1*(1–2), 76–98.

Riera, M., & Iborra, M. (2017). Corporate social irresponsibility: Review and conceptual boundaries. *European Journal of Management and Business Economics, 26*(2), 146–162. Emerald Publishing Limited 2444-8451. https://doi.org/10.1108/EJMBE-07-2017-009

Robbins, P. (2014). Cries along the chain of accumulation. *Geoforum, 54*, 233–235.

Robinson, A. C., Demšar, U., Moore, A. B., Buckley, A., Jiang, B., Field, K., Kraak, M. J., Camboim, S. P., & Sluter, C. R. (2017). Geospatial big data and cartography: Research challenges and opportunities for making maps that matter. *International Journal of Cartography, 3*, 32–60.

Rodríguez, I., & Inturias, M. L. (2018). Conflict transformation in indigenous peoples' territories: Doing environmental justice with a 'decolonial turn'. *Development Studies Research, 5*(1), 90–105.

Roy, B., & Martínez-Alier, J. M. (2017). Blockadia por la justicia climática. *Ecología Política, 53*, 90–93.

Saes, B., Del Bene, D., Neyra, R., Wagner, L., & Martínez-Alier, J. (forthcoming). *Environmental justice and corporate social irresponsibility: The case of the mining company Vale S.A*. Ambiente e Sociedade.

Scheidel, A., Temper, L., Demaria, F., & Martínez-Alier, J. (2018). Ecological distribution conflicts as forces for sustainability: An overview and conceptual framework. *Sustainability Science, 13*, 585–598.

Scheidel, A., Del Bene, D., Liu, J., Navas, G., Mingorría, S., Demaria, F., Avila, S., Roy, B., Ertör, I., Temper, L., & Martínez-Alier, J. (2020). Environmental conflicts and defenders: A global overview. *Global Environmental Change, 63*, 102104.

Schlosberg, D. (2013). Theorising environmental justice: The expanding sphere of a discourse. *Environmental Politics, 22*, 37–55.

Sikor, T., & Newell, P. (2014). Globalizing environmental justice? *Geoforum, 54*, 151–157.

Špirić, J. (2018). Ecological distribution conflicts and sustainability: Lessons from the post-socialist European semi-periphery. *Sustainability Science, 13*, 661–676.

Stein, R. (2004). *New perspectives on environmental justice: Gender, sexuality, and activism*. Rutgers University Press.

Temper, L., & Del Bene, D. (2016). Transforming knowledge creation for environmental and epistemic justice. *Current Opinion in Environment Sustainability, 20*, 41–49.

Temper, L., Del Bene, D., & Martinez-Alier, J. (2015). Mapping the frontiers and front lines of global environmental justice: The EJAtlas. *Journal of Political Ecology, 22*, 256.

Temper, L., Demaria, F., Scheidel, A., Del Bene, D., & Martinez-Alier, J. (2018). The global environmental justice atlas (EJAtlas): Ecological distribution conflicts as forces for sustainability. *Sustainability Science, 13*, 1–12.

Temper, L., McGarry, D., & Weber, L. (2019). From academic to political rigour: Insights from the 'Tarot' of transgressive research. *Ecological Economics, 164*, 106379.

Temper, L., Avila, S., Bene, D., Daniela Gobby, J., Kosoy, N., Billon, L., Philippe Martinez-Alier, J., Perkins, P., Roy, B., Scheidel, A., & Walter, M. (2020). Movements shaping climate futures: A systematic mapping of protests against fossil fuel and low-carbon energy projects. *Environmental Research Letters, 15*, 123004.

Urkidi, L. (2010). A glocal environmental movement against gold mining: Pascua-Lama in Chile. *Ecological Economics, 70*, 219–227.

Veuthey, S., & Gerber, J. F. (2012). Accumulation by dispossession in coastal Ecuador: Shrimp farming, local resistance and the gender structure of mobilizations. *Global Environmental Change, 22*, 611–622.

Wainwright, J., & Bryan, J. (2009). Postcolonial reflections on indigenous counter-mapping in Nicaragua and Belize. *Cultural Geographies, 16*, 153–178.

Walker, G. (2009). Beyond distribution and proximity: Exploring the multiple spatialities of environmental justice. *Antipode, 41*, 614–636.

Walter, M., & Wagner, L. (2021). Mining struggles in Argentina. The keys of a successful story of mobilisation. *The Extractive Industries and Society, 8*, 100940.

Walter, M., Weber, L., & Temper, L. (2020). Learning and teaching through the online environmental justice atlas: From empowering activists to motivating students. *New Directions for Teaching and Learning, 2020*, 101–121.

Whyte, K. P. (2016). Indigenous experience, environmental justice and settler colonialism. In B. Bannon (Ed.), *Nature and experience: Phenomenology and the environment* (pp. 157–174). Rowman and Littlefield.

Wolch, J. R., Byrne, J., & Newell, P. J. (2014). Urban green space, public health, and environmental justice: The challenge of making cities 'just green enough'. *Landscape and Urban Planning, 125*, 234–244. https://doi.org/10.1016/j.landurbplan.2014.01.017

Open Access This chapter is licensed under the terms of the Creative Commons Attribution 4.0 International License (http://creativecommons.org/licenses/by/4.0/), which permits use, sharing, adaptation, distribution and reproduction in any medium or format, as long as you give appropriate credit to the original author(s) and the source, provide a link to the Creative Commons license and indicate if changes were made.

The images or other third party material in this chapter are included in the chapter's Creative Commons license, unless indicated otherwise in a credit line to the material. If material is not included in the chapter's Creative Commons license and your intended use is not permitted by statutory regulation or exceeds the permitted use, you will need to obtain permission directly from the copyright holder.

Chapter 18
The EJAtlas: An Unexpected Pedagogical Tool to Teach and Learn About Environmental Social Sciences

Mariana Walter, Lena Weber, and Leah Temper

18.1 Introduction

The Environmental Justice Atlas (EJAtlas) (www.ejatlas.org) was initially developed to make visible and systematize contemporary struggles against environmental injustice worldwide—with and for affected groups (Temper et al., 2015)—and is becoming an attractive interactive tool to teach and learn about environmental and sustainability concepts and trends from an engaged and innovative approach (Walter et al., 2020).

Since its launch in 2012, the EJAtlas has become a research, teaching, networking, and advocacy resource with thousands of daily visits. As of July 2022, it contains 3700 cases worldwide. Strategists, activist organizers, scholars, and teachers are finding many uses for the database, as well as citizens wanting to learn more about the often-invisible conflicts taking place. The map has become a valuable teaching resource for curricula about environmental social sciences, especially about global environmental justice. It has also fulfilled a valuable networking function connecting groups and actors globally. According to a survey responded to by Atlas users (600 responses), over 37% of Atlas survey respondents report using the EJAtlas for teaching and presentations, representing the most common use of the Atlas by visitors.

Different scholars and policy makers are signaling the improvement of environmental education tools to teach about sustainability as a key strategy to address

---

M. Walter (✉) · L. Weber
Institute of Environmental Science and Technology, Autonomous University of Barcelona, Barcelona, Spain

L. Temper
Institute of Environmental Science and Technology. Autonomous University of Barcelona, Barcelona, Spain

McGill University, Quebec, Canada

ongoing environmental crises (Cotton & Winter, 2010), and the need to further develop adequate tools to address this complex issue (Besong & Holland, 2015). Yet, amongst this broader improvement and development of tools for sustainability education, there is a noted lack of resources for and inclusion of *environmental justice* curricula that is attendant to issues of racism, poverty, capitalism, and inequalities (Garibay et al., 2016), reflective of the historical exclusion of marginalized communities from mainstream environmentalism (Garibay & Vincent, 2018).

It has been pointed out that there is a notable lack of diverse voices in mainstream environmental education programs and course materials (Garibay et al., 2016; Garibay & Vincent, 2018; Bonta, 2008), which risks reinforcing the invisibilisation of power dynamics and social roots behind our current environmental challenges. Lack of engaged attention to tangible environmental justice issues can cause a sense of demotivation and what Val Plumwood (2002) calls epistemic remoteness amongst students—a central characteristic of the 'crisis of relationships' behind socio-ecological destruction (Weber & Hermanson, 2015). Incorporation of environmental justice in higher education can help better equip future leaders in sustainability to address both environmental *and* social aspects of pressing socio-ecological challenges (Garibay et al., 2016).

Considering the uses of the EJAtlas online platform as a pedagogical tool responds to the call to expand research on environmental justice education, and can also contribute to different debates in learning studies, regarding, for instance, the features, strengths, and weaknesses of emergent new technologies (Cotton & Winter, 2010; Mayes et al., 2015) or Virtual Learning Environments (e.g. Dillenbourg et al., 2002) in teaching. Moreover, an important motivation is in the challenge to find solutions that are sustainable and that have an impact beyond the specific and immediate context. In this regard, the EJAtlas is an innovative platform for students. The process of studying and developing cases in the Atlas is a powerful experience for students that can engage in ongoing environmental justice struggles.

Moreover, as a collaborative learning platform for researchers and activists, the Atlas can contribute to an emerging field that explores the pedagogical opportunities inherent within social movements and activist work, a research field that scholars signal as promising (Clover, 2002; Hall, 2009; Lowan-Trudeau, 2017).

Environmental justice movements can result in diverse intentional and unintentional learning outcomes for participants and observers (Clover, 2002; Hall, 2009; Lowan-Trudeau, 2017). Hall (2009, 46) highlights the individual, collective, spontaneous, and (re)generative pedagogical nature of social movements and outlines three common forms of learning related to social movements: (1) informal learning occurring by persons who are part of any social movement; (2) intentional learning that is stimulated by organized educational efforts of the social movements themselves; and (3) formal and informal learning that takes place amongst the broad public, the citizens, as a result of the activities undertaken by a given social movement. The latter avenue could be considered for the EJAtlas.

This chapter presents the initial results of an ongoing systematization and analysis of the pedagogical uses of the Environmental Justice Atlas (www.ejatlas.org). This research aims to explore why and how the EJAtlas is used for teaching/learning environmental justice and sustainability and discuss some challenges involved. After this introduction, we present the methods, results, and discuss the main findings of the research, we then conclude.

18.2 Methods: Exploring Contexts, Experiences, and Users

Different data sources were reviewed to study the contexts in which the EJAtlas is used for teaching (country, type of course, level of teaching), the course content and the learning objectives (which concepts, theories, etc.), the pedagogical approaches (lesson plans, exercises, length, etc.) and the key challenges and lessons (including idiomatic, technological, cultural, etc.). We analyzed a range of teaching experiences from China (Northwest Agriculture and Forestry University), Argentina (Universidad Nacional Jauretche), Bolivia (Universidad Nur), Mexico (Universidad Nacional Autónoma de México), Spain (Autonomous University of Barcelona, Pompeu Fabra), Turkey (Bogacizi University), the UK (University of East Anglia) and the USA (Colorado, Michigan, Bishop State Universities) where the EJAtlas was used as a pedagogical tool both in presential and online courses. We conducted five in-depth semi-structured interviews and about a dozen informal conversations with professors that used the EJAtlas. We reviewed teaching plans and the cases added by students to the EJAtlas during seven courses, including five conducted by the first author of the chapter. Students' comments from an international online course on environmental justice (MOOC, University of East Anglia, 2018) that used the EJAltas were also reviewed to assess students' views on the platform. We also considered the results of an online survey responded to by 600 EJAtlas visitors (from 2015 to April 2018). The survey allowed us to examine the profile of the visitors (i.e. students, activists, academics, journalists, etc.), the reason for visiting, and comments regarding the platform.

18.3 Results

The EJAtlas is used in the context of undergraduate and graduate courses in a wide range of countries around the world (e.g. Latin America, North America, Europe, Asia, Africa, Middle East, Australia). The online platform is mainly used in English-speaking courses; however, there are also examples of Spanish-language courses (many cases in Latin America are in Spanish).

The platform is mainly used in environment and sustainability-related courses (e.g., ecological economics, political ecology, environmental sociology, environment and development, environmental justice, etc.), but also as part of courses on ethics, human rights, political economy, and public administration. Some of the key concepts studied with this tool are: environmental justice, environmental conflicts, ecological unequal exchange, commodity chain, extractivism, ethics or business, and human rights.

Users signaled that the EJAtlas offered a large and diverse number of detailed environmental conflicts aiding professors to: (a) choose cases or build tailored maps to explain concepts and illustrate ongoing trends to support their classes and/or; (b) build practical exercises to guide students in the exploration and comparison of cases of EJ struggles at national and international levels, as well as with diverse thematic foci (e.g. mining, land-grabbing, oil, plantations, etc.). (For more information on these pedagogical approaches, check: Walter et al., 2020).

The survey provided some examples of the motivations of professors visiting the EJAtlas:

> *I am a social sciences teacher. I am interested in finding out about territorial conflicts to teach my students about them.* (Colombian professor, EJATLAS survey)
>
> *The Map is useful to getting global perspective and easy to find project level examples.* (US professor, EJAtlas survey)
>
> *Looking for case studies for public school and business school teaching on ethics, sustainability and other subjects.* (South African professor, EjAtlas survey)

Interviews, the survey, and the MOOC reflected that the use of the EJAtlas creates surprise, interest, and motivation among students. Students value its interactivity, its connection with ongoing processes, that it is useful for activism and social change, and some express their interest to keep exploring the platform after class and become active in adding new cases or sharing with their networks:

> Wonderful resource. I have been exploring all the sites in South Africa and share it with my network via social media. (MOOC student)
>
> Many local activist groups, researchers and conservation practitioners are mobilising against these expansions. I am proud to be involved in this struggle for environmental justice for the local people of ___, and will definitely be working with my friends and colleagues there to register this case in the EJ atlas! (MOOC student)
>
> If I use the Atlas it's to find places where I could volunteer and for me, a map full of former battles is not really useful. (Student from unidentified location, EJAtlas survey)

The potential of the EJAtlas to motivate in a context of classes of "disaffected" students has been explicitly raised in a comment:

> (the EJAtlas is) perfect for showing class of disaffected students all the instances of resistance and creativity that are currently going on. very empowering for them I think. (Masters student from California, USA, EJAtlas survey)

However, during their course in China, Scheidel et al. (2018) found that some students had strong emotional reactions to the cases examined and signaled that:

> As teachers of Political Ecology, we have realized how important it is to show not only the destructive sides of conflicts (e.g., environmental degradation, tensions and sometimes also violence between stakeholders, etc.), but also the productive and creative parts of conflicts when seen as spaces of transformation, where injustices and unsustainabilities are exposed and politicized and where alternative ways to development as usual are explored and put forward (…). As probably many lecturers do, we believe that teaching can make a strong impact in the lives of young students. This impact should be, despite of the heavy topics at times, a positive one that doesn't take away students' hope, but rather motivates them to

further unpack and confront problematic issues. In this context, Ecological Economics and Political Ecology as teaching subjects may have an important future in China. As one student said, *"I think this is not the end of exploration to environmental problems!"* (Scheidel et al., 2018, 12)

18.4 Discussion

The EJAtlas, as a worldwide structured repository of environmental justice struggles, allows teachers to work on complex themes with their classes (social movements theory, environmental justice theory, etc.) with real case studies with which students could connect. The exploration of nearby and faraway environmental struggles, as well as their differences and similarities, allowed teachers to examine EJ concepts and struggles, their roots, power dynamics, and reach. Moreover, the possibility of adding cases, as part—or not—of class, offered students the possibility to become activists and actors of change (Scheidel et al., 2018).

> The Environmental Justice Atlas turned out to be a useful teaching tool to provide concrete empirical case material, on which basis theoretical concepts from Political Ecology and Ecological Economics were discussed. Based on illustrative cases from both China and outside, students could realize the connections between society's material and energy use and the frequently unequal distribution of environmental benefits and burdens across different actors and scales. Thanks to the EJAtlas, students could also see that several environmental problems are not limited to a few single cases, or apply only to countries like China. Rather, they are a systemic feature of those places around the globe where intensive resource extraction and processing is taking place. (Examination of China's teaching experience by Scheidel et al. (2018, 5))

While further research is needed, the initial results suggest that the EJAtlas has the potential to address some key concerns emerging in sustainability studies in higher education- the demotivating 'remoteness' students might feel from tangible, on-the-ground issues and activism, the lack of diverse voices present in course material (Garibay et al., 2016; Garibay & Vincent, 2018) (particularly voices from the frontlines of environmental injustices and resistance movements), and the difficult balance to strike between theory and practice (Weber & Hermanson, 2015). The Atlas offers a platform that students and educators can use to help bridge these gaps- by providing a way for students to tangibly engage with important environmental resistance movements, visibilizing diverse, frontline voices and experiences, and connecting the theoretical to the practical via a range of opportunities for promoting environmental justice work outside of the classroom including advocacy, documentation, networking, and solidarity-building (Weber & Hermanson, 2015). In this vein, Osborne (2017, 852) describes the atlas as a tool for what she terms "Public Political Ecology," in that it "builds a *community of praxis* by using theories of environmental justice and Participatory Action Research methodologies to unite scientists, activist organizations, and policymakers around issues of ecological distribution while rendering resource struggles visible to broader publics."

However, since the atlas was not designed for teaching purposes, but as a research and activist collaborative learning platform, this creates some challenges, such as the tension between timelines for reviewing cases for teachers and the atlas team or the requirements for case study selection or the data form development (e.g. data quality, referencing, narrative voice). Nevertheless, the EJAtlas is a showcase of the pedagogical opportunities inherent within social movements and activist work (Clover, 2002; Hall, 2009; Lowan-Trudeau, 2017). As such, we claim that the EJAtlas is a tool that allows students to learn from and engage with the global Environmental Justice movement (Martinez-Alier et al., 2016; Temper & Del Bene, 2016).

The EJAtlas interactive online functionalities that allow users to search, classify, and add cases also offer a tool for professors to structure dynamic teaching lessons on Environmental Justice and sustainability-related issues. However, we claim that the main strength of the EJAltas as a technological development used for teaching is not only grounded in its interactive functionalities (Mayes et al., 2015), but in its live connection with real-world processes and its capacity to connect with emotions and inspire students inside and outside the classroom.

The experience of Scheidel et al. (2018) also suggests that the deep examination of concepts and real cases of environmental injustice can also be a source of feelings of despair. In this vein, they point to the need to counterbalance the examination of the negative trends and sad stories with the positive processes of social mobilization and the transformations these produce.

18.5 Conclusion

The EJAtlas was developed for and by activists and action research scholars to make visible and connect worldwide EJ struggles, as well as to improve understanding and research in this field. From an empirical perspective, we examined how the EJAtlas was used in different geographical and learning contexts to teach environmental justice and sustainability themes. The analysis of learning and teaching experiences showed that the EJAtlas, as a tool developed for research, social action, and transformation, allowed educators to develop short or long exercises that can motivate students. The possibility of adding a relevant case, which is a more complex, time- and effort-consuming exercise, is attracting interest. The platform offered a pre-built interactive form that allowed for structuring a learning process on understanding the key components of an environmental struggle. Moreover, the result of this exercise can transcend classroom walls and contribute to a wider EJ worldwide collaborative work. Our examination has, however, pointed to some needs and challenges, such as the limited capacity of the current EJAtlas team and the need to improve the involvement of teachers in the final moderation of cases.

From a conceptual approach, the Environmental Justice Atlas contributes to the understanding of the global EJ movement, its claims, and how environmental justice organizations are working around the globe against environmental injustices. The interactive functions of the online platform also allow a wide range of uses and explorations by teachers and students. These features address some of the challenges of learning (with and without new technologies) (Dillenbourg et al., 2002).

Moreover, the EJAtlas has the potential to address some key concerns emerging in sustainability studies in higher education—the demotivating 'remoteness' students might feel from tangible, on-the-ground issues and activism, the lack of diverse voices present in course material (Garibay et al., 2016; Garibay & Vincent, 2018), and the difficult balance to strike between theory and practice (Weber & Hermanson, 2015). Furthermore, while the debate on the use of new technologies for teaching has focused on the ability of these interactive tools to develop capabilities and improve productivity (Mayes et al., 2015), the EJAtlas showcases the potential to inspire and motivate students of interactive technological tools developed with/for social movements.

Finally, we would like to signal that the results of this ongoing research are feeding the discussion regarding how to improve the Atlas and the way we work with educators and students.

References

Besong, F., & Holland, C. (2015). The Dispositions, Abilities and Behaviours (DAB) framework for profiling learners' sustainability competencies in higher education. *Journal of Teacher Education for Sustainability, 17*(1), 5–22.

Bonta, M. (2008, January 3). How do we diversify? *Grist.* http://grist.org/article/how-to-diversify-environmentalism

Clover, D. (2002). Traversing the gap: Concientizacion, educative-activism in environmental adult education. *Environmental Education Research, 8*(3), 315–322.

Cotton, D., & Winter, J. (2010). It's not just bits of paper and light bulbs. A review of sustainability pedagogies and their potential for use in higher education. In *Sustainability education: Perspectives and practice across higher education* (Vol. 1, pp. 39–54). Earthscan.

Dillenbourg, P., Schneider, D., & Synteta, P. (2002). Virtual learning environments. In A. Dimitracopoulou (Ed.), *Information & communication technologies in education* (pp. 3–18). Kastaniotis Editions.

Garibay, C., & Vincent, S. (2018). Racially inclusive climates within degree programs and increasing student of color enrollment: An examination of environmental/sustainability programs. *Journal of Diversity in Higher Education, 11*(2), 201–220.

Garibay, C., Ong, P., & Vincent, S. (2016). Program and institutional predictors of environmental justice inclusion in U.S. post-secondary environmental and sustainability curricula. *Environmental Education Research, 22*(7), 919–942. https://doi.org/10.1080/13504622.2015.1054263

Hall, B. (2009). A river of life: Learning and environmental social movements. *Interface, 1*(1), 46–78.

Lowan-Trudeau, G. (2017). Protest as pedagogy: Exploring teaching and learning in indigenous environmental movements. *Journal of Environmental Education, 48*(2), 96–108. https://doi.org/10.1080/00958964.2016.1171197

Martinez-Alier, J., Temper, L., Del Bene, D., & Scheidel, A. (2016). Is there a global environmental justice movement? *Journal of Peasant Studies, 43*(3), 731–755. https://doi.org/10.1080/03066150.2016.1141198

Mayes, R., Natividad, G., & Spector, M. (2015). Challenges for educational technologists in the 21st century. *Education Sciences, 5*(3), 221–237. https://doi.org/10.3390/educsci5030221

Osborne, T. (2017). Public political ecology: A community of praxis for earth stewardship. *Journal of Political Ecology, 24*(1), 843–860.

Plumwood, V. (2002). *Environmental culture: The ecological crisis of reason*. Routledge.

Scheidel, A., Navas, G., & Liu, J. (2018). Enseñando ecología política en China. *Ecología Política, 56*, 8–13.

Temper, L., & Del Bene, D. (2016). Transforming knowledge creation for environmental and epistemic justice. *Current Opinion in Environmental Sustainability, 20*, 41–49. https://doi.org/10.1016/j.cosust.2016.05.004

Temper, L., Del Bene, D., & Martinez-Alier, J. (2015). Mapping the frontiers and frontlines of global environmental justice: The EJAtlas. *Journal of Political Ecology, 22*(1), 255–278.

University of East Anglia. (2018). *EJAtlas tutorial*. MOOC on Environmental Justice. https://www.futurelearn.com/courses/environmental-justice/0/steps/37211. Accessed 10 July 2018.

Walter, M., Weber, L., & Temper, L. (2020). Learning and teaching through the online Environmental Justice Atlas. From empowering activists to motivating students. *New directions for teaching and learning* (Special issue "Teaching about sustainability across higher education coursework", Vol. 161, pp. 101–122).

Weber, L., & Hermanson, A. (2015). *Anti-oppression and academia. Applying critical methodologies to study identity and student experiences in university settings*. Master's thesis, Lund University. Accessed at: http://lup.lub.lu.se/luur/download?func=downloadFile&recordOId=5387481&fileOId=5471369

Open Access This chapter is licensed under the terms of the Creative Commons Attribution 4.0 International License (http://creativecommons.org/licenses/by/4.0/), which permits use, sharing, adaptation, distribution and reproduction in any medium or format, as long as you give appropriate credit to the original author(s) and the source, provide a link to the Creative Commons license and indicate if changes were made.

The images or other third party material in this chapter are included in the chapter's Creative Commons license, unless indicated otherwise in a credit line to the material. If material is not included in the chapter's Creative Commons license and your intended use is not permitted by statutory regulation or exceeds the permitted use, you will need to obtain permission directly from the copyright holder.

Chapter 19
Commons Regimes at the Crossroads: Environmental Justice Movements and Commoning

Sergio Villamayor-Tomas, Gustavo García-López, and Giacomo D'Alisa

19.1 Introduction

The most widespread body of knowledge around the commons comes from the theory of the commons. The theory, which has also been associated to the idea of community-based natural resource management (CBNRM), has traditionally relied on economic theory to understand whether and how local communities are able to design and change rules that promote cooperation and collective management of shared resources (i.e., commons). The theory has become one of the best-known theories of governance within ecological economics.[1] Over time, however, the theory has also received critiques and co-evolved with them. Relevant for this chapter are the critiques raised for its relative inattention to how historically shaped patterns of power, conflict, the "state", and the broader political-economic context shape the access to and uses of common resources, and CBNRM regimes (Johnson, 2004; Saunders, 2014).

The "critical commons" literature has focused on the political nature of commons initiatives as solutions to ecological distribution conflicts and their entanglements with environmental justice movements (for a review of different approaches Villamayor-Tomas & García-López, 2021); as well as on exploring how the commons can develop a path of emancipation from capitalism by building an alternative mode of production to the state and the market (Caffentzis & Federici, 2014). This

---

[1] The theory has also been called common-pool resource (CPR) theory and institutional or collective action theory of the commons.

S. Villamayor-Tomas (✉)
UAB, Barcelona, Spain

G. García-López · G. D'Alisa
University of Coimbra, Coimbra, Portugal

literature has questioned the managerial emphasis of the theory of the commons, pointing instead to how state's policies led by private and corporate interests have often tried to privatize (i.e., enclose) commons to grab the value commoners produce and boost the capital accumulation process; and how marginalized and resource-dependent populations have mobilized to defend Commons and livelihoods around them.

While the former line of research builds on Marxist analysis of enclosure (De Angelis, 2001), the latter connects with Joan Martínez Alier's materialist approach to environmental conservation. Since the beginning of the 2000s, Martínez-Alier has challenged the post-materialist hypothesis according to which environmental protection and values emerge only among those people that have already secured their high material standard of life (Inglehart, 2000). He contrasted this environmentalism of the wealthy North with that of marginalized, particularly indigenous and peasant communities in the global South, who struggle to defend the environment, not for its own sake, but as their source of material livelihood, health, and identity. Here, we build and expand this approach to make our case for commons movements.

In the next section, we elaborate on how Martínez Alier's environmentalism of the poor thesis was not far at all from that of early commons scholars, and Elinor Ostrom in particular. In Sect. 19.3, we first introduce the Barcelona School's approach to the commons and elaborate what we identify as its three main themes: commons and movements (Sect. 19.3.1), urban commons in crisis/transitions (Sect. 19.3.2), processes of commoning (Sect. 19.3.3), and commons and degrowth (Sect. 19.3.4). In Sect. 19.4, we conclude with some thoughts on the future directions of this approach and themes.

19.2 Communalism and Commons: Tangential Parkours

Part of the work of Martínez-Alier can be traced back to his experience in Peru in the 1970s during the agrarian reform years. Building on his contributions to the Agrarian Archives, he wrote "Los Huacchilleros del Perú" (1973) on the resistance of pastoral communities to being dispossessed from their lands. Influential during these years were works like Florencia Mallon's *The Defence of Community* in Peru's Central Highlands as well as writings from Russian intellectuals around the pre-revolutionary Narodnik movement, which advocated for an agrarian socialism around the autonomy of local communities. Thus, Martínez-Alier was not alien to the "defense of the commons", understood as the struggles of local agricultural communities to defend their lands, and epitomized early experiences of organized agrarianism, or communalism, that followed the Zapatista revolution in Latin America or the early anarchist movement in Spain (*personal communication*). From this perspective, the real tragedy of the commons was not overuse, but that of enclosures, or the encroachment and accumulation of land by big landowners with the sponsoring of governments.

The empirical evidence used both by early commons scholars like Ostrom and political ecologists like Martínez-Alier was quite similar and included historical and anthropological accounts of the self-organization experiences of relatively autonomous local communities around for the management and defense of their resources (see for example the works by Kurien (1991) on community-based fishing organizations and their resistance to the encroachment of their fishing grounds by commercial trawlers). More importantly, both groups of scholars understood the intricate connections between the material well-being of resource-dependent communities and their stakes and capacity to collectively manage and defend those resources (Villamayor-Tomas & García-López, 2018). In a way, they looked at two sides of the same coin.

Ostrom, like other early commons scholars, was concerned about justice issues. In this sense, it could be argued that they were also political ecologists *ante litteram*. Much of the work around the theory of the commons aimed to demystify traditional economic theories that advocated for state control or privatization under the assumption of unavoidably uncooperative local natural resource user groups. Accordingly, the theory of the commons and Ostrom's core legacy in particular is based on countless empirical examples of the ability of natural resource-dependent communities to overcome rivalry and create rules for sustainable management (Forsyth & Johnson, 2014). In turn, Martínez-Alier showed how these same commoners had self-organized to defend their lands, livelihoods, and health against fight exclusionary policies and development and conservation schemes that dispossessed them from their resources.

19.3 The Barcelona School: An Agenda Around the Commons

In this section, we present four themes and associated initiatives that are representative of the Barcelona School approach to the commons: commons movements, urban commons, commoning, and degrowth. These are linked to various projects that have been (co)led by Barcelona School scholars, of which we highlight three: the Environmental Justice Atlas (EJ Atlas and Transform-EJ projects), the Barcelona the European Network of Political Ecology (ENTITLE), the Research & Degrowth (R&D) Collective and the PROCOMÚ (PROCOMMON) project. The EJ Atlas is a collaborative mapping of environmental conflicts co-coordinated by J Martínez-Alier, along with Leah Temper, Daniela Del Bene, Arnim Scheidel, Sara Mingorría, Brototi Roy, Marta Conde, Mariana Walter, and Grettel Navas, among others (see Temper et al., 2018).[2] Members have carried out important research on the relation between EJ movements and commons (see Villamayor-Tomas et al., 2022 for a recent compilation). ENTITLE is a Marie Curie Training Network (ITN) that ran

---

[2] See www.ejatlas.org

until 2015 and has continued in the work of the Undisciplined Environments political ecology blog (previously the ENTITLE blog).[3] The commons were one of the project's key themes, as reflected in one of its main outputs: the *Political Ecology for Civil Society* manual (Beltrán et al., 2015; also Andreucci et al., 2017). The R&D Collective has led discussions connecting commons with degrowth. R&D members' *Degrowth: A Vocabulary for a New Era* (D'Alisa et al., 2015a) included the commons as a keyword in an essay written by commons scholar-activists David Bollier and Silke Helfrich. The collective's Masters Program in Political Ecology also features the commons as a central topic.[4] The PROCOMÚ project, co-led by ICTA and the Institute for Government and Public Policy (IGOP-UAB), has focused on developing an inventory and better categorizing the myriad of citizen-based initiatives that have emerged over the last decades and since the 2008 economic crisis in Barcelona. Finally, the COMOVE project focused on unveiling the synergies and trade-offs between social movements and commons management projects, resulting in a workshop and special issue (Villamayor-Tomas et al., 2022) and several other publications (Villamayor-Tomas & García-López, 2018, 2021).

19.3.1 Commons Movements

In the last decade, scholars with good understanding of both the institutional and critical theories of the commons, have been documenting how the contentious politics of resource users and their allies contribute to advance rights to commons (e.g., Becker et al., 2017; Kashwan, 2017). To some extent, they have continued a thread started by political ecologists working in the interface of anti-extractive resistance movements and autonomism, self-management, and communalism, reflected in concepts such as "environmentalism of the poor" (Martínez-Alier, 2003; Guha & Martínez-Alier, 2013), "grassroots livelihood" movements (Peet & Watts, 1996), "ecological resistance movements" and "popular environmentalism" (Taylor, 1995; Goldman, 1998), "place-based/territorial resistances" (Escobar, 1998), and "local sites of resistance" (Blaikie, 2006).

Scholars associated with the Barcelona School have pioneered and expanded the study of the above "commons movements" as a distinctive type of social movement and commons-making experience (see Villamayor-Tomas & García-López, 2021 for an overview, and Navas et al., 2022 for recent applications to working-class communities and enviornmentalism). Common movements are defined as "politically active community projects that scale out within a territory and/or social mobilizations that materialize into practices of communal management, all aiming for a transformation toward a commons-based society" (pp. 513). Underlying this definition is the hypothesis that there is a cyclical relationship where movements provide

---

[3] www.undisciplinedenvironments.org
[4] See https://master.degrowth.org/masters-in-political-ecology/

impetus for institutional defense of the commons and of commons-based alternative projects, while commons are the social fabric through which movements' demands, visions, and agendas for social and environmental justice, direct democracy and sustainability, can be materialized (De Angelis, 2017). This is particularly evident in the context of rural community-rights movements in the global South, as well as in new water and food commons movements and community energy movements in both the global South and North. Tensions and contradictions of commons-movement dynamics reflect trade-offs between diversity versus uniformization and organizational stability versus expansion of discourses and practices (Villamayor-Tomas & García-López, 2021).

Of relevance here is a special issue published on the topic, which compiles a diversity of cases all reflecting the prominence gained by the commons frame and the diversity of cases illustrating connections with social mobilization in both the rural and urban contexts (Villamayor-Tomas et al., 2022). As pointed out in the introduction of that Issue, there is a history of movements fighting to "reclaim the commons" since the alter-globalization struggles of the 1990s (see Klein, 2001); and a number of recent events including the World Social Forum, the European Commons Assembly, the "indignados" and "occupy" movements in Spain, Greece, and Turkey, or the water commons movements in Italy, Bolivia and elsewhere across the world, have reclaimed the commons as part of their overall agenda and/or grievances. In parallel, more practitioners have placed in the commons the hope to fulfill societal transformations for more democratic, equitable, and ecological lives, resulting in a variety of new agro-ecological food producers and consumer groups, integral cooperatives, urban gardens, community-energy projects, and peer-to-peer designs in cities and rural areas with shared practices and agendas; and in commons-based political platforms run by municipalities. In short, the commons are seen as having an "insurgent power" to advance a society that is "free, fair and alive" (Bollier & Helfrich, 2019).

The SI's diversity of cases was complemented with the use of various combinations of established and emergent conceptual tools, ranging from institutional analysis and political ecology commons theories, to social movement approaches of political opportunities, framing, and transnational networks, to new tools such as collective (re)actions, liminal commons, management opportunity structures, and rooted water collectives to describe the type of organizations that embed both community. In a study of water, forest, and fisheries communities in Mexico and Sri Lanka, for example, Villamayor-Tomas et al. (2020) illustrate how movements can ensure the implementation of collective rights and facilitate the organization of local community organizations and federations of them. By the same token, however, mobilization can reinforce pre-existing divisions within CBNRM regimes; and the failure to mobilize or to achieve the goals of mobilization can trigger dynamics of cooperation defection in local commons governance. Dell'Angelo et al. (2021) offer a panoramic of movements that emerge in reaction to agribusinesses-related land acquisition and commons grabbing especially in the global South and analyze the different configurations of socio-environmental impacts, actors, and forms of protest, that emerge across different contexts. As they show, violent collective reactions employ a wide variety of protest strategies at multiple scales. Dupuits et al. (2020) in turn center on the opportunities and risks for local communities of "transnationalizing" their mobilizations. The authors do so through an analysis of the advocacy for international recognition of the rights to water and rights of nature, by the Latin-American Confederation of Community Organizations for Water Services and Sanitation (CLOSAS) and the Coordination of the Indigenous Organizations of the Amazon Basin (COICA). As they point out, communities gain form of recognition of rights by linking their struggles at higher scales, yet this comes also at the cost of commensuration (i.e., of the diversity of local interests) and associated exclusionary tendencies.

Transversal to the above contributions and others included in the special issue is an interest in the coevolution of movements and commons, the role of heterogeneities, and cross-scalar dynamics. They have opened up, for example, new debates about the tension between commons as emergent, open, or "liminal" processes, and as long-lasting and more institutionalized collective projects (e.g, Varvarousis, 2020; Moreira & Morell Fuster, 2020); or the intended and unintended effects of state regulation on the emergence and consolidation of commons movements (Villamayor-Tomas et al., 2020); highlighted that divisions within communities and movements (e.g., around class-based vs. identity-based grievances, gender, or alliance-building strategies) are the norm rather the exception (e.g., Vos et al., 2020; Dell'Angelo et al., 2021; Tyagi & Das, 2020); or illustrated the ubiquity of scaling-up and out strategies and their opportunities and challenges (e.g., Dupuits et al., 2020; Pera, 2020).

19.3.2 Crisis, Urban Prosumer Groups, and Local Governments

Much of the interest in commons in the Barcelona School has also translated into works on "urban commons", particularly in the aftermath of the 2008 crisis in Europe, and the consequent emergence of the 'movements of the squares'. Some of these works are tightly connected with the School's interest in movements. Camps-Calvet et al. (2015), for example, rely on data collected from 27 urban community gardens in Barcelona to illustrate how they contribute to both building community resilience in times of crisis, and articulating forms of resistance to development pressure and commodified urban lifestyles (see also Calvet-Mir & March, 2019). Varvarousis (2020) focuses on commons-making in the context of crisis and broad social mobilizations, using the case study of Athens, Greece. Using the concepts of "liminality", developed by Turner, and "rhizomatic", developed by Deleuze and Guattari. He argues that during these periods, commoning is not just temporary, they do not "disappear" entirely even when it appears so. Rather, they are "liminal", a rite of passage of sorts where commoning creates new social ties and commons projects, while also becoming disseminated across existing social networks. This expansion is "rhizomatic": they can "facilitate transitions and may transform into or give rise to other, more stable, forms of commoning in their wake". (Varvarousis, 2020: 5). Asara (2020) analyzes the commons as one of the elements of the "radical imaginary" of the Indignados in Madrid – together with autonomism and ecologism. And in comparing the experiences in Greece and Spain, Varvarousis et al. (2021) conclude that the commons were a prefigurative outcome from the movements of the squares.

Apostolopoulou and Kotsila (2021) also explore the making of commons in post-crisis and mobilization contexts in Hellinikon, Greece. Drawing on the theories of critical urban geographers and political ecologists (Harvey, Lefebvre, Smith, Heynen, Kaika, Swyngedouw), they argue that urban "guerrilla" gardening, as a process of commons-making, can gesture towards autogestion, which can embed and foment radical grassroots resistance for the right to the city, against neoliberalization of urban spaces and natures. Moreira and Morell Fuster (2020), for their part, focus on the life cycle and institutional arrangements of a food network that emerged from Porto's solidarity economy movement. They use this to inquire about the nature of the new, post-2008 crisis commons in Portugal. As they show, the preexistence of an ecosystem of local collectives and their leadership were key in the formation of the network and its organization around democratic values, the rejection of food as a commodity, and its openness regarding resources and knowledge.

Beyond connections with movements, others have more genuinely aimed at better conceptualizing urban commons and their relationship with local governments. By building on both institutional, critical, and Marxist scholarship, Ferreri et al. (forthcoming) propose a framework to distinguish urban commons from other social and solidarity economy initiatives, as prosumer groups with strong social and/or environmental transformation agendas and the ambition of constituting alternatives to state and market provision of services. Maestre et al. (forthcoming), in turn, point to the relative diversity of the more than 400 commons initiatives by Barcelona. As illustrated, the 5 clusters of commons initiatives, can be understood by looking at the sector whether they unfold, their social vs. environmental transformation ambition, and their connections with historical experiences of associationism in the city. Connections with local governments show that urban commons do not emerge in a vacuum of governance is well illustrated in Pera (2020) and her study of community socio-cultural centers reclaimed and then managed by local residents in Barcelona in the last decade. As she illustrates, there is a double-edged sword of local government's support: even though local policies that promote and protect the commons represent an opportunity for them to flourish, the agreements established with the city council can limit the capacity for the commons to become alternative spaces for reinventing the city. Popartan et al. (2020) show how the discourses about water as commons (linked to rights, life, and democracy), emerged from the Water is Life movement in Barcelona, aligned with anti-privatization struggles in Latin America and the Indignados movement, and then became embedded in the municipal government's left-populist identity. And Calvet-Mir & March, (2019), other Barcelona School scholars show the contradictory positions on how to govern water between different political actors within the municipal government, leading to a deadlock.

19.3.3 Performative Commons, Commoning and Becoming a Commoner

Barcelona School scholars have also been also involved in rethinking the commons. These scholars engage with a diversity of thinkers[5] to connect commons to other keywords such as praxis, counter-hegemony, performativity, prefigurative politics, (re)subjectification, liminality, insurgency, commonwealth, autonomy, self-management (autogestion), working-class or "commons" environmentalism, and communitarian weavings. These concepts contribute to expanding notions of commons beyond biophysical (material) or intangible "resources" or "goods". Instead, they propose that "everything is a commons" (De Angelis, 2017): simultaneously a social fabric and a principle of the Earth as a shared living space on which we all depend, which is always collectively re/produced, always with consequences to others (human or non-human). Rather than commons as a thing, or a set of rules, these scholars emphasize that commons are networks of coevolving structures that connect social and ecological processes (D'Alisa, 2013); and that they are co-produced, made, and reclaimed through everyday practices, relations, subjectivities, and imaginaries of acting and being in common. Commons in this sense are the idea of commoning, which covers a good part of this understanding, is defined as the process of collectively making the commons, making ourselves in the process as commons subjects or "commoners", shifting towards more equitable and ecological forms of relating to our environment. In this sense, commoning is a "performative" act – a practice which seeks to undo dominant relations and imaginaries or common senses, while being embedded within them (García-Lamarca, 2015; García-López et al., 2021; Velicu & García-López, 2018). It is, furthermore, seen as central to social struggles of communities in defense of their territories, as well as social movements seeking societal transformations. The commons, in other words, become the praxis and political vision of an equitable, deeply democratic, and ecologically sustainable society, "our horizon of peace, freedom and plenty" (De Angeils, 2017: 172).

The casuistic around commoning cases is growing rapidly (e.g., Bollier & Helfrich, 2015; Clement et al., 2019), also within the School. Caggiano and De Rosa (2015), building on De Angelis's ideas on commoning and commons movements / social movement dynamics, illustrate the strategic alliances of environmental activists and social cooperatives in Napoli in their struggle to reclaim and reuse waste disposal lands from the Mafia's in peri-urban areas. They highlight how, by creating new agricultural projects, they not only re-appropriate these land as a commons, but they start to *"make community"*, creating new social ties, cultural

---

[5] Including among others Massimo De Angelis, George Caffentzis, Silvia Federici, Stavros Stavrides, David Bollier, Silke Helfrich, Neera Singh, Andrea Nightingale, Valerie Fournier, Raquel Gutierrez, Judith Butler, Miriam Tola, Jacques Ranciere, Michael Hardt, Antonio Negri, Karl Marx, Antonio Gramsci, Andre Gorz, Cornelius Castoriadis, Henri Lefebvre, Gregory Bateson, and Giles Deleuze and Felix Guattari.

practices, and institutions as part of the struggle to shift from a mafia economy to a social-ecological economy. They thus provide an analysis of strategies and limits for a symbolic and practical project of commoning within environmental justice movements, as part of strategies of remaking their territory.[6]

García-López et al. (2017) use Butler's ideas of performativity and Antonio Gramsci's ideas about counter-hegemony and common sense to analyze the movement and autogestion initiative from Casa Pueblo in the forests of Adjuntas, Puerto Rico. They argue that this initiative entails a process of changing dominant relations and common senses regarding democracy, forests, and community, and creating new "commons senses": a praxis of democratic deliberation where "the people" decide, new economies that provide local livelihoods, the forest as a site of collective care and well-being to be managed communally for the common good, and new forms of community that build trans-local networks of solidarity.

Mingorría (2021) mobilizes Guitierrez's concept of communitarian weavings, as well as Caffentzis, Federici, and De Angelis' ideas on commoning, to analyze agrarian commons in the Maya-Q'eqchi' communities in Guatemala. Such relations can be defined as having relative autonomy from capitalism, reproducing essential needs for "lives worth living. Mingorría identifies the communal relations around agriculture as a permanent form of agrarian commons, but also points to two types of "temporary commons" that intersect with those: "encuentros campesinos" (peasant gatherings), where participants enact daily collective practices; and "land occupations". These show how commons are not fixed nor isolated, and supersede local community through weavings of communities in movement.

Scholars linked to the Barcelona School have also analyzed commoning processes within urban struggles against neoliberalization and for the right to the city (see also Sect. 19.3.2).

Finally, the Undisciplined Environment's and FLOWs collaborative blog series on commoning water (Leonardelli et al., 2021)[7] has also collected a number of insights about the connections of commons-making with struggles against water privatization, for water justice and for self-management and direct democracy in water governance around the world (e.g., Bresnihan, 2020; Olivera & Archidiacono, 2021). Contributors have also mobilized ideas of multi-species commons, including human and more-than-human actors (salmon, beavers, algae, etc.) co-influencing each other (e.g., Woelfle-Erskine, 2020), and the making of commons through daily practices such as swimming (Hurst, 2020).

---

[6] See also De Rosa (2018) who elaborates on the connections of this case to the processes of "territorialization".

[7] See https://undisciplinedenvironments.org/category/series/reimagining-remembering-and-reclaiming-water/ for other contributions.

19.3.4 Commons and Degrowth

In the last decade, scholars from the School have been pushing for an alliance between the degrowth movement (i.e., from the Global North) and the environmental justice movement (from the rural South) to counterbalance the pervasiveness and continuous expansion of the industrial and neoliberal model of development (Martínez-Alier, 2012). As the editors of a recent SI on the topic argued, both movements are materialist and more-than-materialist, stress the contradictions between capitalist accumulation and social reproduction, promote justice and the reconfiguration of the economy, and complement each other's deficits (a broader theoretical frame for environmental justice, and connections to wider social movements for degrowth) (Akbulut et al., 2019). Furthermore, many environmental justice movements represent degrowth claims in practice, even if those movements do not use the term degrowth.

The commons are also part of the above discourse. Rodríguez-Labajos et al. (2019) find important differences across the two movements in terms of values, ideology, strategies, and terminology, but also point to the commons as a way to connect both movements., Martínez-Alier (2020) shows how commons projects have become an important component of the environmental justice movement's repertoire of contentions. Similarly, Velicu (2019) shows how degrowth can be also a source of inspiration for local communities that struggle to defend local natural resources through local democratic practices and alternative economies. In turn, commons is one of the core signifiers of the degrowth imaginary. Most of the grassroots initiatives that degrowthers highlight as alternative development pathways involve commoning processes centered on caring for human and non-human beings (D'Alisa et al., 2015b). As pointed by Helfrich and Bollier (2015), commons and degrowth complement each other and can together trump the growth and neoliberal imaginaries by illustrating ways of doing together and successfully combining well-being, justice, and environmental sustainability. This "social form" of the commons avoids growth compulsion through practices based on voluntariness, autonomy, and needs satisfaction (Euler, 2019).[8]

For degrowthers, commons and commoning practices are pivotal to societies that prosper without growth. A degrowth's primary political strategy is to support commons-based initiatives and associated common senses (of 'being together'), that change dominant culture and slowly debunk the growth hegemony (Kallis et al., 2020). However, some degrowth scholars emphasize that for such societal transformations to occur, commons cannot remain only small-scale initiatives beyond or against the state. Using a Gramscian approach to the state and Wright's insights on theories of change, D'Alisa and Kallis (2020) criticize the lack of focus on

---

[8] It is also worth noting that not all commons scholars associate the development of alternative imaginaries to well-being and environmental sustainability and justice. Most digital commoners are generally optimistic about the ability of commons-based production initiatives to overcome ecological problems and foster economic growth (Fuster Morell et al., 2015).

commons on the transformation of state apparatuses. They maintain that the institutionalization of commons-based initiatives can enforce, spread, and promote further such initiatives, making possible revolutionary transformations that break with socially unfair and ecologically unsustainable capitalist forces (D'Alisa & Kallis, 2020). This is a contested vision in degrowth scholarship that diverges from the Narodnik tradition that Martínez-Alier and other environmental justice activists have followed. It is also probably an essential difference with commons movements activists and scholars, which emphasize the need for autonomy of commons (Euler & Gauditz, 2016) projects or the risks of being co-opted by the governments (e.g., Pera, 2020; Bianchi et al., 2022; see also Sect. 19.3.2).

19.4 Conclusion

In this chapter, we have aimed to highlight projects, authors, and contributions around the commons that can be associated with the Barcelona School. In that process, we have also aimed to give some suggestions for an emerging agenda for this burgeoning field of practice and reflection.

Contributions from the School offer a genuine mix of sensitivities and knowledges around the commons, but they all share an interest in their role as sustainability transformation actors. The commons can not only break through the status quo and "incrementalism" of policy (as social movements would do) but also prefigure, perform and scale up and out alternatives for more socially just, ecologically sustainable worlds. This understanding of the commons as both instances of self-governance and activism can be traced back to the institutional and environmental justice traditions and the works of Elinor Ostrom and Martínez-Alier in particular. They both saw in local communities' environmental and justice concerns a genuine concern about social and ecological sustainability and the seeds of new ways of understanding governance and human-environmental interactions.

The four agenda threads highlighted here (commons movements, urban prosumer initiatives, performative commons, and commons & degrowth) are not independent of each other. As highlighted in Sect. 19.3, the urban commons literature has both built on and contributed to the commons movement agenda through accounts of the emergence of urban commons in the aftermath of the 2008 economic crisis; and the degrowth imaginary and its connections with the commons and movements finds in urban commons initiatives one of its main exponents of the way to go.

Although not explicitly discussed above, it is worth mentioning also the epistemologically eclectic, non-exclusionary approach reflected in the contributions reviewed here. They display a great deal of methodological diversity, ranging from the use of participatory action and case study research to meta-analyses and large-n statistical analyses. We believe this is not random and responds to a belief in the promise of mixed political ecology and (environmental) science methods (Zimmerer, 2015). Also, contributions have not only embraced the critical commons literature but also moved beyond and generated new theory in different degrees, ranging from typological work to new concepts and hypotheses. In the future, we envision a study of the commons that continues to deepen collaborations of multiple actors, to bridge disciplinary frontiers as well as between the divisions of researchers and practitioners.

References

Akbulut, B., Demaria, F., Gerber, J. F., & Martínez-Alier, J. (2019). Who promotes sustainability? Five theses on the relationships between the degrowth and the environmental justice movements. *Ecological Economics, 165*, 106418.

Andreucci, D., Beltrán, M. J., Velicu, I., & Zografos, C. (2017). Capital accumulation, hegemony and socio-ecological struggles: Insights from the entitle project. *Capitalism Nature Socialism, 28*(3), 18–27.

Apostolopoulou, E., & Kotsila, P. (2021). Community gardening in Hellinikon as a resistance struggle against neoliberal urbanism: Spatial autogestion and the right to the city in post-crisis Athens, Greece. *Urban Geography*, 1–27.

Asara, V. (2020). Untangling the radical imaginaries of the indignados' movement: Commons, autonomy and ecologism. *Environmental Politics*, 1–25.

Becker, S., Naumann, M., & Moss, T. (2017). Between coproduction and commons: Understanding initiatives to reclaim urban energy provision in Berlin and Hamburg. *Urban Research and Practice, 10*(1), 63–85.

Beltrán, M. J., Kotsila, P., García López, G., Velegrakis, G., & Velicu, I. (2015). *Political ecology for civil society*, 1–220.

Bianchi, I., Pera, M., Calvet-Mir, L., Villamayor, S., Ferreri, M., Reguero, N., & Maestre Andrés, S. (2022). Urban commons and the local state: co-production between enhancement and co-optation. *Territory, Politics, Governance*, 1–20.

Blaikie, P. (2006). Is small really beautiful? Community-based natural resource management in Malawi and Botswana. *World Development, 34*(11), 1942–1957.

Bollier, D., & Helfrich, S. (Eds.). (2015). *Patterns of commoning*. Commons Strategy Group and Off the Common Press.

Bollier, D., & Helfrich, S. (2019). *Free, fair, and alive: The insurgent power of the commons*. New Society Publishers.

Bresnihan, P. (2020, Sept 15). Rural and urban, green and red, against eco-austerity. *Undisciplined Environments*. https://undisciplinedenvironments.org/2020/09/15/19570/

Caffentzis, G., & Federici, S. (2014). Commons against and beyond capitalism. *Community Development Journal, 49*(suppl 1), i92–i105.

Caggiano, M., & De Rosa, S. P. (2015). Social economy as antidote to criminal economy: How social cooperation is reclaiming commons in the context of Campania's environmental conflicts. *Partecipazione e Conflitto, 8*(2), 530–554.

Calvet-Mir, L., & March, H. (2019). Crisis and post-crisis urban gardening initiatives from a Southern European perspective: The case of Barcelona. *European Urban and Regional Studies, 26*(1), 97–112.

Camps-Calvet, M., Langemeyer, J., Calvet-Mir, L., Gómez-Baggethun, E., & March, H. (2015). Sowing resilience and contestation in times of crises: The case of urban gardening movements in Barcelona.

Clement, F., Harcourt, W., Joshi, D., & Sato, C. (2019). Feminist political ecologies of the commons and commoning (editorial to the special feature). *International Journal of the Commons, 13*(1).

D'Alisa, G. (2013). Bienes comunes: las estructuras que conectan. *Ecología política, 45*, 30–41.

D'Alisa, G., & Kallis, G. (2020). Degrowth and the state. *Ecological Economics, 169*(C).

D'Alisa, G., Demaria, F., & Kallis, G. (Eds.). (2015a). *Degrowth. A vocabulary for a new era*. Routledge.
D'Alisa, G., Forno, F., & Maurano, S. (2015b). Mapping the redundancy of collective actions. *Partecipazione e Conflitto, 8*(2).
De Angelis, M. (2001, September). Marx and primitive accumulation: The continuous character of capital's "enclosures". *The Commoner N.2*. Available at https://libcom.org/files/4_02deangelis.pdf. Accessed on 4 Jan 2022.
De Angelis, M. (2017). *Omnia sunt communia: On the commons and the transformation to post-capitalism*. Zed Books.
De Rosa, S. P. (2018). A political geography of 'waste wars' in Campania (Italy): Competing territorialisations and socio-environmental conflicts. *Political Geography, 67*, 46–55.
Dell'Angelo, J., Navas, G., Witteman, M., D'Alisa, G., Scheidel, A., & Temper, L. (2021). Commons grabbing and agribusiness: Violence, resistance and social mobilization. *Ecological Economics, 184*, 107004.
Dupuits, E., Baud, M., Boelens, R., de Castro, F., & Hogenboom, B. (2020). Scaling up but losing out? Water commons' dilemmas between transnational movements and grassroots struggles in Latin America. *Ecological Economics, 172*, 106625.
Escobar, A. (1998). Whose knowledge, whose nature? Biodiversity, conservation, and the political ecology of social movements. *Journal of Political Ecology, 5*(1), 53–82.
Euler, J. (2019). The commons: A social form that allows for degrowth and sustainability. *Capitalism, Nature and Socialism, 30*(2), 158–175.
Euler, J., & Gauditz, L. (2016). *Commons movements: Self-organized (re)production as a socio-ecological transformation*. Available at https://commons-institut.org/wp-content/uploads/DIM_Commons.pdf. Accessed on 5 Jan 2022.
Forsyth, T., & Johnson, C. (2014). Elinor Ostrom's legacy: Governing the commons, and the rational choice controversy. *Development and Change, 45*(5), 1093–1110.
Fuster Morell, M., Subirats, J., Berlinguer, M., Martínez, R., & Salcedo, J. (2015). *Procomún digital y cultura libre¿ Hacia un cambio de época*. Icaria Editoriales.
García-Lamarca, M. (2015). Insurgent acts of being-in-common and housing in Spain: Making urban commons. In M. Dellenbaugh et al. (Eds.), *Urban commons: Moving beyond state and market*. Birkhäuser.
García-López, G. A., Velicu, I., & D'Alisa, G. (2017). Performing counter-hegemonic common (s) senses: Rearticulating democracy, community and forests in Puerto Rico. *Capitalism Nature Socialism, 28*(3), 88–107.
García-López, G. A., Lang, U., & Singh, N. (2021). Commons, commoning and co-becoming: Nurturing life-in-common and post-capitalist futures (an introduction to the theme issue). *Environment and Planning E: Nature and Space, 4*(4), 1199–1216.
Goldman, M. (1998). *Privatizing nature: Political struggles for the global commons*. Pluto Press.
Guha, R., & Martínez-Alier, J. M. (2013). *Varieties of environmentalism: Essays North and South*. Routledge.
Helfrich, S., & Bollier, D. (2015). Commons. In G. D'Alisa et al. (Eds.), *Degrowth. A vocabulary for a new era*. Routledge.
Hurst, E. (2020, August 18) A swimming commons. *Undisciplined Environments*. https://undisciplinedenvironments.org/2020/08/18/a-swimming-commons/
Inglehart, R. (2000). Globalization and postmodern values. *The Washington Quarterly, 23*(1), 215–228. https://doi.org/10.1162/016366000560665
Johnson, C. (2004). Uncommon ground: The 'poverty of history' in common property discourse. *Development and Change, 35*(3), 407–434.
Kallis, G., Paulson, S., D'Alisa, G., & Demaria, F. (2020). *The case for degrowth*. Polity Press.
Kashwan, P. (2017). Inequality, democracy, and the environment: A cross-national analysis. *Ecological Economics, 131*, 139–151.
Klein, N. (2001). Reclaiming the commons. *New Left Review, 9*, 81.
Kurien, J. (1991). *Ruining the commons and the responses of the commoners*. UNRISD.

Leonardelli, I., García-López, G. A., & Fantini, E. (2021, May 27). Reimagining, remembering and reclaiming water. *Undisciplined Environments*. URL: https://undisciplinedenvironments. org/category/series/reimagining-remembering-and-reclaiming-water/

Martínez-Alier, J. (2003). *The environmentalism of the poor: A study of ecological conflicts and valuation*. Edward Elgar Publishing.

Martínez-Alier, J. (2012). Environmental justice and economic degrowth: An alliance between two movements. *Capitalism, Nature, Socialism, 23*(1), 51–73. https://doi.org/10.1080/10455752.2011.648839

Martínez-Alier, J. (2020). A global environmental justice movement: Mapping ecological distribution conflicts. *Disjuntiva, 1*(2), 83–128. https://doi.org/10.14198/DISJUNTIVA2020.1.2.6

Mingorría, S. (2021). The communitarian weavings: Agrarian commons of the Maya-Q'eqchi' against the expansion of monocultures in the Polochic Valley, Guatemala. *Latin American and Caribbean Ethnic Studies*. https://doi.org/10.1080/17442222.2021.1877876

Moreira, S., & Morell, M. F. (2020). Food networks as urban commons: case study of a Portuguese "prosumers" group. *Ecological Economics, 177*, 106777.

Navas, G., D'Alisa, G., & Martínez-Alier, J. (2022). The role of working-class communities and the slow violence of toxic pollution in environmental health conflicts: A global perspective. *Global Environmental Change, 73*, 102474.

Olivera, M., & Archidiacono, S. (2021, Oct 19) Autogestión, reclaiming the right to self-management of water. *Undisciplined Environments*. URL: https://undisciplinedenvironments. org/2021/10/19/autogestion-reclaiming-the-right-to-self-management-of-water/

Peet, R., & Watts, M. (1996). *Liberation ecologies: environment, development, social movements*. Routledge.

Pera, M. (2020). Potential benefits and challenges of the relationship between social movements and the commons in the city of Barcelona. *Ecological Economics, 174*, 106670.

Popartan, L. A., Ungureanu, C., Velicu, I., Amores, M. J., & Poch, M. (2020). Splitting urban waters: The politicisation of water in Barcelona between populism and anti-populism. *Antipode, 52*(5), 1413–1433.

Rodríguez-Labajos, B., Yánez, I., Bond, P., Greyl, L., Munguti, S., Ojo, G. U., & Overbeek, W. (2019). Not so natural an alliance? Degrowth and environmental justice movements in the global south. *Ecological Economics, 157*, 175–184. (este paper está muy bueno y trabaja sobre los commons como uno de los temas).

Saunders, F. (2014). The promise of common pool resource theory and the reality of commons projects. *International Journal of the Commons, 8*(2).

Taylor, B. R. (1995). *Ecological resistance movements: The global emergence of radical and popular environmentalism*. SUNY Press.

Temper, L., Demaria, F., Scheidel, A., Del Bene, D., & Martínez-Alier, J. (2018, May 20). The Global Environmental Justice Atlas (EJAtlas): Ecological distribution conflicts as forces for sustainability. *Sustainability Science, 13*(3), 573–584.

Tyagi, N., & Das, S. (2020). Standing up for forest: A case study on Baiga women's mobilization in community governed forests in Central India. *Ecological Economics, 178*, 106812.

Varvarousis, A. (2020). The rhizomatic expansion of commoning through social movements. *Ecological Economics, 171*, 106596.

Varvarousis, A., Asara, V., & Akbulut, B. (2021). Commons: A social outcome of the movement of the squares. *Social Movement Studies, 20*(3), 292–311.

Velicu, I. (2019). De-growing environmental justice: Reflections from anti-mining movements in Eastern Europe. *Ecological Economics, 159*, 271–278.

Velicu, I., & García-López, G. (2018). Thinking the commons through Ostrom and Butler: Boundedness and vulnerability. *Theory, Culture & Society, 35*(6), 55–73.

Villamayor-Tomas, S., & García-López, G. (2018). Social movements as key actors in governing the commons: Evidence from community-based resource management cases across the world. *Global Environmental Change, 53*, 114–126.

Villamayor-Tomas, S., & García-López, G. A. (2021). Commons movements: Old and new trends in rural and urban contexts. *Annual Review of Environment and Resources, 46*, 511–543.

Villamayor-Tomas, S., García-López, G., & Scholtens, J. (2020). Do commons management and movements reinforce each other? Comparative insights from Mexico and Sri Lanka. *Ecological Economics, 173*, 106627.

Villamayor-Tomas, S., García-López, G., & D'Alisa, G. (2022). Social movements and commons: In theory and in practice. *Ecological Economics, 194*, 107328.

Vos, J., Boelens, R., Venot, J. P., & Kuper, M. (2020). Rooted water collectives: Towards an analytical framework. *Ecological Economics, 173*, 106651.

Woelfle-Erskine, C. A. (2020, Oct 19). How imaginaries shift in places: Native and settler politics of water and salmon. *Undisciplined Environments*. URL: https://undisciplinedenvironments.org/2021/04/09/how-imaginaries-shift-in-places-native-and-settler-politics-of-water-an/

Zimmerer, K. S. (2015). Methods and environmental science in political ecology. In T. Perreault, G. Bridge, & J. McCarthy (Eds.), *The Routledge handbook of political ecology* (pp. 172–190). Routledge.

Open Access This chapter is licensed under the terms of the Creative Commons Attribution 4.0 International License (http://creativecommons.org/licenses/by/4.0/), which permits use, sharing, adaptation, distribution and reproduction in any medium or format, as long as you give appropriate credit to the original author(s) and the source, provide a link to the Creative Commons license and indicate if changes were made.

The images or other third party material in this chapter are included in the chapter's Creative Commons license, unless indicated otherwise in a credit line to the material. If material is not included in the chapter's Creative Commons license and your intended use is not permitted by statutory regulation or exceeds the permitted use, you will need to obtain permission directly from the copyright holder.

Chapter 20
(In)Justice in Urban Greening and Green Gentrification

Isabelle Anguelovski

20.1 Introduction

This chapter starts with the argument that the association of urban redevelopment with greening creates a paradox (Anguelovski et al., 2018) and examines the production of inequalities as a result of greening projects. Even while greening certainly provides economic, ecological, health, and social benefits to many (Immergluck & Balan, 2018; Baró et al., 2014; Triguero-Mas et al., 2017; Gascon et al., 2016; Wolch et al., 2014; Connolly et al., 2013; Anguelovski, 2014; Wachsmuth & Angelo, 2018), it may create new and deeper vulnerabilities for historically marginalized residents – working-class groups, minorities, and immigrants – even in the many cases where interventions are meant to redress historic inequalities in the provision of parks or green spaces (Landry & Chakraborty, 2009; Heynen et al., 2006b; Hastings, 2007; Park & Pellow, 2011; Dahmann et al., 2010; Grove et al., 2018).

During design and implementation, many greening projects – parks, gardens, greenways, green climate-resilience infrastructure, cleaned-up waterfronts – tend to remain indeed blind to social vulnerabilities (Pearsall & Pierce, 2010) and new affordability issues (Pearsall, 2010; Checker, 2011), and can create what is known as green gentrification: new or intensified urban socio-spatial inequities produced by urban greening agendas and interventions (Gould & Lewis, 2017; Anguelovski et al., 2019, 2022). We refer here to projects such as the Boston Rose Kennedy Greenway, the New York High Line, the Philadelphia Rail Park, or the redeveloped waterfront in Bayview Hunters Point, San Francisco (Pearsall, 2018a; Loughran, 2014; Dillon, 2014).

I. Anguelovski (✉)
UAB (Universitat Autònoma de Barcelona) and ICREA
(Catalan Institution for Research and Advanced Studies), Barcelona, Spain
e-mail: Isabelle.Anguelovski@uab.cat

There are indeed many cases of urban greening and neighborhood redevelopment where real estate developers leverage rezoning ordinances and tax incentives to redevelop vacant land, which they transform into high-end residences adjacent to green spaces (Bunce, 2009; Immergluck, 2009; Quastel, 2009; Dillon, 2014). Urban greening inequalities are thus particularly acute because of what can be defined as "green gaps" upon which municipalities, private investors, and privileged residents capture a "green rent" through new commercial and residential investments (Anguelovski et al., 2018). The term "green gap," builds on Smith's rent gap (Smith, 1987) and extends the concept of an environmental rent gap (Bryson, 2013) to describe how those urban stakeholders find new potential "green rents" from greening projects, couching them under discourses of win-win benefits and public goods for all.

As a result, as I show in this chapter, urban greening interventions targeting lower-income, minority, and immigrant neighborhoods risk being increasingly associated with a GreenLULU or green Locally Unwanted Land Use (Anguelovski, 2016) by socially vulnerable groups because they create enclaves of green privilege for upper-class and racialized privileged residents rather than secured public goods. Such interventions are illustrative of broader and newer trends of urban environmental injustices and are evidence of the "uneven and often debilitating and damaging socio-natural relations of power work together through the urbanization of nature," as political ecological and ecological economists have previously argued (Anguelovski & Martínez-Alier, 2014, p 168; Heynen et al., 2006a). In response, faced with this new green space paradox (Faber & Kimelberg, 2014), marginalized residents and activists are organizing to contest the social effects of greening projects as a central part of efforts to create a just green city (Pearsall & Anguelovski, 2016; Connolly, 2018a).

In the body of this chapter, mostly drawing from established research in North America while bringing in some examples from European and global South cities, I analyze historic inequities in access to urban green amenities and green infrastructure; distill growing trends over green gentrification in different contexts; and examine civic responses to them. I close this chapter with a broader discussion around the need to repoliticize urban greening practices.

20.2 A Historic Lack of Equitable "Access" to Green Space and Amenities

Research on the spatial distribution of green space in cities has found that working-class and immigrant inner-city neighborhoods tend to more often have access to under-maintained, lower quality, less numerous, and smaller parks and public gardens in comparison with more affluent and white neighborhoods (Heynen et al., 2006b; Dahmann et al., 2010; Pham et al., 2012; Wolch et al., 2005; Boone et al., 2009; Connnolly & Anguelovski, 2021). In contrast, wealthier and white communities are in a position of environmental privileges (Park & Pellow, 2011) through the greater presence of nearby parks, coasts, and other open spaces in their area (Landry & Chakraborty, 2009; Heynen et al., 2006b; Hastings, 2007). For example, a 2018 study found that US cities with higher median incomes and lower percentages of Latino and Non-Hispanic Black residents have higher ParkScores [quality park systems] than others (Rigolon et al., 2018).

Such inequalities in access to green space are often explained by uneven urban development, unfair urban planning decisions and regulations, and ensuing inequitable housing tenure (Perkins et al., 2004; Rigolon & Németh, 2018a). In the case of the US, urban neighborhood associations have historically played an active role in promoting new green infrastructure like tree planting. In Milwaukee, for example, these associations were influential in leveraging reforestation program funding towards owner-occupied (i.e., higher-income) urban neighborhoods (Perkins et al., 2004). Similarly, in early XXth century America, those same associations put in place restrictive covenants and fought for segregation ordinances to reserve properties for white homeowners, which then brought a disproportionate share of trees in cities like Baltimore to higher-income white neighborhoods (Boone et al., 2010). These dynamics were sometimes formally officially codified in city policy, as in Austin, TX, where early city plans revealed the creation of separate black spaces in areas that were underserved with parks (Busch, 2017).

However, the historic association between social groups and greening is not always linear. Postwar segregation practices in the United States, which mostly saw whites moving out of city centers, meant that Black residents who moved to formerly white neighborhoods inherited many central city green spaces. In cities like Baltimore, these spaces were often underfunded and included mostly smaller and more crowded parks (Boone et al., 2009). Despite this unequal legacy many city newcomers inherited, communities of color began organizing in the 1980s for the creation of many new green spaces in historically non-white neighborhoods (Anguelovski, 2014). Given this complex, non-linear history, green space distributional inequities – and fights against them – must be clearly connected with long-term exclusionary processes embedded both in the political economy of development and in the (re)creation of urban nature.

In addition, not all groups hold a positive connection to nature and green spaces. Many residents of color associate green space with a traumatic history of disinvestment, racial violence and lynching, and exclusion (Finney, 2014; Brownlow, 2006). For instance, parks might feel more insecure than smaller and closer pocket urban gardens when those larger spaces are in high-crime areas (Anguelovski, 2014). In other cases, such as Los Angeles, Latino residents face ethno-racial and nativist exclusion in parks linked to the predominance of white park users, a lack of minorities in adjacent neighborhoods, fears of aggression, and direct discrimination (Byrne, 2012). Here, the combination of socio-environmental and cultural history creates oppressive and unsafe experiences, anxiety and chronic stress, socio-spatial segregation, and overall poor access to protective green spaces for immigrant, minority and working-class residents.

Another form of exclusion faced by historically marginalized groups emerges when green space planners and designers are unable or unwilling to address issues related to residents' perception, interactions, and use of green spaces (Checker, 2011; Kabisch & Haase, 2014; Haase et al., 2017). Indeed, from a procedural justice standpoint (Schlosberg, 2007), when parks are being designed, if local residents are not involved and incorporated into decisions, their needs, languages, identities, and uses are more likely to be overlooked in the final "product" (Kabisch & Haase, 2014; Byrne, 2012). In contrast, if green spaces can be co-designed and co-production with residents, this process can help them feel more recognized and strengthen their attachment to place and their individual and group identity (Anguelovski, 2014; Scannell & Gifford, 2010), with greater opportunity for strong interpersonal relations (Kabisch & Haase, 2014; Connolly et al., 2013). This is the case of the Parc del Centre de Nou Barris in Barcelona, in which residents' mobilization around the initial park development and design and their ongoing use and community building practices in the park have created strong relational wellbeing for children and families (Del Pulgar et al. 2020).

20.3 Emerging Concerns Over Green Gentrification

Urban green inequalities are not only historical. Since the late 2000s, new studies have examined the social and racial impact of new or restored environmental amenities such as parks, gardens, greenways, or playgrounds (Dooling, 2009; Hagerman, 2007; Quastel, 2009; Tretter, 2013) (Hagerman, 2007), or the clean-up and redevelopment of hazardous or contaminated sites into green and more livable neighborhoods (Gould & Lewis, 2017; Pearsall & Pierce, 2010; Pearsall, 2013; Curran & Hamilton, 2012; Dillon, 2014) – that is green gentrification. Most recently, green infrastructure built to address climate threats and impacts has also been shown to be providing greater security to more privileged, gentrifying and White residents rather than Latinos and Black residents, as exemplified in the case of Philadelphia (Shokry et al., 2020). In other cases of climate infrastructure, such as the Medellin Green Belt, new projects can erase residents' long-term green practices and lead to sociocultural losses of vernacular uses of green space and traditional relationships to nature within "marginal" or informal land.

Overall, much of the green gentrification scholarship has focused on exposing the relationship between the creation or restoration of urban environmental amenities, subsequent demographic changes, and real estate price increases. The core argument is here that new green infrastructure enhances the desirability of a neighborhood – even before their construction – and eventually contributes to increases in property values (Conway et al., 2010; Sander & Polasky, 2009; Immergluck, 2009) and high-end housing constructions. In New York, for instance, the restoration of the Marcus Garvey Park (Harlem) has been accompanied by luxury condominium developments priced well above the historic average (Checker, 2011) benefiting developers and upper-class residents. In Atlanta, housing values have increased by 18% and 27% between 2011 and 2015 for homes located within 0.8 km of the Atlanta's Beltine greenbelt project (Immergluck & Balan, 2018). All in all, the widely agreed-upon conclusion (except in a few studies, see Eckerd, 2011) is that greener neighborhoods in large cities become pricier for vulnerable residents eventually unable to capture the benefits of environmental clean-up, restoration, and green space creation (Checker, 2011; Gould & Lewis, 2017).

Furthermore, understanding the unfolding of green gentrification beyond a few North American or European countries (Pearsall, 2018b), where much of the green gentrification research is situated, leads to insightful findings on urban greening practices in the context of the global smart, sustainable, and resilient city planning orthodoxy (Connolly, 2018a). As cities increasingly sell urban greening (and resilience) as an international brand, the equity implications of land use projects deployed for instance to address "climate resilience" issues – such as flooding in New Orleans or Jakarta or sea level rise in Boston or Manila – and other environmental risks are coming to light. They also reveal the importance of understanding the diverse manifestations of social vulnerability and risk in planning for climate resilience (Connolly, 2018b). As the construction of a green belt in Medellin reveals, thousands of rural migrants escaping the armed conflict have been affected by new large-scale green infrastructure which further illegalizes their land "occupation" and uses, while, at the same time, overlooking other illegal land practices by wealthier residents and real estate developers benefiting from "landscapes of pleasure and privilege".

In other words, green gentrification research contributes to exposing the relationship between environmental change and gentrification, and its implications for residential segregation and economic development dynamics. It reveals that questions of urban greening, secured access over time, and urban equity are all but part of the same equation for socially vulnerable groups and that the green paradox is more alive than ever (Fig. 20.1).

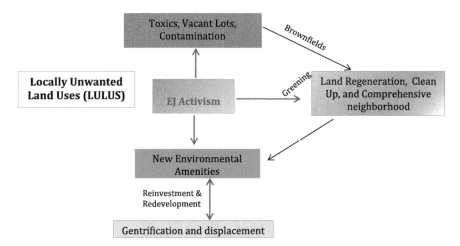

Fig. 20.1 Fighting multiple urban environmental agendas

20.4 Activists vs Green Gentrification

In response to concerns over green gentrification, community activists are organizing at the neighborhood or city level to contest the uneven social impacts of urban greening interventions and to what Martínez-Alier and myself previously called "undeterred processes of development, growth, and speculation" (Anguelovski & Martínez-Alier, 2014, p. 172). While some of the resistance relies on the strategies and tactics of traditional environmental justice movements and illustrates ecological distribution conflicts (Pearsall & Anguelovski, 2016; Anguelovski & Martínez-Alier, 2014), it also demonstrates classic dynamics of collective action at the neighborhood level (Schuetze & Chelleri, 2015; Pearsall, 2013; Tretter, 2013); community organizing through an alliance between EJ groups and community development organizations (Scally, 2012), and direct action tactics (Anguelovski, 2015; Rosol, 2013). In Seoul, for example, stakeholders involved in the planning of a Green Corridor to be part of the city's "Urban Renaissance Master Plan" articulated a vocal opposition against the proposal's top-down approach and lack of concern for traditional small-scale urbanization patterns (Schuetze & Chelleri, 2015). In addition, resistance to green gentrification includes leveraging environmental policies and regulations (Sandberg, 2014; Pearsall, 2013), participating actively in neighborhood planning exercises, and building alliances with progressive gentrifiers (Curran & Hamilton, 2012). Tactics also include advocating for complementary policy tools to ensure the right to housing (Thompson, 2015; Wolch et al., 2014; Ngom et al., 2016).

One of the most commented-upon frames of community resistance has been the "Just Green Enough" strategy (Curran & Hamilton, 2012; Wolch et al., 2014), through which residents, such as those in Greenpoint, Brooklyn, have mobilized for the clean-up of contamination and the incorporation of the neighborhood industrial fabric into redevelopment schemes. Yet, some recent research point at the gentrification potential of re-industrialization strategies (Checker, in press). In general, the long-term ability of "green compromises" to fulfill equity goals remains uncertain and contested (Faber & Kimelberg, 2014). Much attention needs to be placed on risks of cooptation, demotivation over time, and competing goals and conflicts between social organizations and environmental groups (Checker, 2011). Many activists indeed face the risks that their anti-green gentrification grievances might be used by planners and developers to justify siting urban greening interventions in more privileged and white neighborhoods, away from immigrant, working-class, or minority areas.

As well, there are important questions about the lack of true commitment or ability of traditional environmental movements to push for green equity, as many park nonprofits, for instance, consider that it is not "their business" (Rigolon & Németh, 2018b). While many environmental groups articulate a no- (or neutral) growth agenda (Layzer, 2015), urban greening (and green gentrification) does occur in the context of urban growth politics. Given the inherent impact of greening on urban growth, it is not clear whether mainstream environmental movements advocating for urban sustainability are ready to go back on their growth positions and put environmental equity at the center of their advocacy. And, as we have pointed out before, historically, environmental movements and EJ movements have not been allies (Pellow & Brulle, 2005; Anguelovski & Martínez-Alier, 2014). In our 2014 piece, Martínez-Alier and myself already called for possible bridges and alliances between the different varieties of environmentalism, including the Cult of Wilderness (and pristine nature), the Gospel of Eco-Efficiency (in the context of smart growth and sustainability planning), and the Mantra of Environmental Justice and environmentalism of the poor more broadly, in ways that could protect urban territories, defend place-making and identities, and strengthen the right to stay of historically marginalized groups (Anguelovski & Martínez-Alier, 2014).

In return, the role of municipal decision-makers and public agencies in addressing or preventing inequities in greening cannot be left aside: There is still room for cities to offer transformational green interventions that both respond to deep climate agendas and ensure housing equity and anti-displacement. Several municipalities are indeed putting tools in place that ensure a greater right to housing access and affordability, including rent control in Berlin, housing cooperatives in Germany, community land trust in the US, inclusionary zoning in Spain, or social housing construction in France, and that also promotes more inclusive greening practices (Oscilowicz et al., 2022). Figure 20.2 summarizes and analyzes some of community-led or municipality-led green equity tools.

20.5 Conclusion: Reinserting the Political in Planning for Greener Cities

In this chapter, I have shown that some residents face the double circumstance of (a) having historically been excluded from large and quality green spaces (Heynen et al., 2006b) and (b), today, seeing their neighborhoods becoming grabbed, "greened," and rebranded as livable, sustainable, low-carbon, resilient, and/or green at their (future) expense in the context of a new green planning orthodoxy (Davidson & Iveson, 2015). More broadly, the green space paradox makes the clean-up of industrial centers a zero-sum equation where with every gain for a few comes potential losses for (many) others. In view of these questions, reasserting the political into urban greening as a strand of the sustainability agenda is of great urgency, as political ecologists call for (Swyngedouw, 2007).

First, the need to bring the political back into municipal urban greening comes from the urgency to demystify the claim that urban greening is a public good for all. Rather, greening will have a mix of social and ecological effects requiring a deeper analysis of the conditions and pathways through which green projects can help to bring about "just sustainabilities" (Agyeman, 2013). As greening becomes a communication and selling tool for cities, it is indeed increasingly monetized and financialized and can create speculation and rent capture. Many cities indeeed actively

Tool	Definition	Strengths	Limitations	Example
Policy or Planning tool				
Rent Control	Regulations that support renters in obtaining, maintaining, and keeping affordable housing. Traditionally a government regulation that places a ceiling on rent for designated units.	Reduces burden from housing costs on renters allowing them to remain in their homes and gain benefit from economic development in their neighborhood. Supports mixed-income communities	Some real estate economists warn that limiting rent to below market rates will cause over-consumption of controlled apartments and therefore create counterproductive results by constraining housing supply	In Vienna, Austria, limited-Profit Housing Associations are subsidized housing indirectly controlled by the municipality. These units are developed by private developers through a city-regulated process and protected by the Limited-Profit Housing Act which sets parameters for setting fair rent prices and managing repairs and quality.
Inclusionary Zoning	Planning ordinance in which developers set aside a given share of new housing to be affordable for people with low to moderate incomes. Tends to range from 8-12% of units set aside in US cities to 30% or more in Europe (up to 56% in the city of Nantes in France)	Creates Mixed Income Communities. Utilizes private developer funds	Difficult to enforce with complex requirements and a lack of uniform oversight for developers. Reliance on the private market makes IZ policies more suitable to cities with a high demand for housing, they may be less successful in cities with less competitive real-estate markets.	In 2005 the city of Atlanta, GA began construction of the Atlanta BeltLine. In order to mitigate dramatic rises in housing prices around the BeltLine, the city of Atlanta developed the BeltLine Overlay District (roughly ½ mile of the BeltLine Corridor) where standalone or mixed-used multi-unit buildings, with at least 10 units must include 15 percent affordable units.
Defense of Single Family Homes	'Single-Family Home' is the housing typology of a detached unit intended for a single household. Single-family home land use policies, such as Single-Family Home zoning designation prevent communities from building any type of housing in a given area aside from a single-family home.	Creates less dense cities and therefore facilitates a different feel of urban life. Preserves smaller, more affordable homes that in many cases hold cultural significance to communities. Deters the conversion of smaller homes into multi-housing condo complexes, which tend to be built for gentrifying residents	Single-family zoning means *less* housing, and less housing supply may lead to more expensive housing	When land in the La Bajada neighborhood of West Dallas, TX began attracting real estate investment, neighborhood organizations became concerned that the character of the neighborhood, a largely Hispanic neighborhood of small single-family cottage style homes, would be lost to development. The neighborhood was approved as a Neighborhood Stabilization Overlay (NSO) meaning the only type of development that can be built within the area are single family homes, that are no taller than 27 feet and facilities that support the community.
Community-driven initiatives				
Formal Recognition of Right to Stay or Return	Formal recognition of the right to stay or right to return policies have been campaigns led by community activists designed to ensure that the original residents and those with long-term connections to gentrifying neighborhoods are able to receive affordable housing within their communities. These strategies include assistance to renters, home buyers, and existing homeowners provided by municipal agencies.	Formally recognize the importance of honoring generational and emotional connection to place.	Does not address the challenge residents will face as they continue to live in their changing neighborhoods.	The N/NE Neighborhood Housing Strategy in Portland, OR includes loan assistance for home repair, down payment assistance for first time homebuyers, and the creation of new affordable housing for black residents who have been impacted by redlining, racial zoning, and eminent domain.
Community Land Trusts	Community Land Trusts are non-profit, community-based organizations designed to ensure community control and ownership of land.	Guarantees affordable units to members of the Trust for generations. Trusts typically aid tenets to avoid foreclosure by providing technical and real estate assistance	Some groups voiced concern that residents of CLT units are limited to addressing inter-generational wealth gaps (especially between whites and residents of color) due to the inability to sell their homes for full equity.	The Atlanta Land Trust (ALT) has accumulated land in targeted areas of Atlanta, including near the BeltLine, to make them permanently affordable. They are also actively engaged with tenants/members to help them maintain their mortgage payments. ALT is also involved in creating a strong environment for CLT development by promoting congruous public policy, by engaging the community and by fundraising.

Fig. 20.2 Strategies for green equity

brand themselves as being the most livable green city to attract investment and creative class residents in the current trend of competitive urbanism (Garcia Lamarca et al., 2021). New analysis needs to shed light on how urban greening contributes to invisibilizing the environmental and social practices of long-term residents, by rebranding neighborhoods and cities as green, smart, and resilient; by flattening their historical and ecological landscapes; and by erasing their sense of belonging and combined relationship to their neighborhood and to the local nature. Green gentrification does indeed operate both through physical displacement and through social, cultural, and mental displacement and dispossession.

Here, the political ecology literature on land grabbing, green frontier-driven value capture, and accumulation by green dispossession (Safransky, 2014, 2016) highlights the process of (community) losses and (private, concentrated) green wealth capture. In the future, unpacking financial actors, their intermediaries, and economic beneficiaries – both in the global North and South – is an important next step for green gentrification research. New financial instruments and tools – from green bonds to property-assessed clean energy programs – are indeed being mobilized to fund urban greening, in many cases connecting future green urban development to future value and resource creation (Knuth, 2016; Garcia Lamarca & Ullstrom, 2022).

In sum, this chapter on urban greening as both "underrepresented" asset and unwanted greenLULU in historically marginalized neighborhoods aims at repoliticizing current discourses and practices around the green city, and the associated claims that greening brings win-win benefits to everyone in the city. It calls for greater alliances between varieties of environmentalism and for transformational planning practice, whereby racial and social equity would be at the center of greening projects rather than an afterthought or abstract goal flagged in urban plans and interventions. This transformation would allow urban greening to move from being a green privilege and utopia to an environmental good for all.

References

Agyeman, J. (2013). *Introducing just sustainabilities*. Zed Books.
Anguelovski, I. (2014). *Neighborhood as Refuge: Environmental justice, community reconstruction, and place-remaking in the city*. MIT Press.
Anguelovski, I. (2015). Tactical developments for achieving just and sustainable neighborhoods: The role of community-based coalitions and bottom-to-bottom networks in street, technical, and funder activism. *Environment and Planning C: Government and Policy, 33*, 703–725.
Anguelovski, I. (2016). From toxic sites to parks as (Green) LULUs? New challenges of inequity, privilege, gentrification, and exclusion for urban environmental justice. *Journal of Planning Literature*, 1–14.
Anguelovski, I., & Martínez-Alier, J. (2014). The 'environmentalism of the Poor'revisited: Territory and place in disconnected glocal struggles. *Ecological Economics, 102*, 167–176.
Anguelovski, I., Connolly, J., & Brand, A. L. (2018). From landscapes of utopia to the margins of the green urban life: For whom is the new green city? *City, 22*, 417.
Anguelovski, I., Connolly, J., Garcia Lamarca, M., Cole, H., & Pearsall, H. (2019). New scholarly pathways on green gentrification: What does the urban "green turn" mean and where is it going? *Progress in Human Geography, 43*(6), 1064–1086.
Anguelovski, I., Connolly, J. J., H, C., Garcia Lamarca, M., Triguero-Mas, M., Baró, F., Martin, N., Conesa, D., Shokry, G., Pérez del Pulgar, C., Arguelles Ramos, L., Matheney, A., Gallez, E., Oscilowicz, E., López Máñez, J., Sarzo, B., Beltrán, M. A., & Martínez Minaya, J. (2022). *Green gentrification in European and North American Cities Nature Communications, 13*.

Baró, F., Chaparro, L., Gómez-Baggethun, E., et al. (2014). Contribution of ecosystem services to air quality and climate change mitigation policies: The case of urban forests in Barcelona, Spain. *Ambio, 43*, 466–479.

Boone, C., Buckley, G., Grove, M., et al. (2009). Parks and people: An environmental justice inquiry in Baltimore, Maryland. *Annals of the Association of American Geographers, 99*, 767–787.

Boone, C. G., Cadenasso, M. L., Grove, J. M., et al. (2010). Landscape, vegetation characteristics, and group identity in an urban and suburban watershed: Why the 60s matter. *Urban Ecosystem, 13*, 255–271.

Brownlow, A. (2006). An archaeology of fear and environmental change in Philadelphia. *Geoforum, 37*, 227–245.

Bryson, J. (2013). The nature of gentrification. *Geography Compass, 7*, 578–587.

Bunce, S. (2009). Developing sustainability: Sustainability policy and gentrification on Toronto's waterfront. *Local Environment, 14*, 651–667.

Busch, A. (2017). *City in a garden: Environmental transformations and racial justice in twentieth-century Austin, Texas*. UNC Press Books.

Byrne, J. (2012). When green is white: The cultural politics of race, nature and social exclusion in a Los Angeles urban national park. *Geoforum, 43*, 595–611.

Checker, M. (2011). Wiped out by the "Greenwave": Environmental gentrification and the paradoxical politics of urban sustainability. *City & Society, 23*, 210–229.

Checker, M. (in press). Industrial gentrification and the dynamics of sacrifice in New York City. In P. Lewis & M. Greenberg (Eds.), *The city is the factory*. Cornell University Press.

Connolly, J. J. (2018a). From Jacobs to the Just City: A foundation for challenging the green planning orthodoxy. *Cities*.

Connolly, J. J. (2018b). From systems thinking to systemic action: Social vulnerability and the institutional challenge of urban resilience. *City & Community, 17*, 8–11.

Connolly, J. J., & Anguelovski, I. (2021). Three histories of greening and whiteness in American cities. *Frontiers in Ecology and Evolution, 9*, 101.

Connolly, J. J., Svendsen, E. S., Fisher, D. R., et al. (2013). Organizing urban ecosystem services through environmental stewardship governance in New York City. *Landscape and Urban Planning, 109*, 76–84.

Conway, D., Li, C. Q., Wolch, J., et al. (2010). A spatial autocorrelation approach for examining the effects of urban greenspace on residential property values. *The Journal of Real Estate Finance and Economics, 41*, 150–169.

Curran, W., & Hamilton, T. (2012). Just green enough: Contesting environmental gentrification in Greenpoint, Brooklyn. *Local Environnment, 17*, 1027–1042.

Dahmann, N., Wolch, J., Joassart-Marcelli, P., et al. (2010). The active city? Disparities in provision of urban public recreation resources. *Health & Place, 16*, 431–445.

Davidson, M., & Iveson, K. (2015). Recovering the politics of the city: From the 'post-political city' to a 'method of equality' for critical urban geography. *Progress in Human Geography, 39*, 543–559.

Dillon, L. (2014). Race, waste, and space: Brownfield redevelopment and environmental justice at the hunters point shipyard. *Antipode, 46*, 1205–1221.

Dooling, S. (2009). Ecological gentrification: A research agenda exploring justice in the city. *International Journal of Urban and Regional Research, 33*, 621–639.

Eckerd, A. (2011). Cleaning up without clearing out? A spatial assessment of environmental gentrification. *Urban Affairs Review, 47*, 31–59.

Faber, D., & Kimelberg, S. (2014). Sustainable urban development and environmental gentrification: The paradox confronting the U.S. environmental justice movement. In H. R. Hall, C. Robinson, & A. Kohli (Eds.), *Uprooting urban America: Multidisciplinary perspectives on race, class & gentrification* (pp. 77–92). Peter Lang Publishers.

Finney, C. (2014). *Black faces, white spaces: Reimagining the relationship of African Americans to the great outdoors*. UNC Press Books.

García-Lamarca, M., & Ullström, S. (2022). "Everyone wants this market to grow": The affective post-politics of municipal green bonds. *Environment and Planning E: Nature and Space, 5*(1), 207–224.

Garcia Lamarca, M., Anguelovski, I., Cole, H., Connolly, J. J. T., Arguelles, L., Baró, F., Perez del Pulgar, C., & Shokry, G. (2021). Urban green boosterism and city affordability: For whom is the 'branded' green city? *Urban Studies*.

Gascon, M., Triguero-Mas, M., Martínez, D., et al. (2016). Residential green spaces and mortality: A systematic review. *Environment International, 86*, 60–67.

Gould, K. A., & Lewis, T. L. (2017). *Green gentrification: Urban sustainability and the struggle for environmental justice*. Routledge.

Grove, M., Ogden, L., Pickett, S., Boone, C., Buckley, G., Locke, D. H., Lord, C., & Hall, B. (2018). The legacy effect: Understanding how segregation and environmental injustice unfold over time in Baltimore. *Annals of the American Association of Geographers, 108*(2), 524–537.

Haase, D., Kabisch, S., Haase, A., et al. (2017). Greening cities–to be socially inclusive? About the alleged paradox of society and ecology in cities. *Habitat International, 64*, 41–48.

Hagerman, C. (2007). Shaping neighborhoods and nature: Urban political ecologies of urban waterfront transformations in Portland, Oregon. *Cities, 24*, 285–297.

Hastings, A. (2007). Territorial justice and neighbourhood environmental services: A comparison of provision to deprived and better-off neighbourhoods in the UK. *Environment and Planning C, 25*, 896–917.

Heynen, N. C., Kaika, M., & Swyngedouw, E. (2006a). *In the nature of cities: Urban political ecology and the politics of urban metabolism* (Vol. 3). Taylor & Francis US.

Heynen, N., Perkins, H., & Roy, P. (2006b). The political ecology of uneven urban green space. *Urban Affairs Review, 42*, 3–25.

Immergluck, D. (2009). Large redevelopment initiatives, housing values and gentrification: The case of the Atlanta Beltline. *Urban Studies, 46*, 1723–1745.

Immergluck, D., & Balan, T. (2018). Sustainable for whom? Green urban development, environmental gentrification, and the Atlanta Beltline. *Urban Geography, 39*(4), 546–562.

Kabisch, N., & Haase, D. (2014). Green justice or just green? Provision of urban green spaces in Berlin, Germany. *Landscape and Urban Planning, 122*, 129–139.

Knuth, S. (2016). Seeing Green in San Francisco: City as resource frontier. *Antipode, 48*, 626–644.

Landry, S., & Chakraborty, J. (2009). Street trees and equity: Evaluating the spatial distribution of an urban amenity. *Environment and Planning A, 41*, 2651–2670.

Layzer, J. A. (2015). *The environmental case*. Sage.

Loughran, K. (2014). Parks for profit: The high line, growth machines, and the uneven development of urban public spaces. *City & Community, 13*, 49–68.

Ngom, R., Gosselin, P., & Blais, C. (2016). Reduction of disparities in access to green spaces: Their geographic insertion and recreational functions matter. *Applied Geography, 66*, 35–51.

Oscilowicz, E., Anguelovski, I., Triguero-Mas, M., García-Lamarca, M., Baró, F., & Cole, H. V. (2022). Green justice through policy and practice: A call for further research into tools that foster healthy green cities for all. *Cities & Health*, 1–16.

Park, L. S.-H., & Pellow, D. (2011). *The slums of Aspen: Immigrants vs. The environment in America's Eden*. New York University Press.

Pearsall, H. (2010). From brown to green? Assessing social vulnerability to environmental gentrification in New York City. *Environment and Planning C, 28*, 872–886.

Pearsall, H. (2013). Superfund me: A study of resistance to gentrification in New York City. *Urban Studies, 50*, 2293–2310.

Pearsall, H. (2018a). The contested future of Philadelphia's reading viaduct: Blight, neighborhood amenity, or global attraction? In W. Curran & T. Hamilton (Eds.), *Just green enough* (pp. 197–208). Routledge.

Pearsall, H. (2018b). New directions in urban environmental/green gentrification research. In L. Lees & M. Phillips (Eds.), *Handbook of gentrification studies* (pp. 329–345). Edward Elgar Publishing.

Pearsall, H., & Anguelovski, I. (2016). Contesting and resisting environmental gentrification: Responses to new paradoxes and challenges for urban environmental justice. *Sociological Research Online, 21*, 6.

Pearsall, H., & Pierce, J. (2010). Urban sustainability and environmental justice: Evaluating the linkages in public planning/policy discourse. *Local Environment, 15*, 569–580.

Pellow, D., & Brulle, R. J. (2005). *Power, justice, and the environment : A critical appraisal of the environmental justice movement*. MIT.

Pérez Del Pulgar, C., Anguelovski, I., & Connolly, J. J. (2020). Towards a green and playful city: Understanding the social and political production of children's relational wellbeing in Barcelona. *Cities*.

Perkins, H. A., Heynen, N., & Wilson, J. (2004). Inequitable access to urban reforestation: The impact of urban political economy on housing tenure and urban forests. *Cities, 21*, 291–299.

Pham, T.-T.-H., Apparicio, P., Séguin, A.-M., et al. (2012). Spatial distribution of vegetation in Montreal: An uneven distribution or environmental inequity? *Landscape and Urban Planning, 107*, 214–224.

Quastel, N. (2009). Political ecologies of gentrification. *Urban Geography, 30*, 694–725.

Rigolon, A., & Németh, J. (2018a). What shapes uneven access to urban amenities? Thick injustice and the legacy of racial discrimination in Denver's parks. *Journal of Planning Education and Research*, 0739456X18789251.

Rigolon, A., & Németh, J. (2018b). "We're not in the business of housing:" environmental gentrification and the nonprofitization of green infrastructure projects. *Cities, 81*, 71.

Rigolon, A., Browning, M., & Jennings, V. (2018). Inequities in the quality of urban park systems: An environmental justice investigation of cities in the United States. *Landscape and Urban Planning, 178*, 156–169.

Rosol, M. (2013). Vancouver's "EcoDensity" planning initiative: A struggle over Hegemony? *Urban Studies, 50*, 2238–2255.

Safransky, S. (2014). Greening the urban frontier: Race, property, and resettlement in Detroit. *Geoforum, 56*, 237–248.

Safransky, S. (2016). Rethinking land struggle in the Postindustrial City. *Antipode*.

Sandberg, L. A. (2014). Environmental gentrification in a post-industrial landscape: The case of the Limhamn quarry, Malmö, Sweden. *Local Environment, 19*, 1068–1085.

Sander, H. A., & Polasky, S. (2009). The value of views and open space: Estimates from a hedonic pricing model for Ramsey County, Minnesota, USA. *Land Use Policy, 26*, 837–845.

Scally, C. P. (2012). Community development corporations, policy networks, and the rescaling of community development advocacy. *Environment and Planning C, 30*, 712.

Scannell, L., & Gifford, R. (2010). The relations between natural and civic place attachment and pro-environmental behavior. *Journal of Environmental Psychology, 30*, 289–297.

Schlosberg, D. (2007). *Defining environmental justice : Theories, movements, and nature*. Oxford University Press.

Schuetze, T., & Chelleri, L. (2015). Urban sustainability versus Green-washing—Fallacy and reality of urban regeneration in downtown Seoul. *Sustainability, 8*, 33.

Shokry, G., Connolly, J. J. T., & Anguelovski, I. (2020). Understanding climate gentrification and shifting landscapes of protection and vulnerability in green resilient Philaldephia. *Urban Climate*.

Smith, N. (1987). Gentrification and the rent-gap. *Annals of the Association of American Geographers, 77*, 462–465.

Swyngedouw, E. (2007). Impossible sustainability and the postpolitical condition. In R. Krueger & D. C. Gibbs (Eds.), *The sustainable development paradox: Urban political economy in the United States and Europe* (pp. 13–40). Guilford Press.

Thompson, M. (2015). Between boundaries: From commoning and guerrilla gardening to community land trust development in Liverpool. *Antipode, 47*, 1021–1042.

Tretter, E. M. (2013). Contesting sustainability: 'SMART Growth' and the redevelopment of Austin's Eastside. *International Journal of Urban and Regional Research, 37*, 297–310.

Triguero-Mas, M., Donaire-Gonzalez, D., Seto, E., Valentín, A., Martínez, D., Smith, G., Hurst, G., Carrasco-Turigas, G., Masterson, D., & van den Berg, M. (2017). Natural outdoor environments and mental health: Stress as a possible mechanism. *Environmental Research, 159*, 629–638.

Wachsmuth, D., & Angelo, H. (2018). Green and gray: New ideologies of nature in urban sustainability policy. *Annals of the American Association of Geographers, 108*(4), 1038–1056.

Wolch, J., Wilson, J. P., & Fehrenbach, J. (2005). Parks and park funding in Los Angeles: An equity-mapping analysis. *Urban Geography, 26*, 4–35.

Wolch, J., Byrne, J., & Newell, J. (2014). Urban green space, public health, and environmental justice: The challenge of making cities 'just green enough'. *Landscape and Urban Planning, 125*, 234–244.

Open Access This chapter is licensed under the terms of the Creative Commons Attribution 4.0 International License (http://creativecommons.org/licenses/by/4.0/), which permits use, sharing, adaptation, distribution and reproduction in any medium or format, as long as you give appropriate credit to the original author(s) and the source, provide a link to the Creative Commons license and indicate if changes were made.

The images or other third party material in this chapter are included in the chapter's Creative Commons license, unless indicated otherwise in a credit line to the material. If material is not included in the chapter's Creative Commons license and your intended use is not permitted by statutory regulation or exceeds the permitted use, you will need to obtain permission directly from the copyright holder.

Chapter 21
From the Soil to the Soul: Fragments of a Theory of Economic Conflicts

Julien-François Gerber

21.1 Introduction

Environmental conflicts and movements are a core concern of the Barcelona School of Political Ecology and Ecological Economics. Inspired by the work of Joan Martínez-Alier, many scholar-activists at or around the Institute of Environmental Science and Technology ("ICTA") have helped deploy concepts like the environmentalism of the poor, contested metabolisms, activist epistemologies, conflicting languages of valuation, or degrowth. I have myself engaged with several of these concepts since my doctoral years in Barcelona. One of my main interests since then has been to try to strengthen the School's institutional analyses – focussing on ownership and debt – as well as its psychological dimension – mobilizing psychoanalytic and ecopsychological insights.

This chapter draws on recent research along those lines and seeks to expand the School's understanding of environmental struggles. After outlining an overview of the various types of conflicts over market economies, I will suggest that the prominent points of contention of today's neoliberal capitalism are related to its metabolism (ecological dimension) and to its debts (institutional dimension) and that both dimensions as intimately linked with each other. Furthermore, I will argue that capitalism also generates 'inner conflicts' rooted in alienation (psychological dimension). Alienation devitalizes, isolates, and disorients. It disconnects people from themselves, their communities, and their ecologies (Rosa, 2019). To address alienation is politically as important as contesting the immediate causes or the structural causes of economic conflicts. Fortunately, as we can expect, the current state of affairs is generating its countermovements and a new blend of radical ideas is emerging in degrowth and similar 'commonist' movements.

J.-F. Gerber (✉)
International Institute of Social Studies (ISS), Erasmus University Rotterdam, Rotterdam, The Netherlands

Throughout the chapter, I will rely on a broad definition of the economy seen as the various ways by which humans organize their sustenance (Gerber & Steppacher, 2014). The economy ranges from the ecological to the existential and encompasses the fundamental processes of appropriation, extraction, production, distribution, consumption, care, flourishing/alienation, and 'excretion' (waste).

The chapter is organized as follows. Section 21.2 outlines a broad historical framework that systematizes the various struggles over market economies, while Sect. 21.3 delves into the contemporary neoliberal era and its particular combination of conflicts. Section 21.4 then proposes a way of analyzing the triggers and targets of these movements, and Sect. 21.5 illustrates how radical change can only take place via an expansion of consciousness, using debt and degrowth as examples. Section 21.6 offers some concluding remarks.

21.2 Points of Conflicts Over the Economy: A *Longue Durée* Perspective

Different ways of organizing the economy have generated different kinds of conflicts, which have themselves generated different radical ideologies supporting alternatives. But since the birth of market economies in the Fertile Crescent some 5000 years ago, I suggest that it is possible to identify five major sets of economic relations that are especially conflict-prone: land, tax, labour, debt, and ecological resources (Fig. 21.1). These five categories of conflict have varied in combination and intensity over time and space, and they articulate in different ways to class, gender, and race, which is another constant source of fundamental conflicts over the economy.

I will briefly illustrate these ideas with a few examples taken from pre-capitalist Europe, before devoting more space to the capitalist 'early modern', 'modern', and 'neoliberal' eras, respectively. In classical Antiquity, debt and land were arguably the key points of economic conflict. De Ste. Croix (1981: 298) noted that "[t]he programme of Greek revolutionaries seems largely to have centred in two demands: redistribution of land, cancellation of debts" (see also Finley, 1983). Linked to these struggles, the anti-systemic ideologies of the time were often demands for more democratic forms of government (Graeber, 2011). In the Middle Ages, the perennial cause of revolts was associated with the various types of taxes in cash, kind, or labour. Countless peasant rebellions started around tax issues (Burg, 2004). Facing the demands of élites typically justifying themselves on religious grounds, the radical ideologies of the time drew on alternative socio-political understandings of religion. In medieval Europe, there were many anti-systemic movements, such as those around radical theologians like John Ball, Jan Hus or Thomas Müntzer, that experimented with alternative ways of organizing society.

But as the transition to capitalism took place, new economic conflicts and new anti-systemic ideologies appeared. The rise of capitalist social-property relations saw the emergence of new classes of commercial landlords, commercial tenants, merchants, and a nascent working class (Brenner, 1985). For the first time in history, households and workplaces started to be systematically separated, and financial relations began to grow dramatically (Gerber, 2014). Figure 21.1 represents the basic economic relations of market economies (black arrows). The figure includes an ecological dimension: the grey arrows show the flows of materials and energy that run throughout each economic entity, from the input stage (with raw materials, land, and energy) to the output stage (with waste and pollution), following the conventions of a socio-metabolic representation (Gerber & Scheidel, 2018). The stars show the key sites of potential conflicts. This figure systematizes and intertwines the five fundamental categories of conflicts over market economies.

Capitalism requires at least two basic elements without which it cannot function: the separation of the producer from its means of production – through various forms of enclosure and dispossession – and a strong government able to enforce contracts on a large scale. Accordingly, struggles over land (related to dispossession) and taxes (related to the consolidation of the nation-state) can be seen as dominant economic conflicts in many parts of the world witnessing an early strengthening of capitalist dynamics. The corresponding radical ideologies can be sheer escapism from the state, as discussed for example by Scott (2009) in early modern Southeast Asia, or various forms of revolutionary republicanism, like in the French Revolution which started as a tax revolt.

While the basic relations of Fig. 21.1 remain as valid as before, the 'modern era' that started in the early nineteenth century was a period of colossal changes at the world scale. Polanyi (1944) argued that it was the time of a 'Great Transformation' that separated the economic sphere from the rest of society. And after World War II, this era witnessed a 'Great Acceleration', a period characterized by the exponential rise of many ecological-economic indicators such as the use of natural resources and the amount of emissions (Steffen et al., 2011). This 'modern era' is the period

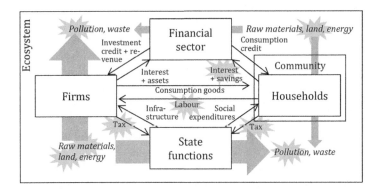

Fig. 21.1 Basic economic relations of market economies. Black arrows represent key socioeconomic relationships which can be further differentiated within an intersectional perspective; grey arrows represent the metabolic flows running throughout each entity; stars represent key sites of struggle

of rapid industrialization linked to waged labour as the strategic site of economic conflict, at least in industrialized countries. Elsewhere, other economic struggles were dominant. In India, for example, debt remained a major cause of rebellion throughout the nineteenth century, and in the colonies more generally tax and land continued to be at the forefront of economic discontent. However, the radical ideology of European modernity was, without a doubt, embodied in the various forms of socialisms.

21.3 Neoliberal Growthism: New Conflicts and New Radical Ideas

Around 1980, the world system can be seen as entering a distinct period, that of neoliberalism. Neoliberalism coincides, among other things, with a massive increase of environmental problems in terms of pollution, biodiversity loss, resources peaks, and climate change. These new circumstances have generated environmental protests to an unprecedented scale in human history. Martínez-Alier (2002: 1) has compared their explosion with the beginning of the socialist movement and the First International. Interlinking a given socio-metabolic configuration with its ecological distribution conflicts has become the hallmark of the Barcelona School and a powerful way of understanding and systematizing environmental struggles.

It is now also possible to link metabolic contestations to their institutional dimension, something that has been relatively less discussed in the Barcelona School. Facing biophysical limits and slower GDP rates, neoliberal capitalism seems to have gradually shifted its strategic form of surplus appropriation from industrial profits to financial rents. The various forms of debt are indeed located at the very heart of the neoliberal project and they have increased to unparalleled levels (Durand, 2017). A new debt-driven proletarianization has taken place since the 1980s, reinforcing the key function of debt as a control and disciplinary device over lower and middle classes, not only specifically over the working class. Unsurprisingly, anti-debt conflicts have exponentially risen since 1980 (Gerber et al., 2021). Class struggle seems to have somehow moved its centre of gravity from the capital/labour relation to the creditor/debtor relation (Lazzarato, 2012).

In sum, ecological conditions and debt represent two most prominent sites of tension over contemporary economies, whether openly or tacitly. Because both are so intimately linked to economic growth, it would not be a surprise if the radical ideologies that emanate from them are articulated around a critique of growth and accumulation. If this is confirmed, degrowth ideas are likely to become central to the radical ideologies of the twenty-first century. This is an important difference with more classic land and labour struggles which are not centred on a critique of growth (Gerber, 2020). While both anti-debt and environmental conflicts can be seen as an attack on growth, they target different 'levels' of it. Anti-debt conflicts contest the 'virtual' financial level, while environmental struggles target what

Martínez-Alier (2009) called the 'real-real' level of economic growth, that is, its metabolic foundation. The 'real' level, for its part, refers to the realm of production and consumption and may not be where the most acute conflicts are taking place right now. This is not to say, of course, that this level is conflict-free; but I suggest that economic struggles have recently expanded to the 'virtual' and the 'real-real' levels with a revenge.

Contemporary capitalism thus presents a unique combination of unsustainability, debt, and addiction to growth. These three elements form a kind of 'complex' that also brings anti-debt and environmental movements together. It turns out for example that many environmental movements are, knowingly or not, targeting the debt economy that pushes companies and indebted countries into predatory extractivism in order to ensure loan repayments (Steppacher & Gerber, 2012).

Another commonality of anti-debt and environmental conflicts is that both often involve different social classes, from workers to entrepreneurs, from landless peasants to wealthy farmers. These conflicts, as a result, usually cannot be understood on the sole basis of traditional class politics. They therefore often lack a clear ideological commitment: their outcome can be radical-revolutionary as well as populist-opportunistic (Gerber et al., 2021). For ecological conflicts, this latter orientation has sometimes been called 'not in my backyard' (NIMBY); for anti-debt struggles, one could call it 'not from my wallet' (NOWA). Having said this, it's important to note that both NIMBY and NOWA movements may represent important starting points for further politicization.

Yet we are left with a perennial puzzle: why do some movements become populist and others radical? How are some protesters able to 'radicalize'? And what makes a radical project viable over the long term? These are crucial questions for the Barcelona School and for any scholar-activist interested in deep transformation. We will examine them in the next section.

21.4 Triggers and Targets: Outline of a Theory

Conflicts over the economy have three different layers of causes and hence also of targets. The first one is concerned with the *immediate impacts* on the protesters' wealth, health, or, following Honneth (1996), recognition. These impacts may for example result in demands for higher wages or equal rights (labour), agrarian reform (land), or the halt of a given polluting industry (ecological resources). Taken together, these demands can be quite radical, but taken individually, they do not really challenge the power structure in place, and in the long run, they may actually reinforce it. Hence, the well-known need for broader answers.

The second layer of causes/targets of conflicts over the economy is thus concerned with the *politico-institutional structure* such as the distribution of ownership or the growth imperative. Within this layer, movements may start with the five fundamental conflict-prone categories of land, tax, labour, debt, and ecological resources, but they typically aim at replacing the entire arrangement represented in Fig. 21.1. Among the structural alternatives put forward, there is, for example, state socialism, radical municipalism, or degrowth. These political projects undoubtedly go beyond a set of top-down policies and do challenge capitalist relationships at their roots. We are so far still in a familiar Marxian terrain – but this is not quite enough yet.

Conflicts over the economy have a third layer of causes and of potential targets: the realm of *consciousness*, namely the protagonists' self- and situational awareness. This is the realm of inner phenomena: why, in terms of collective consciousness, did a given movement start or never started? What are the internalized norms that enable or deter mobilization? How to build a caring community of activists enabling the healthy deployment of a radical ideology? What are the shadow sides of leaders and organizations? To start addressing such questions, one must acknowledge that unconscious forces and emotions are important drivers in the social and political sphere (Gerber, 2022).

This third layer has received much less attention – including from the Barcelona School – and this general neglect has led to too many failed attempts at radical transformation. This is for example true of revolutionary republicanism and communism where, to make it short, one élite was quick to replace another one and reproduce the same old relations. Without a proper psychological theory and praxis, classical Marxism could only coarsely understand the process by which a 'class in itself' ('objective factors') could become a 'class for itself' ('subjective factors') as well as the inner conditions for the long-term viability of socialism. Busy seeking to seize state power, classical Marxism did not emphasize prefigurative politics, that is, the concrete building of emancipated pockets seen as an essential learning ground for further and deeper transformations (Grubačić & Graeber, 2004).

My point is that radical politics requires some kind of 'awakening' work in order to free oneself and heal the multiple potential forms of alienation (Rosa, 2019). The focus should not only be the deeper awareness/healing of the subject's relationship to herself and others, but also to the rest of nature and to the 'underlying reality', be it the unconscious or the divine (Brown, 2017). This work of de-alienation is not an easy one. Recent psychoanalytic work has suggested that the development of capitalism was of course not inevitable, but that "we are, one might say, psychically disposed to invest ourselves in the capitalist system", and this explains its extraordinary grip on many of us (McGowan, 2016: 22). Capitalism's genius, in short, is to postulate that there are solutions for the subject's alienation and emptiness because the resulting desires can be fulfilled by commodities. The contemporary dominant superego commands us to 'enjoy', 'transgress', and even 'self-actualize', but always through consumption. Besides sheer power relations, there are therefore also many unconscious ties which bind us to the system in place. It is thus not enough to rationally criticize capitalism or neoliberalism; it is also essential to identify and come to terms with our own unconscious investments in the system (Gerber, 2022). Next, I will add a few words on how consciousness can be expanded to sustain radical change.

21.5 Expanding Consciousness for Radical Change: The Examples of Debt and Degrowth

Economic relations are 'external' (involving money, contracts, and enforcement) as well as 'internal' (involving values, emotions, and unconscious norms). Many authors have investigated the internal, subjective effects of debt and how these effects tend to hinder resistance. The specific morality of a credit relation is often framed in terms of personal responsibility, that is, of a staged 'mutual trust' between a creditor and a debtor. A loan becomes a bet on whether a particular individual will keep her promise and reputation. To default generates feelings of guilt and shame because lending is intimately linked to values like 'honesty', internalized since a very young age and often projected onto an unjust 'debtfare state' (Soederberg, 2014).

Accordingly, anti-debt mobilizations cannot just be outwardly oriented, but also inwardly, in a consciousness-raising process that requires a "specific kind of subjective conversion […], leaving behind debt morality and the discourse in which it holds us hostage" (Lazzarato, 2012: 164). One way of proceeding, Lazzarato (2015) suggests in a way reminiscent of degrowth, is to collectively retreat from capital's valorization processes by engaging less in the economy, by refusing to work, by consuming less, by organizing autonomously, and by claiming back forms of idleness. Such 'pulling out' would contribute to dissociate our subjectivities from capitalist production and open the time for joint production and for the inner and relational work required for sustaining such a radical project.

In my view, the degrowth movement – at least in its commonist and non-state-centric variants – has already done some work on the third layer of causes/targets I mentioned above. Degrowth goes beyond the ecological critique of capitalist accumulation and includes practical reflections on what constitutes an existentially meaningful mode of relating and coexisting on the planet. One of its core objectives is to overthrow the value system associated with the 'imperial mode of living' and to emphasize instead care as the key relation of (re)production. In short, degrowth is all about reconnecting with what truly matters, away from the noises of capitalist modernity and its 'culture of uncare' (Weintrobe, 2021). Simplicity, conviviality, work-sharing, and commoning become new practices, and outer degrowth opens the door for inner (re)growth (Gerber, 2021; Kaul & Gerber, 2023).

Some elementary psychoanalytic and ecopsychological principles can be helpful along this path. To examine what has been put for us in our superegoic 'laws', to investigate the meaning of our fears and blocks, to seek the guidance of the whole person (the 'soul') instead of merely the ego, to reconnect with our 'ecological unconscious' and with the sacred could all be seen as third layer targets of degrowth as a radical ideology for the twenty-first century. Psychoanalysis advises a humble attitude towards the unconscious: we need to listen to what it seeks to say, as there are potentially destructive energies hidden in it, but also potentially healing and liberating properties. The point is to move from 'ego' to 'eco' and to realize that reason "develops only when the brain and the heart are united, when feeling and thinking are integrated" (Fromm, 1973: 358). Without some inner work, exterior targets alone will be of limited success, especially if state power (at any level) is again taken.

21.6 Concluding Remarks

The problem of indebtedness goes back 5000 years but the prominence of ecological degradations is a very recent phenomenon. As we have seen, these two issues have become central to the current world system. They are intimately associated with the growth addiction of neoliberal capitalism which relentlessly seeks new accumulation opportunities, both virtual and material. These dynamics will intensify in the coming decade as no substantial measures are currently being taken to downsize the global metabolism and limit debt-driven growth. As a result, neoliberalism generates massive ecological and socioeconomic destabilizations that will need to be addressed structurally.

However, this chapter also tried to show that structural 'outer' solutions cannot be the ultimate goal of radical movements. 'Inner' work aimed at addressing alienation and creating caring communities is essential in any preparation or implementation of a revolutionary project. Theodore Roszak – a founder of ecopsychology and an important influence in the degrowth movement – noted that the psychotherapist's role should primarily be "that of raising questions about our standard of sanity. That is an extremely important role, as much for what it might serve to downplay (careerist pressures, money, and status) as for what it might emphasize (our abiding need for wilderness, tranquility, or animal companions)" (Roszak, 1992: 311). For him, "both the therapists and the ecologists offer us a common political agenda for the good of the planet, for the good of the person. It is simply stated: Scale down. Slow down. Democratize. Decentralize" (ibid.). This is in a nutshell the degrowth project. Subjectivity, consciousness, and alienation are research frontiers for the Barcelona School. They have the potential to fortify its scientific and political agenda (Kaul & Gerber, 2023).

References

Brenner, R. (1985). The agrarian roots of European capitalism. In T. Aston & C. Philpin (Eds.), *The Brenner debate* (pp. 213–327). Cambridge University Press.
Brown, A. M. (2017). *Emergent strategy: Shaping change, changing worlds*. AK Press.
Burg, D. (2004). *A world history of tax rebellions*. Routledge.

De Ste. Croix, G. (1981). *The class struggle in the ancient Greek world*. Cornell University Press.
Durand, C. (2017). *Fictitious capital: How finance is appropriating our future*. Verso.
Finley, M. (1983). *Politics in the ancient world*. Cambridge University Press.
Fromm, E. (1973). *The anatomy of human destructiveness*. Holt, Rinehart and Winston.
Gerber, J.-F. (2014). The role of rural indebtedness in the evolution of capitalism. *Journal of Peasant Studies, 41*(5), 729–747.
Gerber, J.-F. (2020). Degrowth and critical agrarian studies. *Journal of Peasant Studies, 47*(2), 235–264.
Gerber, J.-F. (2021). Karl with Carl: Marxism and the Jungian path to the soul. *International Journal of Jungian Studies, 14*(2), 182–203.
Gerber, J.-F. (2022). The psychoanalytic critique of capitalism: Elements for an overview. *Psychotherapy and Politics International, 20*(1–2), 1–23.
Gerber, J.-F., & Scheidel, A. (2018). In search of substantive economics: Comparing today's two major socio-metabolic approaches to the economy – MEFA and MuSIASEM. *Ecological Economics, 144*, 186–194.
Gerber, J.-F., & Steppacher, R. (2014). Some fundamentals of integral economics. *World Futures, 70*(7), 442–463.
Gerber, J.-F., Moreda, T., & Sathyamala, C. (2021). The awkward struggle: A global overview of social conflicts against private debts. *Journal of Rural Studies, 86*, 651–662.
Graeber, D. (2011). *Debt: The first 5000 years*. Melville House.
Grubačić, A. & Graeber, D. (2004, January). Anarchism, or the revolutionary movement of the twenty-first century. *ZNet*.
Honneth, A. (1996). *The struggle for recognition: The moral grammar of social conflicts*. MIT Press.
Kaul, S., & Gerber, J.-F. (2023). Psychoanalysis and degrowth. In L. Eastwood & K. Heron (Eds.), *De Gruyter handbook of degrowth: Propositions and prospects*. De Gruyter, in press.
Lazzarato, M. (2012). *The making of the indebted man: An essay on the neoliberal condition*. MIT Press.
Lazzarato, M. (2015). *Governing by debt*. MIT Press.
Martínez-Alier, J. (2002). *The environmentalism of the poor: A study of ecological conflicts and valuation*. Edward Elgar.
Martínez-Alier, J. (2009). Socially sustainable economic de-growth. *Development and Change, 40*(6), 1099–1119.
McGowan, T. (2016). *Capitalism and desire: The psychic cost of free markets*. Columbia University Press.
Polanyi, K. (1944). *The great transformation*. Beacon.
Rosa, H. (2019). *Resonance: A sociology of our relationship to the world*. Polity.
Roszak, T. (1992). *The voice of the Earth*. Simon & Schuster.
Scott, J. C. (2009). *The art of not being governed: An anarchist history of upland Southeast Asia*. Yale University Press.
Soederberg, S. (2014). *Debtfare states and the poverty industry: Money, discipline and the surplus population*. Routledge.
Steffen, W., Grinevald, J., Crutzen, P., & McNeill, J. (2011). The Anthropocene: Conceptual and historical perspectives. *Philosophical Transactions of the Royal Society A, 369*, 842–867.
Steppacher, R., & Gerber, J.-F. (2012). Meanings and significance of property with reference to today's three major eco-institutional crises. In J.-F. Gerber & R. Steppacher (Eds.), *Towards an integrated paradigm in heterodox economics* (pp. 111–126). Palgrave Macmillan.
Weintrobe, S. (2021). *Psychological roots of the climate crisis: Neoliberal exceptionalism and the culture of uncare*. Bloomsbury.

Open Access This chapter is licensed under the terms of the Creative Commons Attribution 4.0 International License (http://creativecommons.org/licenses/by/4.0/), which permits use, sharing, adaptation, distribution and reproduction in any medium or format, as long as you give appropriate credit to the original author(s) and the source, provide a link to the Creative Commons license and indicate if changes were made.

The images or other third party material in this chapter are included in the chapter's Creative Commons license, unless indicated otherwise in a credit line to the material. If material is not included in the chapter's Creative Commons license and your intended use is not permitted by statutory regulation or exceeds the permitted use, you will need to obtain permission directly from the copyright holder.

Part V
Science and Self-Reflected Activism

Chapter 22
Activism Mobilizing Science Revisited

Marta Conde and Martí Orta-Martínez

The "extraction frontiers" are the place where extraction of natural resources expands geographically, colonizing new lands and territories in search of raw materials (oil, mineral ores, biomass, etc.) to satisfy the increasing demands for materials and energy of industrialized economies (Martínez-Alier et al., 2010; Moore, 2016). This extraction frontier has been advancing since colonial times, in an accelerated rate since the industrial revolution, encroaching and colonizing territories, ways of life and cultures in its wake (Schaffartzik et al., 2014; Krausmann et al., 2009). However, these extraction frontiers do not advance unopposed; on numerous occasions the communities that live near these projects react against the socioenvironmental and cultural impacts on land, water, and ways of life (Martinez Alier, 2003; Conde, 2017; Arsel et al., 2016). Well studied by the BCN school, many of these groups are part of the Environmental Justice (EJ) movement (Martinez-Alier et al., 2016; Martínez-Alier, 2021).

The expansion of the extraction frontier and the resistance movements are marked by an intense controversy between the limits of technology to achieve "sustainable" extraction and the role played in this dispute by scientific knowledge and lay or local knowledge. Scientific knowledge, like all knowledges, is partly socially constructed (Foucault, 1971). Although it depends on observation, experimental- and measurement-based testing is also subject to the interests and the cognitive assumptions of the scientist, social practices, available materials, and, more

---

M. Conde (✉)
GREDS-EMCONET, University Pompeu Fabra, Barcelona, Spain

Universitat de Barcelona, Barcelona, Spain

M. Orta-Martínez
Universitat de Barcelona, Barcelona, Spain

Central University of Catalonia - UVic, Vic, Spain

importantly, to the economic and political interests of the institutions that contribute to its funding, elaboration, and dissemination (Barnes, 1977; Jasanoff, 2004).

Although scientific knowledge has been traditionally used to support the most powerful political forces and actors through the invisible role of expertise, local groups are increasingly engaging in the generation of scientific knowledge. Under terms such as civil science, citizen science, civic science (Bäckstrand, 2003), or advocating for the democratization of science (McCormick, 2009), these groups want to reveal negative socioenvironmental outcomes, bad practices, and/or improve companies' and states' policies and operations. Concerned research groups sensitive to EJ issues are also promoting an alliance between scientists and local affected groups, under the "participatory processes" umbrella. With various degrees of collaboration and participation, we find methods such as Participatory Action Research or the widely used Community-Based Participatory Research where community partners should participate in every step of the process from its inception to the final interpretation of results (Minkler & Wallerstein, 2003; Shepard, 2002). It is important, however, to bear in mind that the degree of participation, the asymmetries of power among actors, and the final use of knowledge for decision-making are controversial factors that can lead to privileged access, a co-optation of the process and the colonization of themes and discourses (Cooke & Kothari, 2001). The underlying idea is that science has political and social implications and that citizens should "have a stake in the science-policy interface" (Bäckstrand, 2003). In this vein, the proposal of a "science without unity" with different forms of knowledge at stake (Irwin 1995) is very relevant for socioenvironmental conflicts. Similarly, Funtowicz and Ravetz (1993) propose the creation of "extended community of equals" to resolve urgent and complex controversies, which must include all knowledgeable people with interests in the subject. Although some authors have questioned the scientific method (Corburn 2005; McCormick, 2009; Funtowicz & Ravetz 1993), many in fact use it to produce alternative policy and action-oriented scientific knowledge.

This is the case in the context of EJ where polluting extractive projects can negatively affect the environment and livelihoods of nearby communities (Nixon, 2011; Martinez Alier, 2003). The usual narrative of the industry is that impacts are nonexistent or that they are not produced by the company but are "natural" due to the geochemical composition of the area, leaving the communities the "burden of proof." If the project has not yet started, communities generally need to challenge the Environmental Impact Assessments (EIAs) submitted by the companies that propose high-technology mitigation and remediation measures that many believe to be false promises or insufficient solutions – an experience gained through previous projects and/or learnt through their alliances (Keck & Sikkink, 1998; Conde, 2017).

Thus, in order to uncover, understand, and denounce present or future impacts, citizens, communities, or local activist groups ally with scientists under the counter-expertise framework (Topçu, 2008, see also Martinez Alier et al., 2011, 2014; Conde, 2015). Through Activism Mobilizing Science (AMS) (Conde, 2014), local groups learn to use measurement tools and scientific language and coproduce new and alternative knowledge that challenges the discourse produced by companies and certain state agencies. Also, under the counter-expertise umbrella, communities whose health is affected by the negative impacts of polluting industries react by carrying out epidemiological studies, measuring the impacts, and quantifying the diseases with their own means. Known as popular epidemiology, communities test the connection between pollution and the impact on their health, report it to gain more support and motivate an "official" epidemiological study (Brown, 1992; San Sebastian & Hurtig, 2005). Similarly, in *Street Science*, Corburn (2005) documented how four groups of marginal populations and/or racial and ethnical minorities from New York self-organized to report impacts on their health based on the coproduction of knowledge. Closely related to the policy-oriented, bottom-up, and coproduction framework of popular epidemiology, street science, and citizen science, AMS goes beyond the analysis and reporting of health impacts to study more carefully the knowledge–activism–scientist nexus and the power differences of this activist strategy.

In order to develop the idea of AMS, three case studies in the frontier of extraction will be analyzed. A short contextualization of the cases is given as follows:

In Niger, Areva (French national nuclear company) has extracted uranium from the underground mines of Somaïr and Cominak since 1968. The towns of Arlit and Akokan were built near the mines to house its workers. Some 60,000 inhabitants live in houses made of clay, iron, and scrap metal. There are two main sources of radioactive hazards: the water distributed by the company for consumption contains radioactive elements at levels 10–100 times higher than OMS recommendations (CRIIRAD, 2008), and the open-air storage of mining waste – that contain 85% of the original radioactivity – as well as the mines' ventilation shafts cause radioactive dust and radon (CRIIRAD, 2008).

In Namibia, Rio Tinto, the big mining giant, has been mining uranium from Rössing mine since 1976, and built the town of Arandis to house its workers. As in Niger, the workers and residents are totally dependent on the mine. The biggest concern for workers is the health impacts; many know of colleagues who have become ill and died but they cannot prove the connection between their illness and their work at the mine (Shindondola Mote, 2008).

In the Peruvian Northern Amazon, two oil concessions known as Block 8 and Block 1AB (now 192) were leased in the late 1960s and early 1970s. First operated by PetroPeru, the Peruvian national oil company, and Occidental Petroleum Corporation (Oxy), a US-based company, were later transferred to a Dutch and Canadian consortium. These oil blocks are the most productive and longest lasting in the Peruvian Amazon (39.2% of total Peruvian production) (Orta-Martínez et al., 2018a). More than 45,000 indigenous people live in the area affected by these oil extraction activities, which have caused serious environmental and health impacts (Orta-Martínez et al., 2007, Orta-Martínez et al. 2018b; Cartró-Sabaté et al., 2019; Yusta-Garcia et al., 2017, O'Callaghan-Gordo et al., 2018, Rosell-Melé et al., 2018). In 2005, the Ministry of Health found that between

98.6% and 66.2% of Achuar children between 2 and 17 years old exceeded the acceptable limits for cadmium and lead in blood (DIGESA, 2006; Orta-Martínez et al., 2007). This risk has been linked to the discharge of 1 million barrels/day of produced water to local rivers and lands (Orta-Martínez et al., 2018b).

22.1 What Is Activism Mobilizing Science?

AMS has two main traits: (i) it is a process driven by activists or local grassroots groups from the areas impacted by extractive projects that engage with scientists to study the impacts of these activities; (ii) in this process, new knowledge is produced merging local knowledge with scientific knowledge in a knowledge coproduction framework. These two traits are examined in detail as follows:

1. Locally driven

In Niger, Almoustapha Alhacen was a worker in the Somaïr mine. In 1999, he saw three colleagues who worked in the uranium concentration division die from causes unknown to them. He contacted CRIIRAD (Commission for Independent Research and Information on Radioactivity), an independent French laboratory specializing in radiation, whose team visited the area in 2003. Together with Alhacen, the team measured radioactivity around the town of Arlit, identifying high levels of radioactivity above WHO-recommended levels in water, air, and soil (CRIIRAD, 2008). To disseminate these results among the local population, Alhacen and the recently created civil society organization, Aghir in'man, organized workshops that included women, journalists, and chiefs of the different tribes.

Spurred by this experience, Bruno Chareyron, director of CRIIRAD, wanted to start a similar process in Gabon where Areva had also carried out uranium mining for 30 years with serious health impacts (CRIIRAD, 2009; Hecht, 2012). However, according to Chareyron, the idea did not come to fruition because there was no strong organization driving the process locally: "We tried to do something [a Geiger counter was sent to the local communities] ... but both parties have to do something." In contrast, in Niger, Aghir in'man leads the process; they take samples, organize workshops, participate in public meetings, give interviews to journalists, and find funds to acquire new equipment.

In Namibia, the AMS process started differently. Two local organizations, LaRRI and Earthlife Namibia, denounced the expansion of mining in Namibia (Kohrs, 2008) and the impact on workers' health based on testimonies from miners of other ongoing mines in Namibia (Shindondola Mote, 2008). Prompted by these studies, Conde conducted her doctoral research in Namibia, inviting Bertchen Kohrs, the director of Earthlife Namibia in an EU-funded project (EJOLT, Environmental Justice Organizations, Liabilities and Trade) as well as CRIIRAD – thus putting them in contact. Was this the start of a participatory research project? This was not Conde's objective. As in Niger, the local organization took the reins of the whole process inviting CRIIRAD to Namibia. Kohrs planned the trip in September 2011 and, once the results were obtained, asked CRIIRAD to come back to present the results, organizing a press conference and meetings with different groups as well as the mining company. The data gathered was shared at an annual general meeting of Rio Tinto, pushing the company to commission a health study.

The AMS process in Peru also has a similar origin to that of Namibia. Invited by indigenous and local organizations (FECONACO – the indigenous organization of the Corrientes River, Grupo de Trabajo Racimos de Ungurahui and Shinai), Orta-Martínez, a PhD candidate at the time, was asked to help mapping the oil spills in the indigenous territory. Shortly after, in January 2006, a team of scientists and activists including Orta-Martínez travelled to the Peruvian Amazon. To their surprise, 13 members of different indigenous communities were waiting for them; the communities had decided they were going to create a community-based monitoring team to map past and new oil spills. During the following weeks the group of scientists and activists trained the "monitors" and at the same time visited the oil-polluted sites to start the monitoring process with photos and their geolocation (Orta-Martínez 2010; Orta-Martínez & Finer, 2010). Indigenous and local organizations have taken over the process looking for their own funding and extending it to other areas and indigenous groups. Orta-Martínez and other scientists have returned almost annually to the area to support the process.

2. Coproduction of knowledge

In these alliances, new scientific knowledge that may refute the data and narratives of the companies is generated. Although in many occasions, local groups know what oil or uranium are and the socioenvironmental impacts they can cause, they ally with scientists to learn how to measure them and, more importantly, to speak the scientific–technical language used by large companies and state agencies. A crucial part of these alliances is the technical training provided by scientists to local groups.

However, scientists could not generate new knowledge alone. They do not understand the local complexity of the area, the local geography, the local and particular impacts on health and environment, and/or the socioeconomic and cultural aspects as well as local practices that are essential to understand and co-interpret the results (Conde & Walter, 2022). Thus, local knowledge is essential to conduct the sampling, understand the exposure routes, and interpret results: What are the local impacts? Where are located the polluted sites and the local sources of pollution (e.g., the fans that expel air from the mines, the mine tailings dams, the oil spills, the dumping sites of produced water)? How can they be accessed to carry out the sampling? What does the company do with the waste products? What drinking water sources are used by the community? Where do livestock feed and drink? What does the local population eat? What are the local customs in relation to their environment?

In Niger, during CRIIRAD's visit in 2003, the coproduction between Alhacen and CRIIRAD's team allowed the group to detect that highly radioactive scrap metal from the mine was sold in the local market. This scrap metal was being used for the construction of houses. Although this practice was, since then, continually denounced by Angir in'man, it was not until 2013 that Areva withdrew all the scrap metal from the market. Similarly, in Namibia, measurements confirmed that groundwater radiation was higher downstream from the mine than upstream, contradicting the company's position that all radiation was natural background radiation and not related to the mining activities (Chareyron, 2014).

In Peru, the company was also forced to improve very poor operational procedures (e.g., in situ burning of oil spills) and remediate a number of oil-polluted sites after the indigenous monitors recorded and denounced such illegal practices (Orta-Martínez et al., 2007, forthcoming). Furthermore, a new-to-science exposure route to local communities was documented by the scientists–locals alliance after the monitors reported that animals they hunted for food were ingesting oil-polluted soils (geophagy) (Orta-Martinez et al., 2018a, b). This motivated research to analyze bioaccumulation of heavy metals in wild game species, a key element of the diet of indigenous people that rely on subsistence hunting for their daily protein intake (Cartró-Sabaté et al., 2019).

22.2 Goals of Activism Mobilizing Science

Local-affected groups want to *learn and understand what is causing the impacts* to their environment and health. In Niger, Alhacen had heard of radioactivity but thought that you had to be in direct contact with uranium for it to affect you. He wanted to understand what radiation was and how it impacted you. With CRIIRAD, he learnt about radiation and, in turn, organized numerous workshops to teach inhabitants in Arlit as well as workmates so that they would take radiation protection measures seriously. For example, he was able to change the practice of taking workplace clothes home for washing over time (Conde, 2014).

A driver of these alliances has been the *coproduction of new knowledge to challenge the state or company discourse*. The measurements taken by CRIIRAD and both teams in Niger and Namibia have allowed to identify high levels of radiation in the nearby areas to the mine. In Niger, for example, they detected radiation 100 times higher than background values, identifying the use of waste rock in the construction of roads as the main source. This was also the case in a parking lot of Rössing uranium mine in Namibia. These bad practices exacerbate the daily radiation exposure for workers or local inhabitants.

The Achuar people were able to challenge the company narrative that argued that high-technology minimized the number of oil spills and other oil-related impacts. They were able to document and prove the existence of numerous oil spills with photos and their geolocation. This was and still is crucial in ongoing efforts to secure an (insufficient) fund for remediation and water treatment plants, to challenge a very limited decommissioning plan – and to ultimately bring the company to the negotiating table and improve its operational practices (Orta-Martínez et al., 2018a, b, forthcoming).

In many occasions, activists and impacted communities already know the impacts and effects of industrial pollution; they directly suffer them or are aware of them from nearby polluting industry projects. Through these alliances they gain scientific prove of the impacts as well as raise public awareness and national and international support; they obtain greater *legitimacy and visibility*.

CRIIRAD already had experience with several nuclear-related projects in France; they would typically carry out visits to the impacted areas, analyze the EIAs, and issue a report and a press release with wide press coverage. In Niger and Namibia, a similar process ensued; after CRIIRAD's visit to the area, a joint press release with the main results and demands was issued. In Namibia the press release was widely distributed giving Kohrs high visibility, with people approaching her on the streets and Earthlife appearing numerous times on the news. At the same time, Kohrs gained legitimacy when discussing "technical" concepts – a language previously only used by company "experts." She was able to talk about radioactivity and its impacts, she participated in the new regulations of the Atomic Regulator and reviewed the legislation for the rehabilitation of mines. In Niger, Alhacen not only learnt about the impacts of radioactivity but as Khors, he started making technical as well as social demands to the company, the press, and government representatives.

For the Achuar monitors in the Amazon, the legitimacy gained through these processes has allowed them to negotiate and speak one-to-one with the oil company managers and state officers using technical terminology. They have placed demands, negotiated settlements, demanded changes in environmental regulations and participated in legal complaints (Orta-Martínez & Finer, 2010; Orta-Martinez et al., 2018a, b, forthcoming).

In all three cases, local groups did not get carried away by the dominant techno-scientific language (Yearley, 1992) but they did learn technical language in order to defend their newly coproduced knowledge and back up their claims for environmental justice.

22.3 Conclusion

Affected communities that engage in alliances with scientists are empowered through the coproduction of new scientific knowledge with which they get more visibility and can challenge the data and discourses used by the companies or state agencies to promote the expansion of the extractive frontier. We want to highlight that their empowerment is also a result of a process of mutual enrichment between local groups and the scientists in terms of knowledge shared. This includes new tools, skills (such as the appropriation of the techno-scientific language), and strategies as well as cultural and place-based knowledges – that many times can be reinforced by friendship and long-term relationships. It is not only the newly produced knowledge that matters but the way it is produced and the dynamics generated.

Although not the scope of this chapter, it is important to pay attention to power differences in these alliances many times related to colonial structures of scientific knowledge production and domination. These alliances might impose scientific knowledge over cultural or local knowledges that can translate in the imposition of certain narratives and demands. Thus, it is important to acknowledge that local knowledge and local demands must be key drivers not only in the coproduction of new scientific knowledge but also in the design of strategies of resistance by local groups. In this regard, AMS has to be understood (and used) as part of a myriad of languages and strategies (Aydin et al., 2017) when confronting wrongdoings of companies and the state, or contesting unwanted projects, in order to achieve changes in decision-making power relations.

References

Arsel, M., Hogenboom, B., & Pellegrini, L. (2016). The extractive imperative and the boom in environmental conflicts at the end of the progressive cycle in Latin America. *The Extractive Industries and Society, 3*(4), 877–879.

Aydin, C. I., Ozkaynak, B., Rodríguez-Labajos, B., & Yenilmez, T. (2017). Network effects in environmental justice struggles: An investigation of conflicts between mining companies and civil society organizations from a network perspective. *PLoS ONE, 12*(7), e0180494.

Bäckstrand, K. (2003). Civic science for sustainability: Reframing the role of experts, policy-makers and citizens in environmental governance. *Global Environmental Politics, 3*, 24–41.

Barnes, B. (1977). *Interests and the growth of knowledge*. Routledge.

Brown, P. (1992). Popular epidemiology and toxic waste contamination: Lay and professional ways of knowing. *Journal of Health and Social Behavior*, 267–281.

Cartró-Sabaté, M., Mayor, P., Orta-Martínez, M., & Rosell-Melé, A. (2019). Anthropogenic lead in Amazonian wildlife. *Nature Sustainability, 2*(8), 702–709. https://doi.org/10.1038/s41893-019-0338-7

Chareyron, B. (2014). *Radiological impact of Rössing Rio Tinto mine* (EJOLT report).

Conde, M. (2014). Activism mobilising science. *Ecological Economics, 105*, 67–77.

Conde, M. (2015). From activism to science and from science to activism in environmental-health justice conflicts. *Journal of Science Communication, 14*(2), C04.

Conde, M. (2017). Resistance to mining. A review. *Ecological Economics, 132*, 80–90.

Conde, M., & Walter, M. (2022). Knowledge co-production in scientific and activist alliances: Unsettling coloniality. *Engaging Science, Technology, and Society, 8*(1), 150–170.

Cooke, B., & Kothari, U. (2001). *Participation: The new tyranny?* Zed Books.

Corburn, J. (2005). *Street science*. MIT Press.

CRIIRAD. (2008). *AREVA : Du discours à la réalité/L'exemple des mines d'uranium du Niger* (Note CRIIRAD 08-02).

CRIIRAD. (2009). *Radiation contamination found in 2009 on the former uranium mining site COMUF – AREVA Mounana (GABON)* (Note CRIIRAD No. 09-118)

DIGESA. (2006). *Evaluación de resultados del monitoreo del río Corrientes y toma de muestras biológicas, en la intervención realizada del 29 de junio al 15 de julio del 2005 Informe No-2006/DEPA-APRHI/DIGESA*. Dirección General de Salud Ambiental (Ministerio de Salud).

Foucault, M. (1971). *The order of things: An archaeology of the human sciences*. Pantheon Books.

Funtowicz, S., & Ravetz, J. (1993). Science for a post-normal age. *Futures, 25*, 735–755.

Hecht, G. (2012). *Being nuclear: Africans and the global uranium trade*. MIT Press.

Irwin, A. (1995). *Citizen science: A study of people, expertise and sustainable development*. Psychology Press.

Jasanoff, S. (Ed.). (2004). *States of knowledge: The co-production of science and the social order*. Routledge.

Keck, M., & Sikkink, K. (1998). *Activists beyond borders*. Cornell University Press.
Kohrs, B. (2008). *Uranium, a blessing or a curse? What you need to know about the uranium industry in Namibia*. Earthlife Namibia. Accessed last 3rd January 2023 at https://www.somo.nl/uranium-a-blessing-or-a-curse/
Krausmann, F., Gingrich, S., Eisenmenger, N., Erb, K. H., Haberl, H., & Fischer-Kowalski, M. (2009). Growth in global materials use: GDP and population during the 20th century. *Ecological Economics, 68*, 2696–2705.
Martinez Alier, J., Kallis, G., Veuthey, S., Walter, M., & Temper, L. (2010). Social metabolism, ecological distribution conflicts, and valuation languages. *Ecological Economics, 70*(2), 153–158.
Martinez Alier, J., Anguelovski, I., Bond, P., Del Bene, D., Demaria, F., Gerber, J.-F., Greyl, L., Haas, W., Healy, H., Marín-Burgos, V., Ojo, G., Firpo Porto, M., Rijnhout, L., Rodríguez-Labajos, B., Spangenberg, J., Temper, L., Warlenius, R., & Yánez, I. (2014). Between activism and science: Grassroots concepts for sustainability coined by Environmental Justice Organizations. *Journal of Political Ecology, 21*, 19–60.
Martinez-Alier, J. (2003). *The environmentalism of the poor: A study of ecological conflicts and valuation*. Edward Elgar Publishing.
Martinez-Alier, J. (2021). Mapping ecological distribution conflicts: The EJAtlas. *The Extractive Industries and Society*.
Martínez-Alier, J., Healy, H., Temper, L., Walter, M., Rodriguez-Labajos, B., Gerber, J. F., & Conde, M. (2011). Between science and activism: Learning and teaching ecological economics with environmental justice organisations. *Local Environment, 16*(1), 17–36.
Martinez-Alier, J., Temper, L., Del Bene, D., & Scheidel, A. (2016). Is there a global environmental justice movement? *The Journal of Peasant Studies, 43*(3), 731–755.
McCormick, S. (2009). *Mobilizing science: Movements, participation, and the remaking of knowledge*. Temple University Press.
Minkler, M., & Wallerstein, N. (Eds.). (2003). *Community based participatory research for health*. Jossey-Bass.
Moore, J. W. (2016). *The rise of cheap nature*. PM Press.
Nixon, R. (2011). *Slow violence and the environmentalism of the poor*. Harvard University Press.
O'Callaghan-Gordo, C., Flores, J. A., Lizárraga, P., Okamoto, T., Papoulias, D. M., Barclay, F., Orta-Martínez, M., Kogevinas, M., & Astete, J. (2018). Oil extraction in the Amazon basin and exposure to metals in indigenous populations. *Environmental Research, 162*, 226–230. https://doi.org/10.1016/j.envres.2018.01.013
Orta-Martínez, M. (2010). *Oil frontiers in the Peruvian Amazon. Impacts of oil extraction for the Achuar of Río Corrientes*. Doctoral thesis. Institut de Ciència i Tecnologia Ambiental (ICTA), Universitat Autònoma de Barcelona.
Orta-Martínez, M., & Finer, M. (2010). Oil frontiers and indigenous resistance in the Peruvian Amazon. *Ecological Economics, 70*(2), 207–218. http://www.sciencedirect.com/science/article/B6VDY-50BDHV3-2/2/13d7fe721d29275f14ca0b85442c538d
Orta-Martinez, M., Napolitano, D. A., MacLennan, G. J., O'Callaghan, C., Ciborowski, S., & Fabregas, X. (2007). Impacts of petroleum activities for the Achuar people of the Peruvian Amazon: Summary of existing evidence and research gaps. *Environmental Research Letters, 2*(4), 45006. http://stacks.iop.org/1748-9326/2/045006
Orta-Martínez, M., Pellegrini, L., & Arsel, M. (2018a). The squeaky wheel gets the grease? The conflict imperative and the slow fight against environmental injustice in northern Peruvian Amazon. *Ecology and Society, 23*(3), art7. https://doi.org/10.5751/ES-10098-230307
Orta-Martínez, M., Rosell-Melé, A., Cartró-Sabaté, C., O'Callaghan-Gordo, C., Moraleda-Cibrián, N., & Mayor, P. (2018b). First evidences of Amazonian wildlife feeding on petroleum-contaminated soils: A new exposure route to petrogenic compounds? *Environmental Research, 160*, 514–517. https://doi.org/10.1016/j.envres.2017.10.009
Orta-Martínez, M., Arsel, M., Pellegrini, L., & Mena, C. (forthcoming). *Barriers to redress of environmental harm from oil extraction in the Amazon*.

Rosell-Melé, A., Moraleda-Cibrián, N., Cartró-Sabaté, M., Colomer-Ventura, F., Mayor, P., & Orta-Martínez, M. (2018). Oil pollution in soils and sediments from the Northern Peruvian Amazon. *Science of the Total Environment, 610–611*, 1010–1019. https://doi.org/10.1016/J.SCITOTENV.2017.07.208

San Sebastián, M., & Hurtig, A. K. (2005). Oil development and health in the Amazon basin of Ecuador: The popular epidemiology process. *Social Science & Medicine, 60*(4), 799–807.

Schaffartzik, A., Mayer, A., Gingrich, S., Eisenmenger, N., Loy, C., & Krausmann, F. (2014). The global metabolic transition: Regional patterns and trends of global material flows, 1950–2010. *Global Environmental Change, 26*, 87–97.

Shepard, P. M. (2002). Advancing environmental justice through community-based participatory research. *Environmental Health Perspectives, 110*, 139.

Shindondola-Mote, H., (2008). *Uranium mining in Namibia. The mystery behind 'low level radiation'*. Labour Resources and Research Institute (LaRRI).

Topçu, S. (2008). Confronting nuclear risks: counter-expertise as politics within the French nuclear energy debate. *Nature and Culture, 3*, 225–245.

Yearley, S. (1992). Green ambivalence about science: Legal–rational authority and the scientific legitimation of a social movement. *The British Journal of Sociology*, 511–532.

Yusta-García, R., Orta-Martínez, M., Mayor, P., González-Crespo, C., & Rosell-Melé, A. (2017). Water contamination from oil extraction activities in Northern Peruvian Amazonian rivers. *Environmental Pollution, 225*, 370–380. https://doi.org/10.1016/J.ENVPOL.2017.02.063

Open Access This chapter is licensed under the terms of the Creative Commons Attribution 4.0 International License (http://creativecommons.org/licenses/by/4.0/), which permits use, sharing, adaptation, distribution and reproduction in any medium or format, as long as you give appropriate credit to the original author(s) and the source, provide a link to the Creative Commons license and indicate if changes were made.

The images or other third party material in this chapter are included in the chapter's Creative Commons license, unless indicated otherwise in a credit line to the material. If material is not included in the chapter's Creative Commons license and your intended use is not permitted by statutory regulation or exceeds the permitted use, you will need to obtain permission directly from the copyright holder.

Chapter 23
Iberian Anarchism in Environmental History

Santiago Gorostiza

23.1 Introduction

Following a renewed interest in anarchism in both social movements and critical academic circles, a growing volume of academic research has vindicated anarchist traditions of thought and increasingly applied an anarchist lens to geography since the early 2010s (see, e.g. Springer et al., 2012; Springer, 2013; Ferretti et al., 2018). As a rich theoretical tradition entangled with praxis, anarchism has the power to illuminate environmental, ethical and political issues faced by our societies today, and inform alternatives. When tracing the historical genealogy of anarchist thought and praxis during the nineteenth and twentieth centuries, Carl Levy pointed to the Paris Commune and "red and black Barcelona" as the most powerful examples of anarchist political action, underlining the role of cities for the transformation of philosophical anarchism into daily practice (Levy, 2018). Despite anarchist organisations in Spain having been repressed during the Francoist dictatorship, anarchist-inspired practices extend to present-day Barcelona. While these have not been central to the research of the Barcelona school of ecological economics and political ecology, several theoretical influences of anarchism and inspiration from local social movements can be identified.

In this chapter I adopt a historical approach to trace and make explicit some of these influences. I start by highlighting the attention devoted to Spanish anarchism and the 1936 revolution by researchers and thinkers interested in the relation between social anarchism and the environment in the 1970s. Next, by focusing on the emergence of political ecology and environmental history in Spain during the 1990s, I examine Eduard Masjuan's research on human ecology and Iberian anarchism, first developed in the journal *Ecología Política*. Masjuan's doctoral research,

S. Gorostiza (✉)
Centre for History, Sciences Po, Paris, France
e-mail: Santiago.Gorostiza@sciencespo.fr

supervised by Joan Martínez-Alier, delved into the rich debates on urbanism and birth control that took place in anarchist circles from Catalonia to Latin America between 1860 and 1937. His writings constitute an essential reference to explore the environmental dimensions of anarchism and have informed degrowth discussions on population and the collective ethics of self-limitation. Yet, despite the impact of Masjuan's research, I argue that an environmental history and political ecology of the 1936 revolution that focuses on practices of self-management, mutual aid and direct democracy is still to be written. I show some examples of work that has been done so far, from urban water management under anarcho-syndicalist principles to collectivised urban agriculture. Finally, I point out that, while not always acknowledged, the influence of anarchist practices can also be found in the research on today's social movements carried out at the Barcelona school of political ecology and ecological economics.

23.2 Environmental Perspectives from Iberian Anarchism

The experiences of Iberian anarchism have shaped radical approaches to social and environmental matters. During the 1960s, Murray Bookchin coined the term "social ecology" in developing an anarchist perspective on environmental problems, arguing that social hierarchies and different types of domination lay at the core of environmental conflicts. Essays like "Ecology and revolutionary thought" (1964) – later included in *Post-Scarcity Anarchism* (Bookchin, 1986 [1971]) – circulated widely in the countercultural movement of the 1960s–1970s. Interested in decentralisation, self-management and mutual aid, Bookchin delved into Spanish anarchism during this period, systematically collecting sources and interviewing exiled anarchists during his visit to Europe in 1967. The resulting book – *The Spanish Anarchists: The Heroic Years* – was a history of the anarchist movement in Spain from 1868 to the 1936 revolution and the beginning of the Spanish Civil War (Bookchin, 1977).

A year after the release of Bookchin's work on Spanish anarchism, the radical journal of geography *Antipode* published a special issue on the relation between social anarchism and the environment, devoting several articles to the 1936 revolution (Breitbart, 1978b). This issue has been referred to as one of the last examples of radical geography scholar engagement with anarchism before a long hiatus during the 1980s and 1990s (Springer et al., 2012). Closely involved in its preparation, together with Myrna Breitbart and Richard Peet, was Maria Dolors García-Ramon, a professor from the Geography Department of the Autonomous University of Barcelona, who pioneered radical geography research in Spain (Albet et al., 2019).

Back in 1978, Myrna Breitbart summarised the importance of the 1936 revolution, stating that "Spain is the only country in the twentieth century where anarcho-communism and anarcho-syndicalism were adopted extensively as revolutionary theories and practices in urban and rural areas" (Breitbart, 1978c: 60). Breitbart's own doctoral research focused on the concept of decentralism in the Spanish revolution (Breitbart, 1978a, d). Other articles in the special issue included a study of the collectivisation of industries in Catalonia during the war (Amsden, 1978), while Maria Dolors García-Ramon examined the theoretical contributions to spatial theory from Spanish anarchists (Garcia-Ramon, 1978). The special issue also included an article on the resurgence of the libertarian movement after the death of the dictator Francisco Franco, a review of Mary Nash's book on the women-only anarchist organisation *Mujeres Libres* (1975), works on Kropotkin and Reclus, reprints of Kropotkin's essay "What geography ought to be" and Bookchin's "Ecology and revolutionary thought".

Anarchist perspectives in geography almost disappeared from international publications during the 1980s and 1990s (Springer, 2013). In Spain, however, the intertwined development of the fields of political ecology and environmental history was a fertile ground for budding research on the socioenvironmental dimensions of Iberian anarchism.

23.3 Iberian Anarchism at the Crossroads of Environmental History and Political Ecology

The foundation of the journal *Ecología Política* (1991) by Joan Martínez-Alier and Anna Monjo has been described as one of the key factors for the coalescence of environmental history in the Catalan context (Martí Escayol, 2019). The vision of environmentalism as "inscribed in a long tradition of emancipatory social struggle" was already explicit in Martínez-Alier's first editorial, thus signalling an interest in history. From Martínez-Alier's viewpoint, environmentalism should not be conceived as a novelty of the countercultural 1970s, because before that "rural and urban social movements that have opposed and oppose exploitation have often been environmentalist movements" (Martínez Alier, 1991: 8–9). Accordingly, during the very same years that *Ecología Política* was taking off, Martínez-Alier also co-edited, together with Manuel González de Molina, one of the first special issues on history and ecology published by a Spanish journal of contemporary history. By examining social conflicts as ecological conflicts, motivated by the inequalities in the access to natural resources and services, Martínez-Alier conceived the contribution of environmental history as a renovation of social history (Martínez Alier, 1993). Such a perspective opened the space to reassess the social dimensions of historical anarchism from an environmental standpoint.

Anarchism was a key part of the political education and intellectual interests of both founders of *Ecología Política*. During the 1980s, Anna Monjo began her doctoral research about the experience of anarchist industrial workers in Barcelona during the Spanish Civil War, presenting her thesis in 1993 (Monjo & Vega, 1986; Monjo, 1993, 2003). Since 1988, Monjo worked at the publishing house Icaria, the publisher of *Ecología Política*, which she eventually came to manage. Martínez-Alier – who refers to himself as a "moderate anarchist" in his memoirs – has acknowledged the political impact of his time spent collaborating with the publishing house Ruedo Ibérico during the 1960s and 1970s (Martínez-Alier, 2019). Ruedo Ibérico was established in 1961 in Paris by Spanish exiles and led by the anarcho-syndicalist José Martínez Guerricabeitia (1921–1986).

During the final years of the Francoist dictatorship, Ruedo Ibérico published dozens of key works from diverse anti-Francoist movements that challenged the official historiography established by the Spanish dictatorship, including Martínez-Alier's own PhD thesis (Martínez-Alier, 1968; Sarría Buil, 2019). Together with José Manuel Naredo, who had a strong influence on the development of ecological economics, Martínez-Alier participated in editing the last issues of the journal *Cuadernos de Ruedo Ibérico* between 1975 and 1979, a period when this publication adopted a more anarchist and environmentalist standpoint (Naredo, 2008; Martínez-Alier, 2019). Until the early 1980s, Martínez-Alier also published regularly in the anarchist periodical *Bicicleta* (1977–1982). While these influences have not developed into an explicit anarchist scholarship, the influence of anarchist perspectives in Martínez-Alier's academic work is evident in his emphasis on peasant agriculture, his hope for disruptive grassroots movements or his distrust of state-based politics. The monumental, collective task of compiling thousands of cases of ecological distribution conflicts in the Environmental Justice Atlas – increasingly portrayed by Martínez-Alier not as an inventory but as an "archive" of environmental justice struggles – is reminiscent of Max Nettlau's efforts to compile a vast collection of documents and archival records to capture the history of anarchism (Gorostiza, 2014).

The interest of *Ecología Política* in exploring grassroots environmentalism and emancipatory social movements made it a good fit for the first insights from Eduard Masjuan's research on human ecology and Iberian anarchism between 1860 and 1937. This was part of his PhD thesis on the topic, supervised by Martínez-Alier (Masjuan, 1993, 1995, 1996, 1998). Masjuan's contribution focused on Catalan anarchism but paid special attention to the international circulation of these ideas, both in and from the Iberian Peninsula and beyond, particularly in Latin America. First, he examined the "organic" tradition of urbanism within Iberian anarchism, connecting local figures to the genealogy of Patrick Geddes and Élisée Reclus. This also involved an analysis of decentralisation and the ideas of "free municipality" (*municipio libre*) and "free commune" (*comuna libre*) as the basic unit of anarchist social organisation, emancipated from the State and federated with other municipalities or communes (Masjuan, 2000: 172–176).

Masjuan's second main contribution was a nuanced discussion of the self-proclaimed "Neo-Malthusian" movement for birth control within Iberian anarchism and in Latin America (Masjuan, 2000). Around the 1870s, despite Thomas Malthus's fervent opposition to the idea of birth control, the notion of "Malthusianism" had ironically become associated with it (Kallis, 2019). Masjuan documented the circulation of "Neo-Malthusian" ideas within the anarchist movement, from Francesc Ferrer i Guàrdia to Emma Goldman. These self-proclaimed "Neo-Malthusians" advocated for birth control or "conscious procreation", explaining poverty not by excess population but by social inequality. Instead of advocating for population policies from above, they preached women's freedom to choose how many children they wanted to have, disseminated contraception measures and challenged religious and state authorities calling for a "womb strike". In supporting "conscious procreation" to prevent the exploitation of women's bodies to produce soldiers and cheap labour, this bottom-up movement was explicitly anti-militaristic and anti-capitalist (Masjuan, 2000).

Published by Icaria – the publishing house directed by Anna Monjo – Masjuan's book *La ecología humana en el anarquismo ibérico* (2000) has been widely cited and constitutes an essential reference to start exploring the socioenvironmental dimensions of Iberian anarchism between 1860 and 1937. Its influence is well apparent in Martínez-Alier's work during the late 1990s and early 2000s (see for instance Martínez-Alier, 1996, 2002: 51–53). Moreover, it has informed degrowth discussions on population and the collective ethics of self-limitation (Kallis & March, 2015; Martínez-Alier, 2015; Kallis, 2019). Nonetheless, Masjuan's focus on the decades preceding the Spanish Civil War leaves an ample space for further historical research. Most of all, while he highlighted the rural and industrial collectivisations of the 1936 revolution as one of the great achievements and legacies of Iberian anarchism, few works from an environmental history or political ecology perspective have delved into this period.

23.4 Anarchism in the City: Barcelona and the 1936 Revolution

The military coup against the government of the Second Spanish Republic on 18 July 1936 marked the beginning of the Spanish Civil War. Aimed at crushing any resistance as fast as possible, the insurrection encountered with the dogged resistance of working-class unions and loyalist forces. In Barcelona, the militants of the main anarchist organisations, the *Confederación Nacional del Trabajo* (CNT) and the *Federación Anarquista Ibérica* (FAI), were at the forefront of the street fights, which ended with the defeat of the insurgent troops after 2 days of combat. The CNT, a confederation of anarcho-syndicalist labour unions, established an antifascist coalition with the rest of Catalan leftist parties. During the first months of the war, the anarchists controlled the regional government in practice, and engaged in a deep transformation of economic and social life, which included the collectivisation of agricultural and industrial activities and many small businesses throughout Catalonia and Aragon. Most of the companies were immediately seized and self-managed by their workers (Castells Durán, 1993; Balcells, 2017). The failed military coup thus ignited a revolution and a civil war that extended until 1939.

In the late 1970s, Bookchin underlined that the work of self-managed anarchist collectives made the Spanish revolution distinct, "challenging [to] popular notions of a libertarian society as an unworkable utopia". Bookchin pointed out how these collectives simultaneously represented the climax and the tragic end of several decades of anarcho-syndicalist tradition in Spain. "To anyone with a concern for novel social forms, the Anarchist collectives of Spain raise many fascinating questions: how were the collective farms and factories established? How well did they work? Did they create any administrative difficulties?" (Bookchin, 1977: 1–2). Since Bookchin posed these questions more than 40 years ago, historical research on anarchism in Spain and anarcho-syndicalist collectives during the Spanish Civil War has significantly expanded (see among many others Bernecker, 1992; Cattini & Santacana, 2002; Ealham, 2005; Castillo, 2016; Balcells, 2017). However, an environmental history and political ecology of the 1936 revolution is still to be written.

Such a task requires examining the anarchist takeover, reorganisation and daily management of industries, agriculture, transport, energy and water supply services. The anarcho-syndicalist current of anarchism, particularly strong in Catalonia, Aragon and Andalusia, and the principles of self-management, mutual aid, direct action and direct democracy are of special interest here. The management of water supply in Barcelona during the war is a case in point. Following the work of both Maria Dolors Garcia-Ramon (1978) and Eduard Masjuan (2000), research on the collectivisation of the Barcelona private water utility has examined the management of a common good under a model that was neither public nor private (Gorostiza et al., 2013). Seized by the CNT during the first days of the war, the private water company *Aigües de Barcelona* was collectivised and self-managed by its workers throughout the conflict. The company launched reforms to increase access to water throughout the city, improved workers' salaries and reduced working hours. The mansion and private gardens of the company's director (today's Parc de les Aigües, in El Guinardó neighbourhood) became a school for the workers' children. Urban water consumption increased in Barcelona throughout the war, even if the collectivised company experienced growing difficulties to replace workers called to the battlefront and maintain water supply under the air raids carried out by fascist forces. These difficulties worsened with the occupation of the Pyrenees hydropower plants by Franco's insurgent troops, which forced the company to resort to the old, coal-powered steam engines to pump and distribute water. The collectivised water company faced increasing costs that obliged them to raise water prices towards the end of the war (Masjuan et al., 2008; Gorostiza et al., 2013; Gorostiza, 2019).

The difficulties experienced by the collectivised water company also bring to light the water–energy–food nexus. The growing demand of water from urban agriculture eventually became a problem for the daily management of water supply. Early after the beginning of the war, CNT's neighbourhood committees occupied monasteries and their adjacent lands in Barcelona and started to cultivate them. With these and other lands in five neighbourhoods, the CNT organised the *Col·lectivitat Agrícola de Barcelona i el seu radi* (Agricultural Collective of Barcelona), employing more than 2000 workers that took care of nearly 850 hectares of land. In addition to this collective, many citizens started cultivating vacant plots as the war progressed and food supply dwindled. These and other gardeners often tapped into the water distribution network, to the extent that the collectivised water company issued warnings to avoid excess use that jeopardised water distribution to houses and hospitals (Gorostiza, 2019; Camps-Calvet et al., 2021, 2022).

The collectivised Barcelona water company or the Agricultural Collective of Barcelona are just two examples of anarchist collectives that can be studied through the lens of environmental history. With more than 75% of the industries collectivised in Barcelona and many other examples of workers' self-management throughout Catalonia, Aragon and Andalusia, there are plenty of cases to be studied to address the questions posed by Bookchin in the late 1970s and explore political ecology approaches in historical perspective. Moreover, and unlike in the 1970s, today the CNT's archival collections are available at the Arxiu Nacional de Catalunya (Sant Cugat) and the International Institute of Social History (Amsterdam) and have been partly digitised.

23.5 Conclusions: Radical Imaginaries from the Past

In this chapter I traced the emergence of the research on Iberian anarchism and the environment intertwined with the establishment of political ecology and environmental history in Spain. The intellectual influence of Eduard Masjuan is visible in many works of ecological economics and political ecology authored by Joan Martínez-Alier and the fast-expanding research on degrowth, among others. Moreover, I have highlighted that there is ample research to be developed on the environmental history of anarchist collectives during the Spanish Civil War. This investigation can contribute to more deeply understanding collective organisation, focusing on the role of self-management, mutual aid and direct democracy in past socio-ecological struggles, and giving insights relevant for the current multiple crisis. Through an anarchist approach that emphasises prefigurative practices, a historical study of these collectives may make it possible "to embed within territorial practices certain organisational functions and structures that are at once effective in building spaces of struggle and developing modes of organisation that prefigure future worlds" (Ince, 2012: 9; cited in Ferretti & García-Álvarez, 2019).

Beyond intellectual influences and gaps in historical research, it is important to point out the connection between researchers at the Barcelona school of ecological economics and political ecology and the city's social movements. Researchers who are directly engaged with (and participated in) local social movements have also been shaped by the political climate of Barcelona. Some have explicitly acknowledged the influences of anarchism or anarchist-connected values in social movements. Claudio Cattaneo and Marc Gavaldà have highlighted the anarchist ideals in the strong urban and rural squatting movement (Cattaneo & Gavaldà, 2010). The squatted *masia* of Can Masdeu, in the hills of Collserola, has frequently hosted degrowth and environmental justice events. Giorgos Kallis and Hug March have pointed out how the Cooperativa Integral Catalana is explicitly inspired by the anarchist tradition and the collectives of the Spanish Civil War (Kallis & March, 2015). In other cases, such influences can be inferred but are not explicitly engaged with. Viviana Asara's research on the radical imaginaries of 15-M (Asara, 2020) has underlined the legacy of workers' associations before the Civil War and pointed to values such as mutual aid, direct participation or solidarity, which directly connect to the anarchist ethos and practices. The "fertile soil" for sustainability-related community initiatives that Filka Sekulova and her co-authors have studied in Barcelona connects to the anarchist past and present of the city. This speaks to the metaphor of "anarchist humus" in which degrowth practices flourish, invoked by Giacomo D'Alisa (Sekulova et al., 2017; D'Alisa, 2019).

These connections to Barcelona could be engaged with much more in environmental justice, political ecology and degrowth research, building both on social movement practices in the city today and the intellectual legacies of anarchism before and during the Spanish Civil War. When unearthing the nuanced anarchist environmental debates in the 1900s, Eduard Masjuan was concerned not only with this past, which was silenced during the Francoist dictatorship and beyond, but also with the relevance of these ideas in the present and future (Masjuan, 2000). Internationally too, anarchist geographers today reclaim the relevance of figures such as Kropotkin or Reclus whilst calling for a view towards the future (Springer, 2013). As put by Marc Dalmau from the cooperative La Ciutat Invisible in the Sants neighbourhood of Barcelona, researching mutual aid practices of the city's past, such as cooperatives, has been "a way of generating reference points relevant for the present, feeding the imagination and providing us with roots which can enable us to build to a future based on greater social justice" (cited in Bibby, 2019; see Dalmau Torvà & Miró i Acedo, 2010).

Acknowledgements The author is grateful to Ekaterina Chertkovskaya for her help in discussing and writing this text. Her comments and those of David Saurí, Borja Nogué, Melissa García-Lamarca, Giorgos Kallis, Sergio Ruiz and Marc Gavaldà greatly improved the initial draft.

References

Albet, A. (Ed.), Monk, J., Prats Ferret, M., & Veleda da Silva, S. M. (2019). *Maria Dolors Garcia-Ramon: geografía y género, disidencia e innovación*. Icaria.
Amsden, J. (1978). Industrial collectivization under workers' control: Catalonia, 1936–1939. *Antipode, 10–11*(3–1), 99–113.
Asara, V. (2020). Untangling the radical imaginaries of the Indignados' movement: Commons, autonomy and ecologism. *Environmental Politics*, 1–25. https://doi.org/10.1080/09644016.2020.1773176. Routledge.
Balcells, A. (2017). Collectivisations in Catalonia and the Region of Valencia during the Spanish Civil War, 1936–1939. *Catalan Historical Review, 10*, 77–92. https://doi.org/10.2436/20.1000.01.133
Bernecker, W. (1992). El anarquismo en la guerra civil española: estado de la cuestión. *Cuadernos de Historia Contemporánea, 14*, 91–115. https://doi.org/10.5209/CHCO.7954
Bibby, A. (2019). *Catalonia celebrates: Exploring the region's co-op heritage, Coop News*. Available at: https://www.thenews.coop/138847/topic/history/catalonia-celebrates-exploring-regions-co-op-heritage/
Bookchin, M. (1971). *Post-scarcity anarchism*. Black Rose Books. https://doi.org/10.4324/9781351319324

Bookchin, M. (1977). *The Spanish anarchists: The heroic years, 1868–1936*. Free Life Editions.

Breitbart, M. M. (1978a). Anarchist decentralism in rural Spain, 1936–1939: The integration of community and environment. *Antipode, 10–11*(3–1), 83–98.

Breitbart, M. M. (1978b). Introduction. *Antipode, 10–11*(3–1), 1–5.

Breitbart, M. M. (1978c). Spanish anarchism: An introductory essay. *Antipode, 10–11*(3–1), 60–70. https://doi.org/10.1111/j.1467-8330.1978.tb00116.x

Breitbart, M. M. (1978d). *The theory & practice of anarchist decentralism in Spain, 1936–1939*. Clark University.

Camps-Calvet, M., Gorostiza, S., & Saurí, D. (2021). Cultivar bajo las bombas. La agricultura urbana y periurbana en Barcelona durante la guerra civil, 1936–1939. *Historia Agraria, 84*, 141–171. https://doi.org/10.26882/histagrar.084e08c

Camps-Calvet, M., Gorostiza, S., & Saurí, D. (2022). Feeding the City and Making the Revolution: Women and Urban Agriculture in Barcelona during the Spanish Civil War (1936–1939). *Antipode*. https://doi.org/10.1111/anti.12819

Castells Durán, A. (1993). *Les col·lectivitzacions a Barcelona, 1936–1939*. Editorial Hacer.

Castillo, A. (2016). Anarchism and the countryside: Old and new stumbling blocks in the study of rural collectivization during the Spanish Civil War. *International Journal of Iberian Studies, 29*(3), 225–239. https://doi.org/10.1386/ijis.29.3.225_1

Cattaneo, C., & Gavaldà, M. (2010). The experience of rurban squats in Collserola, Barcelona: What kind of degrowth? *Journal of Cleaner Production, 18*(6), 581–589. https://doi.org/10.1016/j.jclepro.2010.01.010. Elsevier Ltd.

Cattini, G. C., & Santacana, C. (2002). El anarquismo durante la Guerra Civil. Algunas reflexiones historiográficas. *Ayer, 45*, 197–219.

D'Alisa, G. (2019). The state of degrowth. In E. Chertkovskaya, A. Paulsson, & S. Barca (Eds.), *Towards a political economy of degrowth* (pp. 243–257). Rowman & Littlefield.

Dalmau Torvà, M., & Miró i Acedo, I. (2010). *Les Cooperatives obreres de Sants: autogestió proletària en un barri de Barcelona (1870–1939)*. La Ciutat Invisible.

Ealham, C. (2005). *Class, culture and conflict in Barcelona, 1898–1937, Class, culture and conflict in Barcelona, 1898–1937*. Routledge/Cañada Blanc Centre. https://doi.org/10.4324/9780203493557

Ferretti, F., & García-Álvarez, J. (2019). Anarchist Geopolitics of the Spanish Civil War (1936–1939): Gonzalo de Reparaz and the "Iberian Tragedy". *Geopolitics, 24*(4), 944–968. https://doi.org/10.1080/14650045.2017.1398143. Routledge.

Ferretti, F., et al. (Eds.). (2018). *Historical geographies of anarchism: Early critical geographers and present-day scientific challenges*. Routledge.

Garcia-Ramon, M. D. (1978). The shaping of a rural anarchist landscape: Contributions from Spanish anarchist theory. *Antipode, 10–11*(3–1), 71–82.

Gorostiza, S. (2014). Mapeando conflictos ¿Hacia una nueva ecología política estadística? Entrevista a Joan Martínez Alier. *Ecología Política, 48*, 20–23.

Gorostiza, S. (2019). Critical networks. Urban water supply in Barcelona and Madrid during the Spanish Civil War. In S. Laakkonen et al. (Eds.), *The resilient city in World War II: Urban environmental histories* (pp. 23–46). Palgrave Macmillan. https://doi.org/10.1007/978-3-030-17439-2

Gorostiza, S., March, H., & Sauri, D. (2013). Servicing customers in revolutionary times: The experience of the collectivized Barcelona water company during the Spanish Civil War. *Antipode, 45*(4), 908–925. https://doi.org/10.1111/j.1467-8330.2012.01013.x

Ince, A. (2012). In the shell of the old: Anarchist geographies of territorialisation. *Antipode, 44*(5), 1645–1666. https://doi.org/10.1111/j.1467-8330.2012.01029.x

Kallis, G. (2019). *Limits. Why Malthus was wrong and why environmentalists should care*. Stanford University Press.

Kallis, G., & March, H. (2015). El futuro dialéctico del decrecimiento: ¿ficción distópica o proyecto emancipador? *Revista de Economía Crítica, 19*, 21–33.

Levy, C. (2018). Anarchists and the city: Governance, revolution and the imagination. In F. Ferretti et al. (Eds.), *Historical geographies of anarchism: Early critical geographers and present-day scientific challenges* (pp. 7–24). Routledge. https://doi.org/10.4324/9781315307558

Martí Escayol, M. A. (2019). The environmental history of the Catalan-speaking lands. *Catalan Historical Review, 12*, 43–55. https://doi.org/10.2436/20.1000.01.155

Martínez Alier, J. (1968). *La Estabilidad del latifundismo: análisis de la interdependencia entre relaciones de producción y conciencia social en la agricultura latifundista de la Campiña de Córdoba*. Ruedo Ibérico.

Martínez Alier, J. (1991). Introducción al número 1. *Ecología Política, 1*, 7–9.

Martínez Alier, J. (1993). Temas de historia económico-ecológica. *Ayer, 11*, 19–38.

Martínez-Alier, J. (1996). The failure of ecological planning in Barcelona. *Capitalism Nature Socialism, 7*(2), 113–123. https://doi.org/10.1080/10455759609358685

Martínez-Alier, J. (2002). *The environmentalism of the poor: A study of ecological conflicts and valuation*. Edward Elgar Publishing.

Martínez-Alier, J. (2015). Neo-Malthusians. In G. D'Alisa, F. Demaria, & G. Kallis (Eds.), *Degrowth: A vocabulary for a new era* (pp. 125–128). Routledge.

Martínez-Alier, J. (2019). *Demà serà un altre dia: una vida fent economia ecològica i ecologia política* (Primera ed). Icaria.

Masjuan, E. (1993). Población y recursos en el Anarquismo Ibérico. *Ecología Política, 6*, 129–142.

Masjuan, E. (1995). La ciudad-jardín o ecológica contra la ciudad lineal. *Ecología Política, 10*, 127–140.

Masjuan, E. (1996). Los orígenes del neomalthusianismo ibérico. *Ecología Política, 12*, 19–26.

Masjuan, E. (1998). *El Anarquismo ibérico: sus relaciones con el urbanismo ecologista u 'orgánico', el neomalthusianismo y el naturismo*. Universitat Autònoma de Barcelona.

Masjuan, E. (2000). *La ecología humana en el anarquismo ibérico*. Icaria.

Masjuan, E., et al. (2008). Conflicts and struggles over urban water cycles: The case of Barcelona 1880–2004. *Tijdschrift voor Economische en Sociale Geografie, 99*(4), 426–439. https://doi.org/10.1111/j.1467-9663.2008.00477.x

Monjo, A. (1993). *La CNT durant la II República a Barcelona líders, militants, afiliats*. Universitat de Barcelona. Departament d'Història Contemporània.

Monjo, A. (2003). *Militants: participació i democràcia a la CNT als anys trenta* (1st ed.). Laertes.

Monjo, A., & Vega, C. (1986). *Els treballadors i la guerra civil: història d'una indústria catalana col·lectivitzada*. Empúries.

Naredo, J. M. (2008). *Resumen de mi trayectoria intelectual. Desde la economía y la estadística hacia los recursos naturales y el territorio. Discurso con motivo de la concesión del Premio Internacional de Geocrítica 2008*. Available at: http://www.ub.edu/geocrit/-xcol/naredo.htm

Nash, M. (1975). *Mujeres Libres, España, 1936–1939*. Tusquets.

Sarría Buil, A. (2019). Ruedo Ibérico. In M. Aznar Soler & I. Murga Castro (Eds.), *1939. Exilio republicano español* (pp. 647–650). Ministerio de Justicia, Ministerio de Educación y Formación Profesional.

Sekulova, F., et al. (2017). A "fertile soil" for sustainability-related community initiatives: A new analytical framework. *Environment and Planning A, 49*(10), 2362–2382. https://doi.org/10.1177/0308518X17722167

Springer, S. (2013). Anarchism and geography: A brief genealogy of anarchist geographies. *Geography Compass, 7*(1), 46–60.

Springer, S., et al. (2012). Reanimating anarchist geographies: A new burst of colour. *Antipode, 44*(5), 1591–1604. https://doi.org/10.1111/j.1467-8330.2012.01038.x

Open Access This chapter is licensed under the terms of the Creative Commons Attribution 4.0 International License (http://creativecommons.org/licenses/by/4.0/), which permits use, sharing, adaptation, distribution and reproduction in any medium or format, as long as you give appropriate credit to the original author(s) and the source, provide a link to the Creative Commons license and indicate if changes were made.

The images or other third party material in this chapter are included in the chapter's Creative Commons license, unless indicated otherwise in a credit line to the material. If material is not included in the chapter's Creative Commons license and your intended use is not permitted by statutory regulation or exceeds the permitted use, you will need to obtain permission directly from the copyright holder.

Chapter 24
The Barcelona School of Ecological Economics and Social Movements for Alternative Livelihoods

Claudio Cattaneo

24.1 Introduction

The Barcelona School of Ecological Economics has very strong ties with activist movements. In this chapter, I will show some examples of how it fits with social movements for alternative livelihoods. This is an incredibly valuable characteristic of the Barcelona school that owes much to the political dissidence of Joan Martinez-Alier and of many of his colleagues and disciples.

I moved from Italy to Barcelona to do my Ph.D. in ecological economics in late 2001 and have been living almost entirely in the Can Masdeu community, a rurban squat home to about 25 people and with an active social center, located on the hills of Collserola. One day in January 2002, when I was still more of a tourist than a resident, knew no activists in Barcelona and only had few acquaintances, my classmates Jesus Ramós and Miquel Ortega invited me to join them in a talk on climate change they would have offered a few days later in Can Masdeu. To my joyful surprise – I always had a sympathy for occupied spaces that cost me a reputation among family members of me being a rebel with no serious professional future – I could combine my studying career with an alternative lifestyle and radical activism.

What could explain such coincidence has a lot to do with Joan's legacy with anti-Franco republican activism and his eco-socialist ideas, inspired by nineteenth-century Narodnik anarchist movement and the Barcelona Anarchist government in 1936. The book of his memoirs (Martinez-Alier, 2019) is an excellent source of facts of his life and personal visions.

C. Cattaneo (✉)
Department of Environmental Studies, Faculty of Social Studies, Masaryk University, Brno, Czech Republic
e-mail: Claudio.Cattaneo@uab.cat

The reader can learn about his detention on October 12, 1992, when protesting the 500 years of the European invasion of the American continent – and consequent genocide – or about his smuggling through the French border of books forbidden by Franco, who is indeed the most named person in his book. A clear image of his antiestablishment vision, critical of both political and academic institutions, is contrasted by his sympathy for grassroots movements, indigenous people, peasant struggles, and degrowth ideas.

Barcelona, also known as the Rosa de Foc, is a fertile ground for political dissidence. This is due to its enduring essence as a rebel city; as the capital of the Catalan culture has been for centuries in antagonism with the Spanish central state. This is manifested in many antagonists and social movements that flourish in the city from the grassroots up to Catalan and municipal institutions. The largely unnoticed work of Eduard Masjuan (2000) – one of the first Ph.D. supervised by Joan – offers an excellent historical analysis of the interconnections between anarchist and ecological movements in the Iberian Peninsula. The mediatic relevance of Catalan independentist movements is just the tip of a widespread structural nonconformity across Catalan society. Acts of popular and institutional disobedience are frequent in Catalonia while a portrait of Gandhi has always been hanging on Joan's office wall.

Having shown the legacy of Joan's and the Barcelona school with social movements and grassroots practices, this chapter focuses on how the analysis of alternative livelihoods contributes to ecological economics. I look at those that, by seeking emancipation from the capitalist system and its political establishment, set up self-managed communities and workplaces. Joan's seminal work on the environmentalism of the poor shows how the poor do not necessarily label themselves as environmentalists because they depend on a well-preserved local environment and the services it offers, and they rely on their traditional knowledge rather than technological progress. From an ecological economics perspective, one could argue that the poor are good managers of natural capital even if they do not behave as the growth-obsessed managers of capitalist enterprises of the green economy. In the same vein, the alternative livelihoods here analyzed have a great relevance in ecological economics because they have a lower social metabolism – as shown in my Ph.D. research (Cattaneo, 2008) – and because they constitute examples of grassroots action, self-management, and political activism. Even more, they are relevant for the degrowth branch of the Barcelona school as they constitute visible real-life practices and contribute to the imaginaries of the pluriversal project, as explained by Demaria in his chapter.

Since Joan has used history to track the ecological dimension in economics as far back as the school of the Physiocrats in the eighteenth century, for an essay in his honor I am including the oldest case studies of alternative housing that I could track to offer insight on how alternative communities evolve in the long term.

The first is the earliest case of alternative livelihoods in the Iberian Peninsula, at least since the death of Franco, and is based in Zaragoza. The second is the first squatted eco-community of Barcelona, which I consider the most meaningful example of the legacy of the Barcelona school, with its local activists also inspirational for the more well-known example of Can Masdeu. Finally, I consider another squatted place – to my knowledge among the three oldest examples of political squatting in the Spanish State – which was prompted as an environmental justice conflict at the time of the construction of the Barcelona city bypass.

24.2 Case Studies

24.2.1 E.L.

This is an intentional community that is based on radical principles such as mutual aid, horizontal decision-making, and gender equality. For keeping anonymity, I will refer to it as E.L. The beginning of this communal experience (in 1972) is related to the 1968 movement. Although it did not have a strong resonance such as it did in Paris or Prague, the movement offers with E.L. a case of continuity of nearly half a century, something very rare to witness.

In 1968, far-left communist parties – some of which of Maoist inspiration – and student activists with no party affiliation took control of the Physics Faculty at the University of Zaragoza. They were self-managed and offering seminars on topics forbidden by the Franco dictatorship, such as questioning the mononuclear family, and published a bulletin.[1] Formal teaching was suspended by the State in 1971–1972 and the faculty became self-organized by students: a truly revolutionary spirit was permeating it.

Meanwhile, interest in alternative livelihoods was rising and several student communities were set up, whose political lines got inspiration from communes abroad such as those in Germany. E.L. was established in October 1973 by a group of 11 recently graduated students moving into two apartments – first rented and later bought – in a working-class neighborhood. In contrast to the acratic tendency of the rest of the students' communes – which also did not last too long – E.L. has been since the beginning a highly organized project with a very strict working schedule. It progressively increased the reach of the communal aspect: initially limited to sharing the cost of rent and to mandatory participation in few house chores, it evolved towards a more comprehensive communal organization. Family boundaries were extended and personal income from work collectively shared. Its members were mainly employed as teachers, with a preference for the private sector. In any case they stayed away from the university system – against which they had fought a lot – and that in turn avoided employing former conflictive students.

Their pragmatic capacity to shape ideals into real practice was also evident in the choice of the community location: a working-class district where they could reach more easily the proletarian mass and work towards raising their class conscience. In the beginning they were highly criticized for their anti-conformism with respect

---

[1] See for instance those copies in the deposit of Autonomous University of Barcelona: https://ddd.uab.cat/record/56757 (last retrieved on October 12, 2020).

to socially established gender and family norms. For instance, men and women were equally involved in care tasks. However, thanks to their capacity to create good relationships and to work hard towards the common good, they soon got accepted. Simple actions such as cleaning the corridors and stairs of their block of apartments, actively participating in the creation of an adult school or, in later years, offering care support for the ageing parents of the community members are all examples that have contributed to their social acceptance.

Experimenting with new family relationships could be perceived as one of their most revolutionary acts – the personal is in first place political. In 1974, the first of eight community children was born; they would all be raised collectively. As a matter of fact, E.L. members nowadays still differentiate between their biological and their non-biological children. As the community grew in population, more flats were rented – reaching a maximum of four, all in the same neighborhood – and children would move and live together in one of these apartments from a very early age; between 1978 and 1980, two couples left the project and later one woman joined it. Throughout its history, the project has maintained family arrangements that are explicitly opposed to the bourgeois traditional family, and a recurring debate has dealt with how to combine such a challenge with children education that is not sectarian. For instance, while severely limiting the use of television and convincing children that Mazinger-Z (a popular cartoon) was a fascist and a macho, children were going to schools in the neighborhood. In addition, E.L. did not follow the leaning of intentional communities of the time towards open relationships and no couples.

In 1989, they bought land and built one collective house for the seven adults and five children still living together, which reduced the previous dispersion of the community in different flats. They, however, remained in the neighborhood in which they had already set up good relationships; the children were schooled, played, and had friends in the neighborhood while the adults worked a lot in networking, fomenting adults' schools, and avoiding the elitism that their university degree would have allowed them to generate. Throughout its history the project's mutual aid principles have allowed house members to maintain a good quality of life, with enough free time to invest in local activism and the privilege to still have a coherent lifestyle with the initial principles, 50 years down the way.

24.2.2 Kan Pasqual

In part, this essay is a praise to Marc Gavaldá: among those without a Ph.D., Marc is probably Joan's favorite student, with whom I share some publications and teaching at Universitat Autònoma de Barcelona (UAB), both of us as precarious professors. Marc lives an eclectic life between research, journalism, permacultural house building, and activism (see for instance Gavaldà, 2003).

If the genesis of the Barcelona School of Ecological Economics is associated to the launching of its Ph.D. program in 1997 – with Eduard Masjuan and Fander

Falconi as the first doctors – its gestation can be related to the meeting – organized by Joan – where the idea of the International Society of Ecological Economics and its academic journal were discussed and the consequent launching, in 1992, of the undergraduate degree in environmental science, with Marc Gavaldá in the first cohort of students.

In the heyday of the Barcelona squatter's movement – fall 1996 – Marc witnessed the brutal and spectacular eviction of Cine Princesa (Batista, 2002), a short-lived social center in via Laietana that was using the premises of a former cinema to offer countercultural activities at low cost, provide a visible example of a self-managed alternative to capitalism right in the heart of the city, and serve as the hub of the Barcelona squatter's movement. The movement was like a rising star and posing a serious threat to the city's political establishment. It is in those post-Olympics years that tension between anarchist and anti-nationalist movements on one side and capitalist and institutional ones on the other was culminating into astonishing actions, such as the substitution in the palace of the Catalan Generalitat of the squatter's flag for the Spanish one. However, there were strong protests and political repressions, such as the one against the 500th anniversary, which Joan reminds us very well in his *memoires* (2019).

Few months after the eviction of Cine Princesa, the squatter's movement decided that it was better to decentralize into the neighborhoods and so a group of students in environmental science at UAB joined forces with other activists and decided to occupy an empty farmhouse whose aim was to "open the rural frontier to the squatting practice" (https://radar.squat.net/ca/barcelona/kan-pasqual).

Kan Pasqual is among the five oldest squats in Barcelona and has served as a bridge between urban squatting and rural squatting in the Iberian Peninsula. Until the early 1990s the squatters' movement was more related to young punks who looked for abandoned places to be used for playing their music concerts with no hassles (Jony, 2011), while reclaiming spaces for social interaction and enhancing the right to housing. The earliest squats were mainly located around the districts of Gracia and Guinardó; in working-class belt towns, such a Cornellá or Terrassa, and in Sants – a district with a blue-collar heritage of cooperatives. With Kan Pasqual, opening the rural frontier meant introducing the environmental agenda into countercultural squatting with activists finding a common heritage in the university degree that had been recently created under Joan's leadership.

Concern for the environment did not take too long in turning into action: among the squatters' movement the customary approach to energy supply was, and still is, getting an illegal connection to the electricity grid – granting access to free energy, which is often abused. Contrary to that custom, the Kan Pasqual collective heroically cut the electricity connection and turned candle lights on, spending the early winter months nearly into darkness until the first recycled solar panels and batteries were installed.

A quarter of a century later, Kan Pasqual is still a vibrant radical eco-community (Gavaldà & Cattaneo, forthcoming) that, conserving a strong anti-capitalist spirit, remains disconnected from the electricity grid, inspires new generations of activists, and represents one example for degrowth into practice.

24.2.3 Ateneu Popular de Nou Barris

The origin of Ateneu Popular de Nou Barris (Ateneu 9B) is an early case of environmental justice in the Nou Barris district, Barcelona. In the 1970s, environmental racism awareness was raising in the United States (Bullard, 1983) but not yet in Europe. Nonetheless, the case of Ateneu 9B even if still quite unknown in the environmental justice framing, deserves a mention. In the occasion of its 40th anniversary, a historical recompilation of its trajectory has been produced (Tudela, 2018), here briefly summarized.

Barcelona's population was rapidly expanding at the time, with migrants flowing in from depressed areas of the peninsula and infrastructures being built quickly, often without care for environmental standards. To the north of the city, in a forgotten peripheral neighborhood standing next to the building site of the city bypass, the municipality installed an asphalt plant that immediately raised neighbors' concerns who reclaimed cleaner air for their lungs. In early January 1976, after receiving false promises from the municipality, 200 people invaded the premises of the asphalt plant and sabotaged it. These were times when, after 40 years of dictatorship, the anarchist movement was flowering in Barcelona; meanwhile, the municipal administration was also failing in the provisioning of basic cultural services so that, few months after its sabotage, the neighbors squatted the place, organized a cultural festival that lasted 30 h, and was attended by more than 50,000 people: the sociocultural project of the Ateneu Popular de Nou Barris was born.

After the first 2 years in which a lot of activities and performances were held, both in the space of Ateneu and in the neighborhood streets, participation decreased dramatically in 1979 and 1980 and the first problems arrived. On the one hand, many activists were co-opted by the new institutions of the democratic government; on the other hand, the abuse of heroine destroyed the lives of many young people of these peripheral areas (for an account of the general sociopolitical context of those years, see Tudela & Cattaneo, 2016). With renewed strengths, activities picked up again in 1981. Since then, Ateneu 9B has been characterized by a long and steady trajectory marked by artistic and cultural performances, circus and theatre education, community work in the neighborhood – which remains one of the most marginalized – and participation in local politics and cultural manifestations.

The project was originally managed via informal processes and voluntary work. As it became more established, it formalized its relationship with the municipality, which granted some funding – so that paid work positions could be opened – and in late 1984 conceded access to another building, up until then not used, and where a restaurant and a bar was opened. In late 1985 the reformation of the building begun – which had been reclaimed for a while. In 1989 the project was constituted into a legal entity, more workers were employed, and in the early 1990s, with further reforms to the building, Ateneu de Nou Barris became a core cultural project in the neighborhood first and then in the city. Since then, it has balanced the high-quality cultural activities with community work aimed at improving social conditions in the neighborhood. Its capacity to remain a self-managed institution inspired by anarchist principles, however, is put under constant pressure by its increasing institutionalizaiton and the need to square its financial budget.

24.3 Discussion and Conclusions

These three examples show early cases of movements for alternative livelihoods that have persisted over several decades. One is linked to the Barcelona School of Ecological Economics, one is an early case of environmental justice in Barcelona, and one is an early example of politically inspired alternative livelihoods that is still alive nearly half a century later in Zaragoza.

First, by observing some examples from the early times, we learn that the Barcelona School of Ecological Economics is directly or indirectly related to activism for alternative livelihoods. Its genesis and gestation have coincided, in the 1990s, with the establishment of Kan Pasqual and the heyday of the squatters' movement in Barcelona. The school and Joan's activism are related to the anarchist character of Barcelona and Catalonia, which has its roots in the glorious experience of 1936, as first-hand documented by George Orwell (1938). Zaragoza has a similar but less well-known history, with many grassroots experiments and social movements appeared in the city since 1975 (Vari@s Autor@s, 2009). In both cities, beginning with the 1968 movement and the end of the dictatorship, these new practices and movements could flourish in a social context that favored their development. To this extent, the school is set on the fertile ground of Iberian and in particular Catalan dissidence reinvigorated by the end of 40 years of dictatorship and the centuries-old struggle for Catalan independence and against the Spanish State.

Second, a continuity in politically inspired and alternative livelihoods can be established between the present and 1968, which is a date nearly one decade earlier than the first examples of squatting in the Spanish State (Aguilera et al., 2017; Debelle et al., 2018). E.L. has not been analyzed before and constitutes an example of how ideas travel across borders. Like the first countercultural squatters in the Barcelona of the 1980s got inspiration from the Anglo-Saxon context, early intentional communities connected with the 1968 ideals took inspiration from abroad and then they shaped – in the case of E.L. – their project adapted to the local neighborhood context and the coupling preferences of their members.

Third, these cases have been enduring quite enough to offer valuable insights into how alternative political projects evolve. The case of Ateneu 9B, established in 1977, took the momentum of the early years of the Democratic Transition and succeeded in turning what today could be considered as a terrorist act into a project that, in its fifth decade of existence, is trying to keep a balance between different aims and challenges. One challenge is to keep working as a space at the service of the neighborhood, while at the same time offering a high-level cultural program – sometimes at not-so-popular prices; lately there has been a shift towards the latter and there is an ongoing discussion on how to strike a better balance. The other challenge is of an institutional character and is related to the space being governed by a horizontal assembly open not only to formal members and waged employees but also to common people with no formal link to the project. Within it, the work of waged employees is paired with that of volunteers and activists, but at times conflicts arise between the former who are more empowered than the latter. In turn, the obligations of a formal institution tied to budget agreements with the public administration allow for the project to function as a self-managed workplace that employs several workers; however, the institutionalization of its practices has crowded-out some volunteers in the past years.

E.L. has lasted much more than any other community born with the spirit of 1968 ideals. Although they could break with many of the social norms related to family structure, economic organization, children education, and gender division of care work, they kept a very disciplined work schedule with members contributing equally to common tasks. At times, E.L. had been mocked by other more acratic communes for such a formal organization; however, unlike most of these, it did not collapse when the voluntary distribution of community workload became too unequal and unsustainable in the long term, neither when open relationships between members became unbearable. To a certain extent, E.L. has been able to anticipate what in more recent times eco-communities are achieving, by balancing between sustainability – also known as duration in time – and the number of social norms to break, –which tend to be less ambitious and more realistic than what used to be in the 1970s. For a comparison of how intentional communities have organized family relationships in different decades in Barcelona, see "Familia no nuclear", a video documentary by Lopez Lloret (2019).

Kan Pasqual has managed a long-term supply of electricity from renewable sources and autonomy from monetary expenditure, contributing to the low-impact lifestyle of most of its members, and is the first case connecting the urban squatters' movement with the agroecological practice. It has been a referent in the rural and urban squatters' networks, has inspired the more well-known project of Can Masdeu, and the connection between squatting and gardening – since the early 2000s, many squatted gardens have emerged in the city, constituting now a common phenomenon in the use of social spaces. Open spaces are more visible and offer to the passers-by an easy image of what squatters do, which is not distorted by mass media's disinformation that tend to criminalize them.

In conclusion, this chapter has shown the presence of a legacy between the Barcelona School of Ecological Economics and social movements for alternative livelihoods in Catalunya and Spain. In turn, by analyzing three cases of the most long-lasting ones and still existing up to now, a connection between the 1968 movement and the present has been established and some insights have been offered on how such projects evolve and the impact they have.

Acknowledgments The author wishes to thank Concha for her kind availability in providing information about E.L. community, and Sergio and Roldan for their patience.

References

Aguilera, T., Cattaneo, C., Dee, E. T. C., Martinez, M. A., van der Steen, B., & Warnecke, J. (2017). Mapping the Movement: Producing maps of squatted social centres in Western Europe. *Trespass Journal, 1*, 84–102.

Batista, A. (2002). *Okupes. La mobilització sorprenent*. Rosa dels Vents.

Bullard, R. D. (Ed.). (1983). *Confronting environmental racism: Voices from the grassroots*. South End Press.

Cattaneo, C. (2008). *The ecological economics of urban squatters*. PhD thesis in environmental science. Autonomous University of Barcelona.

Debelle, G., Cattaneo, C., Gonzalez, R., Barranco, O., & Llobet, M. (2018). Squatting cycles in Barcelona: identities, repression and the controversy of institutionalisation. In M. Martinez (Ed.), *The urban politics of squatters' movements* (pp. 51–73). Palgrave.

Gavaldà, M., & Cattaneo, C. (forthcoming). Feeding together: The revolution starts in the kitchen. In S. Stavridis & P. Travlou (Eds.), *Housing as commons*. Zed Books.

Gavaldà. (2003). *La recolonización. Repsol en America Latina: invasión y resistencias*. Icaria Editorial.

Jony, D. (2011). *Que pagui Pujol!: una cronica punk de la Barcelona de los 80*. La Ciutat Invisible.

Lopez Lloret, J. (2019). *Familia no nuclear*. Video documentary. Retrieved on October 12th 2020 at https://www.filmin.es/pelicula/familia-no-nuclear

Martinez-Alier, J. (2019). *Demà serà un altre día*. Icaria Editorial.

Masjuan, E. (2000). *La Ecología Humana en el Anarquismo Ibérico. Urbanismo "orgánico", neomalthusianismo y naturismo social*. Icaria Editorial.

Orwell, G. (1938). *Homage to Catalonia*. Secker and Warburg.

Tudela, E. (2018). *40 anys fent l'Ateneu Popular 9 Barris. Un altre relat de la cultura a Barcelona*. Barcelona. Retrieved on October 6th 2020 at https://issuu.com/ateneupopular9barris/docs/catalogo_nou_barris

Tudela, E., & Cattaneo, C. (2016). Beyond desencanto: the slow emergence of new social youth movements in Spain during the early 1980s. In K. Andersen & B. var den Steen (Eds.), *A European youth revolt* (pp. 127–141). Palgrave Macmillan.

Vari@s Autor@s. (2009). *Zaragoza Rebelde. Guìa de Movimientos Sociales y antagonismos, 1975–2010*. Traficantes de Sueños.

Open Access This chapter is licensed under the terms of the Creative Commons Attribution 4.0 International License (http://creativecommons.org/licenses/by/4.0/), which permits use, sharing, adaptation, distribution and reproduction in any medium or format, as long as you give appropriate credit to the original author(s) and the source, provide a link to the Creative Commons license and indicate if changes were made.

The images or other third party material in this chapter are included in the chapter's Creative Commons license, unless indicated otherwise in a credit line to the material. If material is not included in the chapter's Creative Commons license and your intended use is not permitted by statutory regulation or exceeds the permitted use, you will need to obtain permission directly from the copyright holder.

Chapter 25
The Ups and Downs of Feminist Activist Research: Positional Reflections

Sara Mingorria, Rosa Binimelis, Iliana Monterroso, and Federica Ravera

25.1 Introduction

We are four women activist-researchers, who studied together at the Barcelona School of Ecological Economics, motivated by and interested in contributing to processes of social transformation from the perspective of research and radical and activist feminism. We accepted and undertook the challenge of writing together about our similar experiences in feminist activist-research during the middle of a lockdown for the COVID-19 pandemic. Such systemic crisis, and specifically the lockdown, has highlighted for us, particularly as women researchers, the difficulties of finding time and energy to continue our research and activism, and at the same time moved at the center the caring for life (including that of a newborn) and the care of others (grandparents, children, vulnerable neighbors).

After presenting our personal trajectories and link them to what we learned at the Barcelona School, this chapter will focus on feminist research and Participatory

---

This work was supported by Juan de la Cierva Incorporación Fellowship [Grant Number IJC2020-045451-I] and the Spanish Ministry of Science, Innovation and Universities through the senior fellowship Ramón y Cajal (RyC2018-025958-I).

S. Mingorria (✉) · F. Ravera
Universitat de Girona (UdG), Girona, Spain

R. Binimelis
Arran de terra SCCL, Barcelona, Spain

I. Monterroso
Center for International Forestry Research (CIFOR), Guatemala City, Guatemala

© The Author(s) 2023
S. Villamayor-Tomas, R. Muradian (eds.), *The Barcelona School of Ecological Economics and Political Ecology*, Studies in Ecological Economics 8, https://doi.org/10.1007/978-3-031-22566-6_25

Action Research (PAR)[1] processes in which we have been involved. Additionally, it also provides a reflection on the "positionality" of us as researchers, the challenges and the ups and downs we have experienced, and the mutual support woven into these processes.

25.2 Activist and Feminist Research: Individual Trajectories

We define ourselves as activist-researchers based on Borras's (2016) definition, which refers to those who are mainly in academic institutions and carry out activist work (three of us); those who are based mainly in social movements or political projects and carry out academic activism from within (one of us); and those who are based mainly in independent research institutions and are activist-academics.

While studying at the Barcelona School, we have had access to ongoing discussions on research and political action. Such studies overlap with the questioning of linearity and the search for "integrative sciences," whose epistemological basis is rooted in the weak comparability and incommensurability of values (Martinez Alier et al., 1998), which places ecological economics outside the mainstream of neoclassical economic theory. The PAR processes in which we have been involved have simultaneously included scientific research, training, and political action. They have considered critical analysis, diagnosis of situations, and practice as sources of knowledge (Fals-Borda, 1978). Furthermore, we want to highlight three aspects of PAR that emerge during the process: (a) dealing with research topics of social relevance; (b) the importance of overcoming the subject/object dichotomy in the research process to actively promote knowledge co-elaboration processes; (c) knowledge is mediated by the subjects that produce it; therefore, there is no neutrality in the way of knowing or in the knowledge that is produced.

We learned from the "post-normal science" developed by Funtowicz and Ravetz (1993) that research must recognize and include the plurality of values and knowledge through the participation of the actors in the research. This participation is justified in the first place both by a question of justice in giving a voice to those who experience the greatest impacts of environmental management decisions, and to guarantee the quality of research that supports decision-making. We also learned to be aware that the uncertain context can be analyzed from multiple angles and values; hence the importance of transparency in the research process (Munda, 2004); and the importance of integrating local wisdom, from the knowledge of ancestral peoples and peasants, and from a complex systems perspective, as political agroecology has shown us (Cuéllar & Sevilla, 2009). Such theoretical base we received at the Barcelona School was enriched by reflections from Freire's pedagogy, who recognized that knowledge is only possible through dialogue between the social

---

[1] We will use the term Participatory Action Research (PAR) and Activist Research interchangeably, referring to the same concept.

actors who directly suffer oppressions and the thinkers or activists committed to social change (Freire 1970). In this regard, these definitions are also close to the concept of situated knowledge elaborated by Haraway (1995). She proposes specifying from which point of view one starts, and why that perspective and not another, thus making the political position explicit. In the words of the author: "relativism is a way of not being anywhere while simultaneously pretending to be everywhere; at the same time the neutrality of positioning is a denial of responsibility and a critical search" (Haraway 1995: 329).

Finally, we also rely on the feminist concept of reflexivity, defined as the examination of one's own beliefs, judgments, and practices during the research process. This implies the recognition of a positionality of researchers, based on the legitimate powerful position of science with respect to other systems of knowledge, and also the position of each one of us that influences the relationship within research staff, with other subjects, and the entire research process (England, 1994). We reflect on how our personal and collective research trajectory is not a process alien to our context.

25.2.1 Activist-Research from and in the South

I belong to a small proportion of women and have obtained a university degree in Guatemala. The options for women are limited, although in the years of positive discrimination policies the possibilities have increased. My academic career was marked by scholarship programs and the search to understand the confluence between the analysis of environmental conflicts – from a social perspective. From the beginning, employment options were linked to forestry issues, dominated in Latin America by the perception of being male-dominated fields (Rocheleau & Edmunds, 1997). The role that social movements have had, from indigenous groups in Latin America to women's movements in India and Sub-Saharan Africa (Mies & Shiva, 1997), means that anyone who wishes to do research on this issue is obliged to engage with these mobilization processes. My engagement with the forestry sector was based on training and exchange processes with indigenous and peasant communities on forest lands. The time I spent in Barcelona was fortuitous. The basis of my study had been closer to institutional analysis and studying the commons (Monterroso et al., 2019). However, during my stay in Barcelona I sought to articulate theoretical and methodological development with a deeper problematization of power issues and the reconfiguration of social relations that frames environmental conflicts. Barcelona was key for me in my search for methodological plurality, as it recognizes the importance of addressing different types of knowledge. Perhaps, much more important to me was the recognition of the absence of neutrality in the research process and the importance of promoting a reflective process from the perspective of rights and agency. This was key to understanding my positionality with respect to research in other countries and in other contexts (Monterroso et al., 2019; Gnych et al., 2020). The process was not easy. As a woman in a field where men predominate and the challenges are diverse: not only to ensure that the value of the contributions and proposals are recognized but also to be able to carry out research in isolated places – where the condition of being a woman is a challenging one.

25.2.2 Activist-Feminist Research on Climate Change: An Epistemological Revolution

Being a woman, a mother, and a researcher in the areas of sciences in Spain is not an easy task. According to recent statistics,[2] only 40% of researchers from public universities are women (and this figure has been stagnant since 2009) and only 19% of women are accepted to study in the areas of science and technology. Several feminist authors have provided evidence of the underrepresentation of women, indigenous populations, and Afro-descendants, as well as their knowledge, in environmental studies on climate change, a dominant topic of environmental policies today. In fact, climate change is studied mainly as a biophysical problem, based on predominantly positivist and technocratic knowledge and does not include a sociopolitical dimension. This lacuna motivated me to open my research to the multifaceted, contested, and political topic of adaptation to climate change (Ravera et al., 2016).

During the time of my doctorate at the university in Barcelona, I began to wonder about what role science and scientific knowledge play in complex issues such as climate change. Various interests are at stake and the decisions to be made are urgent, and can have repercussions for some actors more than others, those who normally have no say in those same decisions. Thus, I began to reflect on epistemological issues, legitimacy, and power in the construction of knowledge in dynamic socio-ecological systems. Guided by Roger Strand and Giuseppe Munda, I began reading "post-normal science." Since those early years, PAR allowed me to work on research processes based on an appreciation of transdisciplinary dialogue, creating bridges with other forms of knowledge and applying inclusive methods and tools. Various action–research processes in which I was involved in my PhD and postdoctoral years have also taught me a reciprocity in the relationship with the subjects who participate in the research, as a way to reverse the privileged position of science and scientists regarding local and indigenous communities and knowledge. Along the way, this aspect has been decisive in deepening my reflections on climate change knowledge construction from a feminist perspective.

In 2015, myself and a group of colleagues from Institute for Environmental Science and Technology (ICTA) and from the Autonomous University of Madrid (UAM) began to reflect on the discrimination and violence of the neoliberal academic system in our lives as women, researchers, and activists. We understood that

---

[2] https://www.fundacionaquae.org/mujeres-ciencia-espana/

we first need to recognize where we were discussing these issues, since we were used to working in the academic world. In 2017, we established the collective FRACTAL[3] as a feminist activist-research refugia, which implies, in the first place, appropriating an agency to build another way of doing research, through networks, and taking care of ourselves and the process itself. Moreover, the reading and reflection meetings that we held together allowed me to delve into feminist studies of science and decolonial theories, focusing on how women, indigenous populations, and local knowledge systems have historically been silenced and marginalized in the construction of narratives and representations of global environmental change (Schnabel, 2014). Thus, in recent years I have been collecting life stories of women in the pastoral sector facing global environmental changes, in order to listen to their invisible voices. On many occasions, together with shepherdesses and women livestock managers, we have reflected on multiple individual categories that create multiple relationships of power and violence. Recently, I am also reflecting on how to decolonize our language and knowledge production as well as how to be sure to include silenced voices, such as migrants or laborer. Finally, in the last years, I have been working with various feminist methods of action–research processes, including situated ethnography and work with artists. This allows for the exploration of other forms of knowledge construction through the body and multiple disciplines, as well as to communicate with greater impact the messages about challenges and barriers and overcoming these in power relations at different scales: family, community, sectors, and society.

25.2.3 Activist in Academia

Ever since I can remember, social justice has been an important aspect for me, and many of the paths I have taken are linked to activism. Entering academic life and deciding to do a PhD in Ecological Economics at ICTA was part of this too. I finished the PhD during a period of great changes in my life, and my motivation then was to find a job, and one that would allow me to contribute to social change. I was connected to social movements associated with political agroecology and anti-transgender struggles, food sovereignty, and feminisms, among other issues, and I wanted to continue contributing to them. So, when I went to see Joan Martínez-Alier to ask about my doctorate, he asked me to work with him, and I accepted that right away. Choosing the topic for my doctorate was a collective process. In fact, with colleagues from various social movements involved in political agroecology, we discussed the relationship and have voice within the academic world (similarly to what Heller, 2002 suggests).

At that time, despite the great social controversy around the introduction of agricultural GMOs, the scientific studies that were being published on social, ethical,

---

[3] https://fractalcollective.org

and economic impacts were totally dominated by the analysis of economic implications and very often based solely on mathematical models (Catacora-Vargas et al., 2018). Thus, I decided to work on the ethical, legal, and socio-economic impacts of transgenics both from the perspectives of political ecology and environmental conflicts, as well as the role of uncertainty and complexity in governance and decision-making on GMOs from science and technology studies (Binimelis & Strand, 2009).

Once I had finished my doctorate, my options to continue working in academia in Catalonia were through precarious and short-term contracts, which coincided with the birth of my first daughter. After stringing together several projects, and just when I was thinking of abandoning research altogether, I won a contract scholarship to work in Norway, where I worked for 6 years under working conditions that I would never have had if I had stayed in Catalonia. Even so, after living in different countries for a while, and a second maternity leave, I decided to return to Catalonia where myself and colleagues, who were also doing activist-research, started academic projects with a view to social transformation. I have always tried to combine the more theoretical research with practical perspectives, delving into aspects that are vital but little valued in the academic world, such as the return of research results (almost always paid for with public money) to society that has paid for it. I thus began to work on how to communicate the research using other languages and formats beyond the classical academic ones (which are, on the other hand, the only ones that are valued in a competitive academic "career," something far removed from my values and principles).

I am currently working on projects related to the promotion of fairer and more sustainable agri-food systems, linked to local and community economies. I research from practice, with one foot inside academia and the other with the people: from different spheres, creating networks with the idea of moving towards territorialized food systems.

25.2.4 Activist-Researcher from the Global North in the Global South

I write these reflections from the perspective of my condition as a woman, an activist-researcher, born in the European Global North, specifically in a working-class neighborhood of the city of Madrid (Spain), a fellow during all my years of university education and now also a mother (on maternity leave and without a contract).

My first activist-research reflected on and denounced the logic of human dominance over nature in zoos. I was working as an environmental educator in a zoo and I was taking a course taught by members of the Laboratory on Socio-Ecosystems at UAM, where I heard about ecological economics for the first time.

Subsequently, I received a scholarship to carry out my end-of-degree project at the UAM, and I began my relationship with the university in Barcelona as I requested scholarly advice from Dr. Laura Calvet (at the time a PhD student) and Dr. Victoria Reyes-Garcia, both from ICTA-UAB, on local wisdom and its potential use in the management of natural spaces (Gómez-Baggethun et al., 2010).

It was Reyes-García who endorsed me to continue studying at the school in Barcelona through the Master's in Ecological Economics. The classes and the debates in the master's degree allowed me to theorize and further evaluate some of my own concerns and to have new arguments to question theories that had been given to me as "truths" during my degree. My views broadened after learning about the foundations of "post-normal science." Likewise, I had access to different techniques and research methodologies that I was able to apply during more than 7 years of activist-research, as part of my master's and doctoral thesis in the conflictive region of the Polochic Valley in Guatemala.

The Polochic is one of the regions of Guatemala with the greatest conflicts over land, where the State and oligarchic families use direct violence against indigenous communities to maintain their interests (Mingorría, 2018). Throughout this research, myself and other researchers, organizations, and communities reflected on the dynamics of these conflicts while also attempting to define and/or strengthen new mobilization strategies. I followed the PAR framework and carried out transdisciplinary and multiscale analyses adapting research questions to the changing context (Mingorría, 2016). This experience compelled me to learn to manage the intersections between experiences, activism, and the production of academic knowledge in contexts of direct and structural violence. In addition, I continue to grapple with and reflect on other issues not discussed at the Barcelona School such as the demands and conditions in European neoliberal academia; the changes in roles throughout research; ethics in research; the intersections between my condition of being a young, white, European woman with university degrees in the activist contexts of Latin America and academia of the Global South and Global North. I was also a co-funder of the FRACTAL collective, which was created to provide a safe space for such reflections.

At the end of the doctorate and during the post-doc, in addition to continuing to carrying out activist-research, especially in Indonesia, on environmental conflicts through the EnvJustice[4] project directed by Joan Martinez-Alier, I continue looking for tools to work in academia without neglecting my self-care or my activism. I continue to broaden my views and investigate new schools of thought and approaches such as feminist political ecology or research from the south, or how to create bridges between art, education, activism, and research (Mingorria & Heras, 2019), and I reflect on my own research processes with other academic colleagues, activists, and community members (such as writing in this book with co-authors)[5] (see, e.g., Johnson et al., 2020).

---

[4] http://www.envjustice.org

[5] These reflections are also the result of a series of discussions within the project entitled "Social sciences and organized political subjects. Methodological implications for collaborative work" in which Latin American academics and activists from different disciplines participate.

25.3 Ups and Downs in the PAR Processes

In this section, we reflect on the difficulties and challenges posed by conducting Participatory Action Research (PAR) and the importance of reflecting on the privileges and oppressions we face.

25.3.1 The Demands of Neoliberal Positivist Research Versus Activist-Research

When we refer to the impacts that neoliberal academia has on our lives, we refer to how it promotes a competitive modus operandi, which is based on an intense accounting of results (Mountz et al., 2015), with a focus not on content but on the count (Öhman, 2012). This promotes business management practices in universities, urging the mobility of researchers and that they endure highly precarious working conditions. Women in many cases have suffered impacts on our own bodies, evident in our health, with effects such as stress, shame, a sense of guilt, health problems and exhaustion, and a feeling of isolation and disconnection (Mountz et al., 2015). Many of us have experienced it. The impacts can vary depending on the context of each one, and are heightened in researchers with racialized conditions, are non-English speaking, and/or are socio-economically disadvantaged.

There are also many criticisms of and opposition to how neoliberal research excludes and makes it impossible to develop activist-research, within the academy. We will focus on discussing two elements that were key in our activist-research, and that collided with what was expected/established in the neoliberal academy: (a) the constant change of objectives, reflections, discussions, proposed results, and roles versus the fixed and linear research proposal necessary to maintain the role of a university researcher; and (b) decisions about publications.

PAR requires you to adapt to a constant swing and change of objectives, reflections, discussions, proposals, phases of fieldwork, and publications. In research, as in real life, events are uncertain and complex. In most cases, the research proposal is modified by the participation of other subjects in the research, by the uncertain context itself, by the lack or search for funding, or by the times set by political agendas and the movement to or from the university. These changes in roles that we experience, as well as all those shifts that we have often made in projects, were on many occasions judged negatively by university evaluators or in the request for funding.

Furthermore, one of the great dilemmas that anyone who works as a researcher in the framework of academia has to solve is what, how, and where to publish. In academia, publications are not only the only cover letter in your CV, they are also the main evaluation criterion in the PhD program, how your scientific quality is valued and, according to the research ethics protocols, one of the criteria to judge/evaluate your research ethics. According to the Ethical Commission on Animal and

Human Experimentation (CEEAH in its Spanish acronym), the code of good scientific practices of the Autonomous University of Barcelona,[6] one of the ethical rules is that the person carrying out the research has the obligation to publish all the results. This point is often critical; nevertheless, during an activist-research process it is not an arbitrary topic. Sometimes, much of the information collected cannot be published due to security issues, or a decision of the people involved in the investigation, or the lack of time and prioritization of other activities or forms of communication. Another issue is the format of the publication. Although formal academic centers prioritize indexed publications, these usually require waiting for an advanced process in the investigation with concrete results. However, in practice, social processes require information in the short term to make urgent decisions in highly dynamic contexts. There is nothing better than the example of the COVID-19 pandemic to illustrate this idea, where the need to discuss responses coincides with the urgency of having the necessary information to deal with its impacts, as well as to address the pandemic's deepening social differentiation (Gausman & Langer, 2020).

25.3.2 Intersections as Activist-Researchers

Haraway (1995) proposes specifying from which point of view someone starts the research process and why that and not another, thus making the political position explicit. Thus, by reflecting on our research process, we seek to raise awareness about the multiple intersections that condition privileges and oppressions that we live as activist-researchers as key elements of the research process. According to black feminist thoughts on intersectionality (Davis, 2004; Crenshaw, & Bonis 2005), these intersections have been influenced by multiple categories, such as gender, race, and social class age, and origins and cultural background. Most of us have lived through these intersections as researchers, beginning with the selection interviews to participate in research projects, especially when we were students, for being women and young as well as foreign origin. In these processes, intersections of class, ethnicity, and place of origin also intervene and question the spaces and places where each one decides to investigate.

In addition, our appearance as young women conditioned our research both in activist spaces and in academia itself. In both spaces, we are constantly evaluated and questioned, having to demonstrate the validity of our reflections and contributions. These conditions have prevented us from going into depth in some interviews with members (men) of social organizations and participating more actively in assemblies and meetings, although on other occasions they were useful when dealing with topics considered "adequate" to be discussed with us, such as those related to health. To avoid these biases, we have resorted to various strategies: in some cases, the interviews were repeated by men.

---

[6] www.recerca.uab.es/ceeah

We are also aware of our privileges as academics, as well as those that come from our class and origin. What does it mean to do research as white, highly educated, middle-class women in countries of the Global South? What does it mean for a woman from countries of the Global South to do research in countries in the Global North? In some cases, these privileges may have made it easier for us to obtain information and interviews with institutional representatives with influence and power. However, on the other hand, this condition of privilege also keeps us from understanding and talking about the reality we wanted to approach. In feminism, the suggestion is that by breaking the dichotomies of reason/emotion and body/mind, we can find some window to collaborate in co-producing another type of knowledge. Thus, adding the language of emotions and the body to our research can allow a dialogue to continue doing research to denounce or transform social and environmental injustices.

25.4 Final Thoughts

Ecological economics and political ecology are spaces for reflection that connected us, as disciplinary references and as colleagues. In addition, they allowed us to problematize from the point of view of academic feminism, as other colleagues in different fields and geographies have done (Nightingale, 2011). However, the influence of feminist thought in ecological economics has been limited to a few authors (Mies & Shiva, 1997) and the discussion of gender, feminisms, and rights in ecological economics has also been limited, with the exception of some specific proposals coming from feminism (Hanaček et al., 2020). Additionally, our own experiences have shown that in order to be radically transformative in our research on strong sustainability and social justice, we must first challenge the ways of doing science, deviating from the rules of the neoliberal academy, as activist feminism teaches us. This invites us in the coming years to fill this void of reflections.

Also noteworthy is the final reflection on the research process itself (see, e.g., Johnson et al., 2020), in that it invites us to assess and co-construct initiatives that promote PAR from a feminist approach, in both academic and activist spaces. Some recent examples question the current model of neoliberal academia and also suggest ways in which PAR and feminisms have a place, for example, feminist laboratories, Great Lakes Feminist Geography Collective (Mountz et al., 2015), and a feminist collective FRACTAL. We also propose the facilitation approach (Mindell, 2004) to work on conflicts that are generated at multiple levels during the PAR process and create spaces where we collectively and consciously share remedies, emotions, results, and proposals among those who participate in the process of investigation.

References

Binimelis, R., & Strand, R. (2009). Spain and the European debate on GM moratoria vs coexistence. In A. Guimeraes Pereira & S. Funtowicz (Eds.), *Science for policy* (pp. 120–134). Oxford University Press.

Borras, S. M. (2016). Land politics, agrarian movements and scholar-activism. *Inaugural Lecture, 14*.

Catacora-Vargas, G., Binimelis, R., Wynne, B., & Myhr, A. I. (2018). Socio-economic research on genetically modified crops: A study of the literature. *Agriculture and Human Values, 35*, 489–513.

Crenshaw, K. W., & Bonis, O. O. (2005). Mapping the margins: Intersectionality, identity politics, and violence against women of color. *Cahiers du Genre, 39*(2), 51–82.

Cuéllar, P., & Carmen i Sevilla Guzmán, E. (2009). Aportando a la construcción de la Soberanía Alimentaria desde la Agroecología. *Ecología Política, 38*, 43–51. https://www.ecologiapolitica.info/novaweb2/wp-content/uploads/2016/03/038_Cuellaretal_2009.pdf

Davis, A. (2004). *Mujeres, raza y clase* (Vol. 30). Ediciones Akal.

England, K. V. (1994). Getting personal: Reflexivity, positionality, and feminist research. *The Professional Geographer, 46*(1), 80–89.

Fals-Borda, O. (1978). *El problema de como investigar la realidad para transformarla: por la praxis*. Ediciones Tercer Mundo.

Freire, P. (1970). *Pedagogía del oprimido* (Capítulo I). Siglo XXI Editores.

Funtowicz, S. O., & Ravetz, J. R. (1993). The emergence of post-normal science. In *Science, politics and morality* (pp. 85–123). Springer.

Gausman, J., & Langer, A. (2020). Sex and gender disparities in the COVID-19 pandemic. *Journal of Women's Health, 29*(4), 465–466.

Gnych, S., Lawry, S., McLain, R., Monterroso, I., & Adhikary, A. (2020). Is community tenure facilitating investment in the commons for inclusive and sustainable development? *Forest Policy and Economics, 111*, 102088.

Gómez-Baggethun, E., Mingorria, S., Reyes-García, V., Calvet, L., & Montes, C. (2010). Traditional ecological knowledge trends in the transition to a market economy: Empirical study in the Doñana natural areas. *Conservation Biology, 24*(3), 721–729.

Hanaček, K., Roy, B., Avila, S., & Kallis, G. (2020). Ecological economics and degrowth: Proposing a future research agenda from the margins. *Ecological Economics, 169*, 106495. https://doi.org/10.1016/j.ecolecon.2019.106495

Haraway, D. (1995). Cyborgs and symbionts: Living together in the new world order. In *The cyborg handbook* (pp. 101–118).

Heller, C. (2002). From scientific risk to paysan savoir-faire: Peasant expertise in the French and global debate over GM crops. *Science as Culture, 11*, 5–37.

Johnson, A., Zalik, A., Mollett, S., Sultana, F., Havice, E., Osborne, T., Valdivia, G., Lu, F., & Billo, E. (2020). Extraction, entanglements, and (im)materialities: Reflections on the methods and methodologies of natural resource industries fieldwork. *Environment and Planning E: Nature and Space*, 1–46.

Martinez-Alier, J., Munda, G., & O'Neill, J. (1998). Weak comparability of values as a foundation for ecological economics. *Ecological Economics, 26*(3), 277–286.

Mies, M., & Shiva, V. (1997). *Ecofeminismo. Teoría, crítica y perspectivas* (Colección Antrazyt, 111). Icària editorial.

Mindell, A. (2004). *Sentados en el fuego: Como transformar grandes grupos mediante el conflicto y la diversidad*. Icaria Editorial.

Mingorria, S. (2016). *The Nadies weaving resistance: Oil palm and sugarcane conflicts in the territory, communities and households of the Q'eqchi', Polochic Valley, Guatemala*. PhD thesis. Universitat Autònoma de Barcelona.

Mingorria, S. (2018). Violence and visibility in oil palm and sugarcane conflicts: The case of Polochic Valley, Guatemala. *Journal of Peasant Studies, 47*(7), 1314–1340. https://doi.org/10.1080/03066150.2017.1293046

Mingorria, S., & Heras, M. (2019). *Expresiones Transculturales del Movimiento de Justicia Ambiental, Plan piloto educativo en la Universitat Autònoma de Barcelona* [Transcultural expressions of the environmental justice movement: An educational pilot plan in the Autonomous University of Barcelona]. Fundació Autónoma Solidària. http://docus-ecoeco.net/wpcontent/uploads/2019/01/FichasDidacticas.pdf

Monterroso, I., Cronkleton, P., & Larson, A. M. (2019). *Commons, indigenous rights, and governance*.

Mountz, A., Bonds, A., Mansfield, B., Loyd, J., Hyndman, J., Walton-Roberts, M., Basy, R., Whitson, R., Hawkins, R., Hamilton, T., & Curran, W. (2015). For slow scholarship: A feminist politics of resistance through collective action in the neoliberal university. *ACME: An International Journal for Critical Geographies, 14*(4), 1235–1259.

Munda, G. (2004). Social multi-criteria evaluation: Methodological foundations and operational consequences. *European Journal of Operational Research, 158*(3), 662–677.

Nightingale, A. J. (2011). Bounding difference: Intersectionality and the material production of gender, caste, class and environment in Nepal. *Geoforum, 42*(2), 153–162.

Öhman, A. B. (2012). Leaks and leftovers: Reflections on the practice and politics of style in feminist academic writing. In M. Livholts (Ed.), *Emergent writing methodologies in feminist studies* (pp. 27–40). Routledge.

Ravera, F., Iniesta-Arandia, I., Martín-López, B., Pascual, U., & Bose, P. (2016). Gender perspectives in resilience, vulnerability and adaptation to global environmental change. *Ambio, 45*(3), 235–247.

Rocheleau, D., & Edmunds, D. (1997). Women, men and trees: Gender, power and property in forest and agrarian landscapes. *World Development, 25*(8), 1351–1371.

Schnabel, L. (2014). The question of subjectivity in three emerging feminist science studies frameworks: Feminist postcolonial science studies, new feminist materialisms, and queer ecologies. *Women's Studies International Forum, 44*, 10–16.

Open Access This chapter is licensed under the terms of the Creative Commons Attribution 4.0 International License (http://creativecommons.org/licenses/by/4.0/), which permits use, sharing, adaptation, distribution and reproduction in any medium or format, as long as you give appropriate credit to the original author(s) and the source, provide a link to the Creative Commons license and indicate if changes were made.

The images or other third party material in this chapter are included in the chapter's Creative Commons license, unless indicated otherwise in a credit line to the material. If material is not included in the chapter's Creative Commons license and your intended use is not permitted by statutory regulation or exceeds the permitted use, you will need to obtain permission directly from the copyright holder.

Chapter 26
From the Environmentalism of the Poor and the Indigenous Toward Decolonial Environmental Justice

Brototi Roy and Ksenija Hanaček

Academic articles and textbook on origins of environmental justice and evolution of environmentalism describe the three main varieties of environmentalism:[1] the cult of wilderness, the gospel of eco-efficiency and the environmentalism of the poor (Guha & Martinez-Alier, 1997). Of these three, the term "environmentalism of the poor" was analyzed and popularized by Joan Martinez-Alier and Ramachandra Guha, who started using the phrase since their first meeting in 1988 in Bangalore, India (Martinez-Alier, 2002). Born within the discipline of social history, the term centers on social justice, including claims to recognition and participation, builds on the premise that the fights for human rights and environment are inseparable (Martinez-Alier, 2002, p. 514). It refers to the multiple environmental justice movements where the impoverished, marginalized, and Indigenous communities resist

---

[1] This is not to say that there are no other concepts to understand different forms of environmental struggles in different parts of the world: resigned activism to denote China's quiet environmentalism (Lora-Wainwright, 2017), subaltern environmentalism in the United States (Egan, 2002; Simonian & Pulido, 1996), bourgeois environmentalism that analyzes the role of the heterogenous middle class as actors of environmental justice concerns in India (Baviskar & Ray, 2011; Mawdsley et al., 2009), among others.

B. Roy (✉)
ICTA-UAB, Institut de Ciència i Tecnologia Ambientals, Universitat Autònoma de Barcelona, Barcelona, Spain

Department of Environmental Sciences and Policy, Central European University, Wien, Austria

K. Hanaček
ICTA-UAB, Institut de Ciència i Tecnologia Ambientals, Universitat Autònoma de Barcelona, Barcelona, Spain

Faculty of Social Sciences, Global Development Studies, University of Helsinki, Helsinki, Finland

against state and businesses carrying out projects of resource extraction, waste disposal, and big infrastructure.

However, with mounting evidence of the disproportional impacts of environmental injustices on Indigenous communities around the world, recent writings by Joan Martinez-Alier incorporated a more comprehensive phrasing of the concept, referring to it as environmentalism of the poor and the Indigenous (Martinez-Alier et al., 2016).

Both environmentalism of the poor and the Indigenous and environmental justice are frameworks to understand unjust and unequal distribution of environmental benefits and harms, more often than not, at the expense of historically subaltern communities, such as Indigenous, Women, Peasants, Romani, African and Latin American people. In this regard, the Atlas of Environmental Justice Movements – the EJAtlas – was a tool cocreated with activist–academic collaboration to document and study such movements against socio-environmental injustices.

Mapping such struggles is certainly a first step toward understanding movements against socio-environmental injustices. But is it enough? In this chapter, further, we push forward the decolonial understanding of environmental justice research and what it entails. We do so by providing insights from India and the Arctic as two examples of the Global South . According to de Sousa Santos (2016, pp. 18–19), the South is not a geographical definition but *"rather a metaphor for the human suffering caused by capitalism and colonialism on the global level… and speaks of a South that also exists in the geographic North (Europe and North America), in the form of excluded, silenced and marginalised populations…"*. This is the definition we follow throughout this chapter.

26.1 Our Positionalities

Brototi grew up as an expatriate Bengali Hindu in the state of Jharkhand, from an upper caste, middle class family and currently lives as a precarious person of color in Europe. Ksenija's lived reality is a "label" as Eastern European, as her home country, Croatia, is situated at the periphery of the Western European economic core (Roncevic, 2002) and subordinate by Western way of being, thinking, and knowing (de Sousa Santos, 2016). At the time of writing this chapter, we were both young immigrant women based in Barcelona and navigating between the multiple identities, languages, ideas, and positionalities (Smith, 2012). By the time it will be published, though we would be in Vienna and Helsinki, respectively, pursuing fixed-term post-doc positions. We offer the analytical and empirical insights in this chapter as junior foreign scholars navigating these multiple social relations and ideological agendas, which is often challenging, sometimes contradictory but always profoundly fulfilling.

26.2 New Directions in Environmental Justice Scholarship: Engagement with Decoloniality

In recent years, there has been a distinct interest in environmental justice scholarship to recognize and analyze multiple forms and phases of injustices (Malin & Ryder, 2018). The four pillars of critical environmental justice scholarship as proposed by Pellow (2016) include intersectionality (Crenshaw, 1991), multi-scalarity, anti-authoritarianism, and indispensability, aims to provide a framework to do so. Furthermore, the hegemonic theories of Western environmental justice scholarship have been challenged by proposing newer ones from the margins as an important intent to resist continuous coloniality (Quijano, 2007) of knowledge (Grosfoguel, 2002; Parra Romero, 2016). For example, in Central America, resistances are not only against climate injustices but also against violence of patriarchy and coloniality. In India, the resistances of the marginalized Indigenous and Dalit communities are manifestations of a longer struggle against both external and internal colonialism and/or caste-based discrimination (Martinez-Alier & Roy, 2019). Similarly, the working class, Indigenous population and Romani people have been the racial subjects of dispossession, colonialism, and domination within Europe and can be understood as the subalterns in the North. This in no way diminishes overseas slavery and exploitation in the majority world (Latin America, Africa, and Asia) but rather recognizes the continuous racial othering and domination of some Europeans by other Europeans as well (Robinson, 2000).

Yet, there remains a lack of meaningful engagement with complexities of theories and experiences of environmental injustices as well as engagement with decolonial thought in environmental justice scholarship across geographies (Álvarez & Coolsaet, 2018). A welcomed exception in recent years has been scholarship from South America, which has been putting forward the need of decolonial environmental justice by examining intercultural communication (Escobar, 2011; Rodríguez & Inturias, 2018), the politics of ontology (Blaser, 2013; Escobar, 2016), and decolonization of knowledges and acceptance of multiple worldviews (Rodríguez & Inturias, 2018). This scholarship is establishing an emerging decolonial thinking, which is crucial when conducting research on environmental conflicts and injustices.

The main arguments revolve around colonial imposition as a violent way of invading the earth, subjugating lands, humans, and non-humans to maintain colonial relations in the so-called "post-colonial" present (Escobar, 2011; Quijano, 2007). A colonial worldview that invented a hierarchy between races and different lands of the globe transposing Western ideas and approaches in case studies of Global South without understanding the context and the multiple marginalities that communities face causes a "coloniality of justice" (Álvarez & Coolsaet, 2018; Ferdinand, 2019). Imposition of such concepts and frameworks without contextualization, even if it is well-intended, could be counterproductive and lead to further inequalities and injustices (Mawdsley et al., 2009). That is to say, diverse subaltern environmental struggles must be acknowledged (Pulido & De Lara, 2018).

This is explained using the concepts of the coloniality of power, the coloniality of knowledge, and the coloniality of being (Maldonado-Torres, 2008). A decolonial "switch" against such colonial assumptions of environmental justice combines with critical thinking about race, gender, and class as a contribution to the radical epistemological traditions (Pulido & De Lara, 2018) against the dominant Western World-System (Grosfoguel & Cervantes-Rodríguez, 2002).

This "switch" toward decolonizing environmental justice aims to explore other ways of understanding human–nature relationships, different methodologies involved, processes of resistances, and acknowledgment of multiple lived experiences and worldviews. Because colonial silence separates environmental and colonial thinking and excludes a whole swath of people, a decolonial environmental justice recognizes people's need for justice based on historical and structural injustices related to environment, but functioning within a broader structure of colonialism, racism, casteism, communalism, and patriarchy (Sultana, 2020; Sultana & Loftus, 2012).

The long-lasting history of colonial environmental extraction against communities' well-being is opposed by people who are at the core of social, environmental, and cultural injustices in different geographies and, who call for the decolonization (Escobar, 2008; Maldonado-Torres, 2008) of socio-ecological distribution conflicts (Martinez-Alier, 2002; Temper, 2019), with or without using those words. Their opposition to the continuous domination of modern, colonial, capitalist, and extractive tendencies (Escobar, 2001; Grosfoguel & Cervantes-Rodríguez, 2002; Svampa, 2015), which stem from racial/ethnic marginalization, poverty, gendered discrimination, ageism, rural/urban divides, and many other dynamics, are at the heart of many motivations for resistance against environmental injustices (Kojola & Pellow, 2020).

The domination of Western cultural imaginaries through development and extractive logics explains social and environmental injustices as arising from the project of modernity and economic growth. The decolonization of knowledge, culture, and social relations is one of the key challenges for overcoming the history of oppression and marginalization in development and contributes to decolonizing structures, relations, and ways of being (Grosfoguel & Cervantes-Rodríguez, 2002). Scholars must play a role in decolonizing environmental injustices through a commitment to engage with the structural and historical forces that create marginalization and exclusion in the use of natural resources and territories (Mar & Edmonds, 2010).

Decolonial environmental justice addresses socio-cultural environmental dimensions and responsibilities of a given place, such as traditional knowledge (*knowing*), spirituality, identities (*being*), and different ways of struggles (*transformative power*) (Grosfoguel & Cervantes-Rodríguez, 2002). Many conflicts are experienced with extractive industries such as mining, infrastructure, and intensive agriculture as they involve enormous physical transformation of traditional landscapes, leaving behind the intangible way of feeling about the environment, being part of the environment and knowing about the environment. As Fernández-Giménez and Arturo (2015) and Parra Romero (2016) argue, decolonial shift in the analysis of environmental conflicts includes cultural, economic, and political dynamics as continuous colonial heredities.

In the rest of the chapter, we provide two examples of how and why decolonizing environmental justice is relevant for the Global South, providing evidence from Indigenous communities in India and the Arctic. We finally conclude with some potential research directions toward decolonial environmental justice.

26.3 Indian Adivasi Thinking

The Indigenous population of India, officially called "Scheduled Tribes," comprises more than 700 different communities. According to the last census data in 2011, 8.6% of India's total population (more than 100 million people) are made up of *adivasis*, literally translated as first inhabitants or original dwellers, and are the world's largest population of Indigenous people (Faizi & Nair, 2016). There are different sets of laws depending on the geographical location of the communities in peninsular India or north-eastern India, as fifth and sixth schedules respectively. According to the EJAtlas, more than half of the environmental justice movements in India (57%) have Indigenous people mobilizing, and estimates show that more than 40% of the people affected or displaced as a result of ecological distribution conflicts are *adivasis* (Shrivastava & Kothari, 2012).

Many of the early grassroot resistance to colonial rule in India such as the *Santhal* revolt of 1855 had clear environmental undertones. If we understand environmentalism of the poor and the Indigenous as movements of people fighting for issues beyond environmental safeguard, but protection of a way of life, culture and traditions, and livelihoods, it would not be too much a stretch to see the early adivasi resistances much different from present-day environment of the poor. Similarly, although Birsa Munda is remembered as a tribal hero for the freedom fight against colonialism in the late 1800s, in today's context, his fight for the safeguard of the forests and their resources, as well as Indigenous autonomy over those forests and resources, can be understood also from the lens of environmental justice struggles. Many Indigenous struggles today remember and invoke Birsa's bravery and persistence of fighting against extraction and injustices.

Yet, despite such a rich tradition of fighting for socio-ecological equality, adivasi thinkers are quite marginalized, both in India and globally. In recent years, young adivasi leaders are critical of this position that they are historically put in, "as bodies for the protests, and not minds for the movement" (interview with JK). According to Jacinta Kerketta, an Indigenous poet, journalist, and social activist from the central Indian state of Jharkhand, and belonging to the *Oraon* tribe, this is a form of epistemological (knowing) injustice (as she explains in her words): "*The first fundamental thing to question is this very concept of how one individual or a group of individuals can claim to 'develop' another individual or society. Development for me implies a life of dignity. And that necessarily implies respect and understanding of the Indigenous way of life. You can't develop someone if you consider yourself superior to them, that only leads to oppression."* Jacinta has grown up witnessing and participating in the struggle of the vast adivasi society to preserve their land, forests, rivers, languages, and heritage and culture, which she expresses in her poetry.

Her concerns and frustrations are neither new nor surprising. It resonates completely with the claims of Archana Soreng (AS), who is an Indigenous activist and researcher from the *Khadia adivasi* community in the Eastern Indian state of Odisha, and one of the seven members of the UN Secretary General's Youth Advisory Group on Climate Change. She says it is crucial for incorporation of Indigenous practices and worldviews for issues of biodiversity conservation and climate justice, since for centuries, Indigenous communities have remained the responsible stewards for biodiversity protection, yet they have very little decision-making power, and instead have been faced with forcible displacement and nonconsensual relocation due to large-scale mining and infrastructure projects by states and private corporations (interview with AS).

26.4 Resistance to Coloniality In and Around the Arctic

The Arctic is a colonized territory (Cameron, 2012; Josephson, 2014; Kuokkanen, 2019; Stuhl, 2016) and so are livelihoods, cultures, traditions, languages, and identities of Indigenous peoples. In that regard, Indigenous lands and culture have been fragmented by oil fields, wind-power parks, and mining projects, among others (Naykanchina, 2012). Extractive and industrial activities on traditional Indigenous Arctic lands are both the consequence of colonization including rising global commodity extraction frontiers (Hanaček et al., 2022; John, 2016; Naykanchina, 2012; Tlostanova & Mignolo, 2009). Extractive and industrial colonization of the Arctic are commonly perceived as justified, because states acquire the land and hand it for extractive and industrial purposes (Gritsenko, 2018; Muller-Wille, 1987). Yet, these activities jeopardize and, therefore, continue to marginalize Indigenous people, their lands, identities, and worldviews (Lassila, 2020).

In the process, there is also discrimination and racial prejudice against Indigenous people of the Circumpolar North, which continue to persist in both the private and public sectors in Sweden, Norway, Finland, Russian Federation, Canada, and the United States (Kumpula et al., 2011). The prejudice constrains the opportunities and the rights of people to express their own concerns regarding cultural identity and their colonized lands. This, for example, includes recognizing reindeer herders' use and management of grazing land by identifying cultural practices for Indigenous land use (Naykanchina, 2012). However, marginalization and oppression of Indigenous herders are common when the herders do not follow "modern" industrial development logic, or when they prioritize traditional cultural values and worldviews (Huntington, 2016; Nuttall, 1998). The state ignores the fact that these activities, identities, and human nature relationships are the foundation of local economies and livelihoods (Naykanchina, 2012) .

As of January 2022, there are 1913 cases in the EJAtlas, which reports loss of traditional knowledge, practices, and cultures as one of the social impacts of divergent extractive projects around the world. Given the fact that Indigenous peoples of the Arctic and beyond call for the need of socio-cultural dimensions in environmental questions, and that is, the spiritual foundations of their cultural identities along their (physical) lands (Dorough, 2014). Thus, cultures related to the environment are fundamental in environmental conflicts and injustices studies, which deepen in power relations and coloniality (Rodríguez & Inturias, 2018). By focusing on stories of those on the frontline is important to envision decolonial justice and sustainable future paths (Wiebe, 2019) As Indigenous people put it in the "Our Cultures Our Rights" video for the Cultural Survival (2017) movement, which advocates for Indigenous peoples' rights and supports Indigenous communities' self-determination, cultures, and political resilience:

> We draw upon knowledge given to us by our ancestors to be in spiritual relationship with Mother Earth and all living things, and to appropriately honor and steward the land. We protect, defend, resist, renew with our art and traditions.

Similarly, Indigenous Buryat[2] woman explains in an interview for the Cultural Survival (2019):

> I come from the Buryat Peoples who have lived in Siberia for millennia, on both sides of Lake Baikal, the deepest and largest fresh-water lake. My grandmother would tell me stories which encapsulated the wisdom of our ancestors and have been passed down for generations. I participated in our traditional ceremonies. I still recall the fire, the chants, and the prayers of the women in my community. I grew up with a deep sense of understanding of our lifeways and belongingness to the land, to my people, and a deep love for my culture. It was not until I was 24 when I first encountered the term 'Indigenous Peoples'. It took leaving and living far away to understand the degree of both external and internalized oppression, colonization, and paralysis that my people and other Indigenous Peoples in Russia currently face.

The above stated words by Indigenous peoples bring into focus spiritual and identity relationships as an important angle in decolonial environmental justice research, precisely because traditional cultural significance of the people and the environment in different places strongly oppose to the continuous domination of colonial relations embedded in extractive and industrial tendencies (Escobar, 2008; Maldonado-Torres, 2008). What is important to mention, however, is that these places and stories of marginalization must be seen as spaces of resistance (Tuhiwai Smith, 1999).

---

[2] Buryat, northernmost of the major Mongol people, living south and east of Lake Baikal. By the Treaty of Nerchinsk (1689), their land was ceded by China to the Russian Empire, as an arrangement between the two empires (Chen, 1966); ("Buryat." Encyclopaedia Britannica, December 5, 2018, https://www.britannica.com/topic/Buryat. Accessed 12 January 2022).

26.5 Conclusion and Prospects for Further Research

In this chapter, we have argued that there is an urgent need for decolonial environmental justice research both theoretically and methodologically. We claim that a future research agenda on environmental justice must include multiple drivers and forms of oppression across relevant historical and contemporary social contexts that intersect to control and dominate nature and the communities on the frontline while simultaneously privileging powerful actors in environmental distribution conflicts.

Methodologically, this research agenda must also explore pedagogical aspects for decolonial research. This is crucial for real transformations toward sustainability, and can be achieved when the answers and decisions come *from* the South itself – telling their own stories and theorizing as well as implementing their own alternatives to colonial extractivism, patriarchy, racism, classism, and other forms of oppression.

It is vital to engage and advance different forms of intersectional, interdisciplinary, and international decolonial and feminist inquiries to address ongoing socio-ecological crises. We conclude that the future path of critical political ecology must be paved by engaging with and valuing the scholarship that advances complexities of power, relational privileges, intersectional politics, and epistemological differences by fostering decolonized environmental politics, climate activism, and alliances and solidarities with Indigenous peoples (Sultana, 2020), including engagements with (inter)colonialism. We argued in this chapter that embracing decoloniality in environmental justice research is the way to go about it.

References

Álvarez, L., & Coolsaet, B. (2018). Decolonizing environmental justice studies: A Latin American perspective. *Capitalism Nature Socialism, 5752*. https://doi.org/10.1080/10455752.2018.1558272

Baviskar, A., & Ray, R. (2011). *Elite and everyman: The cultural politics of the Indian middle classes*. Routledge India.

Blaser, M. (2013). Ontological conflicts and the stories of peoples in spite of Europe. *Current Anthropology, 54*, 547–568. https://doi.org/10.1086/672270

Cameron, E. S. (2012). Securing indigenous politics: A critique of the vulnerability and adaptation approach to the human dimensions of climate change in the Canadian Arctic. *Global Environmental Change, 22*, 103–114. https://doi.org/10.1016/j.gloenvcha.2011.11.004

Chen, V. (1966). The Treaty of Nerchinsk. In *Sino-Russian relations in the seventeenth century* (pp. 86–105). Springer. https://doi.org/10.1007/978-94-015-0847-6_9

Crenshaw, K. (1991). Mapping the margins: Intersectionality, identity politics, and violence against women of color. *Stanford Law Review, 43*, 1241. https://doi.org/10.2307/1229039

Cultural Survival. (2017). *Our cultures our rights*.

Cultural Survival. (2019). *Q&A with Cultural Survival's New Executive Director: Galina Angarova*.

de Sousa Santos, B. (2016). Epistemologies of the South and the future. *From Eur South, 1*, 17–29. https://doi.org/10.1080/10570310802636334

Dorough, D. S. (2014). Declaration on the Rights of Indigenous Peoples. In *11th annual conference on the parliamentarians of the Arctic region: Governance models and decision making process*. United Nations Permanent Forum on Indigenous Issues.

Egan, M. (2002). Subaltern environmentalism in the United States: A historiographic review. *Environment and History*. https://doi.org/10.3197/096734002129342585

Escobar, A. (2001). Culture sits in places: Reflections on globalism and subaltern strategies of localization. *Political Geography, 20*, 139–174. https://doi.org/10.1016/S0962-6298(00)00064-0

Escobar, A. (2008). *Territories of difference: Place, movements, life, redes*. Duke University Press.

Escobar, A. (2011). *Encountering development: The making and unmaking of the third world*. Princeton University Press.

Escobar, A. (2016). Thinking-feeling with the Earth: Territorial struggles and the ontological dimension of the epistemologies of the South. https://doi.org/10.11156/aibr.110102e

Faizi, S., & Nair, P. K. (2016). Adivasis: The world's largest population of indigenous people. *Development*. https://doi.org/10.1057/s41301-017-0115-8

Ferdinand, M. (2019). *Une écologie décoloniale – penser l'écologie depuis le monde caribéen*. Le Seuil.

Fernández-Giménez, M. E., & Arturo, E. (2015). Territorios de diferencia: la ontología política de los "derechos al territorio." *Cuadernos de Antropología Social, 20*, 25–38. https://doi.org/10.5751/ES-08054-200429

Gritsenko, D. (2018). Energy development in the Arctic: Resource colonialism revisited. In A. Goldthau, M. F. Keating, & C. Kuzemko (Eds.), *Handbooks of research on international political economy series*. Edward Elgar Publishing.

Grosfoguel, R. (2002). Colonial difference, geopolitics of knowledge, and global coloniality in the modern/colonial capitalist world-system. *Utop Think, 25*, 203–224. https://doi.org/10.2307/40241548

Grosfoguel, R., & Cervantes-Rodríguez, A. M. (2002). *The modern/colonial/capitalist world-system in the twentieth century: Global processes, antisystemic movements, and the geopolitics of knowledge*. Praeger.

Guha, R., & Martinez-Alier, J. (1997). *Varieties of environmentalism. Essays North and South, varieties of environmentalism*. Essays North and South. https://doi.org/10.1016/s0264-8377(97)00047-1

Hanaček, K., Kröger, M., Scheidel, A., Rojas, F., & Martinez-Alier, J. (2022). On thin ice – The Arctic commodity extraction frontier and environmental conflicts. *Ecological Economics, 191*, 107247. https://doi.org/10.1016/j.ecolecon.2021.107247

Huntington, H. P. (2016). The connected Arctic. *Environment*. https://doi.org/10.1080/00139157.2016.1112197

John, A. T. (2016). *Alaska indigenous governance through traditions and cultural values, indigenous governance*. Native Nations Institute, University of Arizona.

Josephson, P. R. (2014). *The conquest of the Russian Arctic*. Harvard University Press.

Kojola, E., & Pellow, D. N. (2020). New directions in environmental justice studies: Examining the state and violence. *Environmental Politics*, 1–19. https://doi.org/10.1080/09644016.2020.1836898

Kumpula, T., Pajunen, A., Kaarlejärvi, E., Forbes, B. C., & Stammler, F. (2011). Land use and land cover change in Arctic Russia: Ecological and social implications of industrial development. *Global Environmental Change, 21*, 550–562. https://doi.org/10.1016/j.gloenvcha.2010.12.010

Kuokkanen, R. (2019). At the intersection of Arctic indigenous governance and extractive industries: A survey of three cases. *The Extractive Industries and Society, 6*, 15–21. https://doi.org/10.1016/j.exis.2018.08.011

Lassila, M. (2020). The Arctic mineral resource rush and the ontological struggle for the Viiankiaapa peatland in Sodankylä, Finland. *Globalizations*, 1–15. https://doi.org/10.1080/14747731.2020.1831818

Lora-Wainwright, A. (2017). *Resigned activism: Living with pollution in rural China*. MIT Press.

Maldonado-Torres, N. (2008). La descolonización y el giro des-colonial. *Tabula Rasa*, 61–72. https://doi.org/10.25058/20112742.339

Malin, S. A., & Ryder, S. S. (2018). Developing deeply intersectional environmental justice scholarship. *Environmental Sociology, 4*, 1–7. https://doi.org/10.1080/23251042.2018.1446711

Mar, T. B., & Edmonds, P. (2010). *Making settler colonial space: Perspectives on race, place and identity*. Palgrave Macmillan.

Martinez-Alier, J. (2002). *The environmentalism of the poor: A study of ecological conflicts and valuation*. Edward Elgar Publishing.

Martinez-Alier, J., & Roy, B. (2019). Editorial: Some insights on the role of violence. *Ecology, Economy and Society, 2*, 27–30.

Martinez-Alier, J., Temper, L., Del Bene, D., & Scheidel, A. (2016). Is there a global environmental justice movement? *Journal of Peasant Studies, 43*, 731–755. https://doi.org/10.1080/03066150.2016.1141198

Mawdsley, E., Mehra, D., & Beazley, K. (2009). Nature lovers, picnickers and bourgeois environmentalism. *Economic and Political Weekly, 11*.

Muller-Wille, L. (1987). Indigenous peoples, land-use conflicts, and economic development in circumpolar lands. *Arctic and Alpine Research*. https://doi.org/10.2307/1551399

Naykanchina, A. (2012). *Indigenous Reindeer Husbandry: The impacts of land use change and climate change on indigenous reindeer herders' livelihoods and land management, and culturally adjusted criteria for indigenous land uses*. https://reindeerherding.org/images/projects/Nomadic_Herders/publications/UNPFII-2012-Reindeer-Husbandry_Final23Nov.pdf

Nuttall, M., 1998. Protecting the Arctic: Indigenous peoples and cultural survival. Routledge Taylor & Fancis Group. 9789057023552.

Parra Romero, A. (2016). ¿Por qué pensar un giro decolonial en el análisis de los conflictos socioambientales en América Latina? *Ecologia política*, 15–20.

Pellow, D. N. (2016). Toward a critical environmental justice studies. *Du Bois Review: Social Science Research on Race, 13*, 221–236. https://doi.org/10.1017/S1742058X1600014X

Pulido, L., & De Lara, J. (2018). Reimagining 'justice' in environmental justice: Radical ecologies, decolonial thought, and the Black Radical Tradition. Plan. E Nat. Sp. https://doi.org/10.1177/2514848618770363

Quijano, A. (2007). Coloniality and modernity/rationality. *Cultural Studies, 21*, 168–178. https://doi.org/10.1080/09502380601164353

Robinson, C. (2000). *Black Marxism: The making of the Black Radical Tradition*. University of North Carolina.

Rodríguez, I., & Inturias, M. L. (2018). Conflict transformation in indigenous peoples' territories: Doing environmental justice with a 'decolonial turn'. *Development Studies Research, 5*, 90–105. https://doi.org/10.1080/21665095.2018.1486220

Roncevic, B. (2002). Path from the (semi) periphery to the core: On the role of socio-cultural factors. *IES Proceedings, 1*, 1–26.

Shrivastava, A., & Kothari, A. (2012). *No churning the earth: The making of global India*. Penguin Books.

Simonian, L., & Pulido, L. (1996). Environmentalism and economic justice: Two Chicano struggles in the southwest. *The Western Historical Quarterly*. https://doi.org/10.2307/970546

Smith, L. T. (2012). Choosing the margins: The role of research in indigenous struggles for social justice. In *Decolonizing methodologies: Research and indigenous peoples*. Zed Books.

Stuhl, A. (2016). *Unfreezing the Arctic: Science, colonialism, and the transformation of Inuit lands*. University of Chicago Press.

Sultana, F. (2020). Political ecology 1: From margins to center. *Progress in Human Geography*. https://doi.org/10.1177/0309132520936751

Sultana, F., & Loftus, A. (2012). The right to water: Prospects and possibilities. In S. Farhana & A. Loftus (Eds.), *The right to water* (pp. 1–18). Earthscan.

Svampa, M. (2015). Commodities consensus: Neoextractivism and enclosure of the commons in Latin America. *South Atlantic Quarterly, 114*, 65–82. https://doi.org/10.1215/00382876-2831290

Temper, L. (2019). Blocking pipelines, unsettling environmental justice: From rights of nature to responsibility to territory. *Local Environment*. https://doi.org/10.1080/13549839.2018.1536698

Tlostanova, M. V., & Mignolo, W. D. (2009). Global coloniality and the decolonial option. *Kult, 6*, 130–147.

Tuhiwai Smith, L. (1999). *Decolonizing methodologies: Research and indigenous peoples* (3rd ed.). Zed Books.

Wiebe, S. M. (2019). "Just" stories or "just stories"? Mixed media storytelling as a prism for environmental justice and decolonial futures. *Engagement School Journal of Community-Engaged Research Teaching, Learning, 5*, 19–35. https://doi.org/10.15402/esj.v5i2.68333

Open Access This chapter is licensed under the terms of the Creative Commons Attribution 4.0 International License (http://creativecommons.org/licenses/by/4.0/), which permits use, sharing, adaptation, distribution and reproduction in any medium or format, as long as you give appropriate credit to the original author(s) and the source, provide a link to the Creative Commons license and indicate if changes were made.

The images or other third party material in this chapter are included in the chapter's Creative Commons license, unless indicated otherwise in a credit line to the material. If material is not included in the chapter's Creative Commons license and your intended use is not permitted by statutory regulation or exceeds the permitted use, you will need to obtain permission directly from the copyright holder.

Part VI
Public Policy Applications

Chapter 27
Agrobiodiversity in Mexican Environmental Policy

Nancy Arizpe and Dario Escobar-Moreno

As an example of a multi-scalar analysis approach from Barcelona School of Ecological Economics, in this chapter we present a case study applied to Mexican agrobiodiversity policies. In Sect. 27.1, we explain the role of agrobiodiversity at different scales, Sect. 27.2 exposes the importance of maize in Mexican agrobiodiversity, and Sect. 27.3 analyzes multi-scalar agrobiodiversity polices in Mexico focusing in maize.

27.1 The Role of Agrobiodiversity in Rural Systems

In countries such as Mexico, different factors affect agricultural systems, such as: (i) international markets, which do not reflect the importance of most of the countries linked to maintaining food security; (ii) the biophysical performance of agriculture, which plays a special role in alleviating poverty; and (iii) the effect of market prices of agricultural products, which underestimate the indirect effects of agricultural growth.

Rural livelihoods have been incorporating new combinations of technological, discursive, commercial, and financial elements in recent decades (Hecht, 2010), along with the fragmentation of working classes and migratory flows in multiple directions between rural and urban, national and international, and in permanent and cyclical modes (Borras, 2009). For example, the agro-industrial model of maize expansion has induced changes in land use and production that generate severe negative impacts (i.e., sociocultural, ecological, biophysical) associated with

N. Arizpe (✉)
Universidad Nacional de Misiones, Posadas, Argentina

D. Escobar-Moreno
Universidad Autonoma de Chapingo, Texcoco, Mexico

malnutrition, migration, poverty, disease, lack of food, among others. There are also changes in monetary flows and energy inputs such as machinery, petroleum products, fertilizers, and transgenic seeds, which also result in negative social and environmental impacts (Arizpe et al., 2011; Holland et al., 2008). In traditional agriculture, both men and women participate in different activities, including what feminist scientists call "care work" (Jochimsen & Knobloch, 1997), "reproductive work" (Biesecker & Hofmeister, 2010), "domestic production," or "subsistence," as well as productive agricultural work. Likewise, traditional agriculture considers the diversity of crops, also called agrobiodiversity.

Agrobiodiversity involves various areas of knowledge (agronomy, anthropology, ecology, botany, etc.) and reflects complex human social relationships, influences conservation policies for cultivated ecosystems, and promotes food security, social inclusion, and sustainable development. In addition to taking into account the cultural processes, knowledge and practices of farmers as key elements in the maintenance of agroecosystems (Bergel, 2017). Agrobiodiversity includes all the variety and variability of animals, plants, and microorganisms that are used directly or indirectly for food and agriculture, including crops, livestock, trees, and fish. Created and managed by farmers, shepherds, fishermen, and forest dwellers, it encompasses the diversity of genetic resources (varieties, races) and species used for food, fodder, fibers, fuel, and medicine. It also includes the diversity of unharvested species that support production (soil microorganisms, predators, pollinators) and those in the broader environment that support agroecosystems (agricultural, pastoral, forestry, and aquatic), as well as the diversity of agroecosystems (FAO/PAR, 2011), which are considered to be constituted by all the biological elements in it, while agrobiodiversity refers to all the components of biodiversity that deal with food and agriculture (Salazar et al., 2015).

27.2 Maize, an Emblematic Case of Agrobiodiversity in Mexico

Mexico is the center of origin and diversification of 15.4% of the species that are used as food sustenance worldwide, 62 languages are also spoken within its territory, and it is the country with the greatest cultural diversity in America. In this regard, the domestication of species is a biocultural event and therefore the conservation of agroecosystems is of great importance (Barrera-Bassols et al., 2011). Among the 130 crops of which Mexico is the center of origin, domestication or genetic diversity of maize, beans, chili, squash, chilacayote, amaranth, tomato, avocado, sweet potato, nopal, tobacco, cocoa, and vanilla stand out, crops that coexist with its wild relatives, for example, 7 species of teosinte have been identified, which are the ancestors of modern maize, 67 species of wild beans, 10 of squash, and 38 of amaranth (Casas & Parra, 2016).

Peasants not only contribute to maintaining the evolution of maize, but the scale at which this is done becomes by itself an irreplaceable evolutionary or ecosystem service. These ecosystem services arise because peasant maize agriculture combines in a single system, three of the main factors that are known to positively affect adaptive evolution: large effective population size, high-level genetic diversity, and environmental change (Bellon et al., 2018).

Peasants have created, or inherited, complex agricultural systems that, for centuries, have helped them meet their subsistence needs, even under very adverse environmental conditions (Altieri, 1999). The fundamental objective of the peasant economy is to satisfy the consumption needs of the family nucleus and to exchange the surpluses. Therefore, the knowledge and comprehensive use of the largest amount of biotic and abiotic resources in their environment is essential.

At least 9 million hectares dedicated to agricultural production in Mexico are found in mountain areas at altitudes above 2000 m above sea level, where it is frequent that many of the cultivated plots are located in hillside areas, on marginal soils, and with irregular rainfall regimes. These are generally poor producers, which have very limited means of production, and therefore make strategic use of the scarce productive resources they possess: labor, land, and seeds. Under the above conditions, maize seeds and the knowledge that producers have about them and their environment are two of the fundamental elements that guarantee the survival of the peasants and their families, which could number around 12 million people. Therefore, maize seeds constitute a vital resource for a very important segment of Mexican rural society.

27.3 Multi-scalar Policies on Agrobiodiversity Issues

Few studies have analyzed how agrobiodiversity is maintained in farmers' fields. Research on the subject, especially from South America and Asia, suggests that there is a connection between the conservation of agrobiodiversity in farmers' fields and the exchange of seeds (Badstue et al., 2007).

In 1983, FAO adopted an International Undertaking on Plant Genetic Resources for Food and Agriculture, which is a voluntary agreement signed by 113 countries, whose purpose is to promote international harmony regarding access to plant genetic diversity. In 2001, the 180 countries of the FAO Conference adopted an International Treaty on Plant Genetic Resources for Food and Agriculture. This Treaty represents a global agreement on a subject of primary importance: the management of the world's agrobiological diversity. The objectives are the conservation and sustainable use of plant genetic resources for food and agriculture, and the fair and equitable sharing of the benefits derived from their use, in harmony with the Convention on Biological Diversity, for sustainable agriculture and food security. The agreement covers all plant genetic resources relevant to food and agriculture.

In Mexico, until very recently, there has not been a policy to promote and conserve biodiversity, nor agrobiodiversity, nor in particular the diversity of its great wealth of native maize. It is enough to point out that agricultural policies since the middle of the past century have been characterized by promoting agricultural productivity per unit area from the use of a set of inputs such as hybrid seeds, fertilizers, pesticides, and agricultural machinery. So, we have had policies that promoted the predominance of a few types of improved varieties with high productivity, which are grown in large areas, and which have undermined the preservation of the diversity of native maize. However, it is surprising that a relatively recent study published by the National Commission for the Knowledge and Use of Biodiversity (CONABIO, 2006) has documented that the diversity of native maize in Mexico is kept under cultivation in practically all of the territory of the country.

The T-MEC, continuation of the North American Free Trade Agreement (NAFTA), regarding the "Protection of New Varieties of Plants" (seeds), has an agreement where Mexico must ratify to the International Convention UPOV–91, which obliges Mexico to continue adhering to UPOV, better known by social organizations that resist GMOs, such as the Monsanto Law, which is nothing more than "privatizing seeds" and concentrating its ownership in transnational monopolies.

The Federal Law for the Promotion and Protection of Native Maize (LFFPMN) seeks to recognize the production, commercialization, consumption, and constant diversification of this species, as a national cultural manifestation. This new Law proposes that the State must guarantee and promote, through all the competent authorities, that the general population has access to the informed consumption of native maize, as well as to derived products, in conditions free of genetically modified organisms and other genetic improvement techniques. The current federal government, through the Rural Development Secretariat (SADER), seeks to promote a "long-term" policy, "more productive, sustainable, and inclusive," with productive and commercial objectives. Two undersecretariats, Agriculture and Food Self-sufficiency, were created. The former is responsible for matters of agriculture productivity and rural development; the latter is in charge of 4 of the 25 strategic programs of the new government, all "socially oriented" and with the objective of paying for the purpose of achieving food self-sufficiency. The appearance of what would appear to be new programs is reduced to objectives and instruments very similar to those that have operated in the last 25 years. Between 72% and 82% of rural maize-growing households use their production for human consumption; that is, they do not market it. According to recent studies, the value of the maize that these households harvest for their subsistence is ten times higher than its price in the market. For the same reason, maize price subsidies do not benefit this majority; on the contrary, they generate adverse conditions, by definition, to their activity. The presumption that such prices will convert this population into surplus producers has no clear grounds. More than incentives to increase their yields, the subsidies create conditions for the concentration of land within their communities. By capitalizing on land rent, the proposed subsidies will be regressive. The big losers of the sector reform could well be the peasants, the rural population and agrobiodiversity, in the social sphere, and the native ecosystems and climate change in the environmental aspect (Dyer et al., 2019).

Although we began to have in Mexico the first laws that address the problem of conservation and promotion of biodiversity, and in particular agrobiodiversity, such as the Federal Law for the Promotion and Protection of Native Maize (DOF, April 13, 2020). In practice, there is no evidence that the legal framework is responsible for the conservation and promotion of the country's biodiversity, quite the contrary, there is clear evidence of how, for decades, "rural development" laws have agrobiodiversity of the country undermined.

In recent years, the Mexican federal government has implemented a set of programs that, at least in their objectives, consider the protection and promotion of agrobiodiversity. Such is the case of the "Sembrando Vida" program, in charge of the Ministry of Welfare, which, in addition to proposing as an objective the reforestation of 1 million hectares with native tree species in 19 states of the country, promotes the sowing of native seeds from each region and locality through the milpa system. The National Council of Science and Technology (CONACYT) has also implemented the National Strategic Research Programs (PRONACEs), including the one aimed at the recovery of Food Sovereignty and which also seeks to rescue native seeds and peasant knowledge. Within the Ministry of Agriculture and Rural Development (SADER), there is a Sub-secretariat for Food Self-Sufficiency that promotes a set of programs that also emphasize the promotion of native seeds and peasant knowledge as a central part of their strategies. Lastly, we can mention the Ministry of the Environment and Natural Resources (SEMARNAT), which also includes a set of programs aimed at protecting, recovering, and promoting agrobiodiversity in Mexico. One of the great challenges for the current federal administration will be to not only implement these policies and programs but also to interconnect them to enhance their effects and achieve the best possible results. However, we will have to wait, at least for the remainder of this government, to start evaluating its effects, it is still too early to assess its impacts.

Regarding local policies, we find that different movements and organizations that promote agrobiodiversity emerge from below, mainly rejecting the entry of transgenics, such as the Regional Organization of Purhépecha Farmers in Defense of Criollo Maize (ORAPDMC). Thus, it gave rise to a sui generis indigenous resistance movement by promulgating a precautionary policy in the face of the possible arrival of GMOs in the open field, with repercussions at the national and international level and by articulating with other resistance movements in the country. For the time being, a joint program began that involved a dozen indigenous and mestizo communities of the basin, academics, non-governmental organizations (NGOs, hereinafter), government agencies, donor foundations, artisans and artists, and the public from the city and the countryside. The above-mentioned program is articulated through several components that allow it to be comprehensive: (i) the revaluation of local agronomic knowledge, including agricultural rituals, the exchange of seeds between producers, and the return of native germplasm to its original places; (ii) the launch of agroecological projects and training together with academics, technicians, and NGOs; (iii) the revaluation of maize as its own cultural icon in the media (radio, press, and television); and (iv) political work based on conferences, regional fairs, and local maize festivals.

In Mexico, we find other movements of a sociopolitical nature that emerge with this cause and that adapt their own cultural and natural contexts, and that in their petition documents have points such as (i) rejection of aggressive agricultural technologies such as biotechnology, (ii) rejection of the local effects of the global market, (iii) rejection of the health effects caused by both, and (iv) defense of food sovereignty against the loss of agrobiodiversity, among others. For example, some of them are The Totonaca Nahuatl Indigenous Unit (UNITONA), The Vicente Guerrero Comprehensive Rural Development Project (GVG), The Union of Organizations of the Sierra Norte de Juárez in Oaxaca (UNOSOJO), among others.

27.4 Conclusions

Contrary to what public policies for agricultural development have been, the native maize of Mexico remains a clear expression of the resistance of indigenous and peasant people, who have been ignored for years by such policies.

Different movements and organizations emerge to revalue or fight for agrobiodiversity. In this chapter, we find that beyond the fact that agrobiodiversity is important for the current Mexican government for its incorporation into its strategic lines, these policies are not yet implemented. However, we find that communities continue to resist along with the support of other national–international organizations, local governments, or experiences of other peasant movements.

In this regard, there is a continuous struggle to preserve agrobiodiversity and associated biocultural processes, both by peasants as well as by urban areas that consume agrobiodiversity products. According to our analysis, local policies are more successful in activities that promote and preserve agrobiodiversity. National policies in many cases are conflicting between programs. It is important that top-down and bottom-up policies are integrated, in addition to considering the entire Mexican territory.

References

Altieri, M. A. (1999). Applying agroecology to enhance productivity of peasant farming systems in Latin America. *Environment, Development and Sustainability, 1*, 197–217.

Arizpe, N., Giampietro, M., & Ramos-Martin, J. (2011). Food security and fossil energy dependence: An international comparison of the use of fossil energy in agriculture (1991–2003). *Critical Reviews in Plant Sciences, 30*, 45–63.

Badstue, L. B., Bellon, M. R., Berthaud, J., Ramírez, A., Flores, D., & Juárez, X. (2007). *The dynamics of seed flow among small-scale maize farmers in the Central Valleys of Oaxaca, Mexico* (p. 29).

Barrera-Bassols, N., Astier M., Orozco, Q., & Boege-Schmidt, E. (2011) Saberes locales y defensa de la agrobiodiversidad. In S. Á. Cantalapiedra (coord.), *Convivir para perdurar. Conflictos ecosociales y sabidurías ecológicas* (pp. 289–310). Icaria.

Bellon, M., Mastretta-Yanes, A., Ponce-Mendoza, A., Ortiz Santamaría, D., Oliveros-Galindo, O., Perales, H., Acevedo, F., & Sarukhán, J. (2018). Evolutionary and food supply implications of ongoing maize domestication by Mexican campesinos. *Proceedings of the Royal Society B: Biological Sciences, 285*, 20181049. https://doi.org/10.1098/rspb.2018.1049

Bergel, D. S. (2017). *La agrobiodiversidad como tema bioético* (Alegatos, num. 96) (pp. 349–364).

Biesecker, A., & Hofmeister, S. (2010). Focus: (Re)productivity. Sustainable relations both between society and nature and between the genders. *Ecological Economics, 69*(8), 1703–1712.

Borras, S. M., Jr. (2009). Agrarian change and peasant studies: Changes, continuities and challenges – An introduction. *Journal of Peasant Studies, 36*(1), 5–31.

Casas, A., & Parra, F. (2016). Capítulo 1. El manejo de recursos naturales y ecosistemas. La sustentabilidad en el manejo de recursos genéticos. In A. Casas, J. Torres-Guevara, & F. Parra (Eds.), *Domesticación y en el Continente Americano. Volumen 1. Manejo de biodiversidad y evolución dirigida por las culturas del Nuevo Mundo* (pp. 25–50). Universidad Nacional Autónoma de México/Universidad Nacional Agraria La Molina.

CONABIO. (2006). *Orígen y diversificación del maíz. Una revisión analítica*. CONABIO. 116p.

DOF, Diario Oficial de la Federación del 13 de abril de 2020. Secretaria de Gobernación, Gobierno de México.

Dyer, G., HernÁndez-Solano, A., Meza-Pale, P., & Yúnez-Naude, A. (2019). Las reformas de mercado, el TLCAN y la política agropecuaria y rural: 1988–2017. In José Luis Calva (Coordinador), *La economía de México en el TLCAN: Balance y perspectivas frente al T-MEC (USMCA)* (pp. 611–642). Juan Pablos Editor. ISBN 978-607-711-530-4.

FAO/PAR. (2011). *Biodiversity for food and agriculture: Contributing to food security and sustainability in a changing world*. Available at http://www.fao.org/3/a-i1980e.pdf

Hecht, S. (2010). The new rurality: Globalization, peasants and the paradoxes of landscapes. *Land Use Policy, 27*(2), 161–169.

Holland, N., Joensen, L., Maeyens, A., Samulon, A., Semino, S., & Sonderegger, R. J. (2008). *The round table on IR-responsible soy. Certifying soy expansion, GM soy and agrofuels*. ASEED Europe, BASEIS, CEO, and Rain Forest Action Network.

Jochimsen, M., & Knobloch, U. (1997). Making the hidden visible: The importance of caring activities and their principles for any economy. *Ecological Economics, 20*(2), 107–112.

Salazar, L., Magaña, M., & Latournerie, L. (2015). Importancia económica y social de la agrobiodiversidad del traspatio en una comunidad rural de Yucatán, México. *Agricultura, Sociedad y Desarrollo, 12*(1), 1–14. Sánchez, G. J. J. 2011.

Open Access This chapter is licensed under the terms of the Creative Commons Attribution 4.0 International License (http://creativecommons.org/licenses/by/4.0/), which permits use, sharing, adaptation, distribution and reproduction in any medium or format, as long as you give appropriate credit to the original author(s) and the source, provide a link to the Creative Commons license and indicate if changes were made.

The images or other third party material in this chapter are included in the chapter's Creative Commons license, unless indicated otherwise in a credit line to the material. If material is not included in the chapter's Creative Commons license and your intended use is not permitted by statutory regulation or exceeds the permitted use, you will need to obtain permission directly from the copyright holder.

Chapter 28
Conventional Climate Change Economics: A Way to Define the Optimal Policy?

Jordi Roca and Emilio Padilla

28.1 Introduction

Among the earliest economic models for the integrated assessment of climate change was the Dynamic Integrated Climate-Economy (DICE) model (Nordhaus, 1993). This models the links between macroeconomic developments, greenhouse gas emissions, climate change, and their economic costs. It was proposed as a model for the world as a whole and has had several subsequent revisions, the last one being in 2018 (Nordhaus, 2018a). This line of research has led to many other integrated assessment models (IAMs) that aim to contrast the economic costs of different policies against climate change with those of not acting.

28.2 The DICE Model and Its Results

The DICE model is a normative model that maximizes an objective function, the present value of the sum of current and future utility of "extended consumption" (i.e., consumption minus the effects of climate change on well-being, valued monetarily). The utility in each moment depends on this consumption according to a function in which the marginal utility of consumption is decreasing. The impact of climate change caused by greenhouse gases is considered to reduce utility, in terms of its direct effects both on well-being and production. A simple quadratic function

J. Roca (✉)
Department of Economics, University of Barcelona, Barcelona, Spain
e-mail: jordiroca@ub.edu

E. Padilla
Department of Applied Economics, Autonomous University of Barcelona, Bellaterra (Cerdanyola del Vallès), Spain

is assumed for the relationship between temperature change and economic damage. In the initial model, the possibility of extreme events was ignored, while later versions assume this possibility assigning probabilities that are deemed "reasonable" or using other ad hoc procedures. However, mitigation policies also have an economic cost, reducing the possibilities of consumption. Adopting this cost–benefit approach, Nordhaus estimates that the climate change "optimal policy" is – in the Pigouvian tradition – to establish a global carbon tax equal to the "marginal social cost of carbon." The model assumes that the economy always moves along a path of investment-consumption that maximizes well-being, and so it calibrates the model, with the exception of an externality – the damage of climate change – that is necessary to consider to achieve the optimal path (Goulder and Williams III, 2012). Nordhaus does not explicitly make the extravagant – though popular in macroeconomics – assumption that human society is equivalent to a representative agent who lives infinitely, but his model only seems consistent with this assumption.

The application of the DICE model in the 1990s recommended little deviation from the scenario without mitigation policy (Nordhaus, 1993, 1994). Nordhaus's results contrasted with the calls to act quickly at the 1992 United Nations Framework Convention on Climate Change, and influenced political decisions that legitimized the inaction of the US administration. Its most current version (DICE 2016) suggested limiting warming to around 3.5 °C by 2100 (Nordhaus, 2018b).

The Economics Nobel Prize award to Nordhaus was announced precisely the same day (October 8, 2018) that a special Intergovernmental Panel on Climate Change (IPCC) report (2018) was released. However, their conclusions regarding the intensity of the actions that need to be taken were radically different, so the so-called "economics of climate change" contrasts sharply with the dominant scientific opinion on what should be done. The report pointed to the need for a rapid and radical reduction in emissions to limit warming to 1.5 °C, to reduce risks, and to facilitate adaptation. Nordhaus' "economic optimum" implied an increase in emissions for approximately 25 years followed by a slow decrease (Nordhaus, 2018a, b). As Pezzey states, when projections such as Nordhaus' are made, "climate scientists typically express disbelief, derision, or dismay" (Pezzey, 2017, p. 3).

It is doubtful whether cost–benefit logic can show us how to tackle climate change. Although Nordhaus highlights an "optimal path", his model can lead to as many optimal paths as choices are made on certain controversial parameters so that by modifying them, virtually any level of reduction can be justified (Padilla, 2004). The quantitative results give an impression of being scientific, but they depend essentially on some nontransparent opinions of "experts" (Pindyck, 2017). In the rest of the chapter, we discuss the main problems with Nordhaus' proposals.

28.3 Discounting the Future

Conventional economic climate change analysis gives less weight to what may happen in the future. The choice of the rate by which the future is discounted ends up determining the level of mitigation that is considered optimal. Nordhaus considers a rate around 4–5% to be adequate, while the influential Stern report (2007) used a rate of 1.4%. This largely explains their different conclusions regarding the appropriate level of mitigation. The choice does not depend on scientific enquiry, but is based on specific assumptions and value judgments.

When using a utilitarian function, as Nordhaus does, there are essentially two arguments that are employed to justify time discounting. The first is based on the assumption that people prioritize present over future consumption. This assumption, called pure time preference (or impatience), is applied to the social discount rate. However, it is not clear that it is the dominant human preference. Most people seem to prefer distributing their resources to sustain – or even improve – their well-being throughout life. It is also questionable as to whether impatience can be considered a rational preference (Strotz, 1956). Even Ramsey (1928), author of some of the first macroeconomic models of intertemporal maximization, wrote that discounting the future is "a practice which is ethically indefensible and arises merely from the weakness of the imagination" (p. 543).

However, the main criticism is that impatience for a person's own consumption should not be transferred to a social preference between the present and the future, when the decision affects other generations (Padilla, 2002). This would only be justified if the society were made up of immortal individuals. Is it ethical that we give more importance to the present at the cost of what happens to future generations? Nordhaus' answer is yes, because the time preference of present generations must be respected. Many authors have opposed this position. Solow (1974, p. 150) states that: "in social decision-making (…) there is no excuse for treating generations unequally (…) we ought to act as if the social rate of time preference were zero." Thus, Cline (1992) and Stern's (2007) climate change models reject pure time preference (except that due to the very small probability of human extinction in the near future).

The second argument for discounting future consumption – which is different from discounting future utility – is based on the decreasing marginal utility of consumption and the optimistic assumption that there will be a continuous increase in per capita consumption. A growth of around 2% per year, as predicted by Nordhaus, means that consumption will multiply by a factor of approximately 7 in 100 years. Despite climate change, the well-being of future generations (as measured by the much-criticized criterion of per capita consumption; Roca Jusmet, 2011) would be well above the current one. So, why should we worry about them? The basis of this argument is the blind belief in future welfare improvements, which would render superfluous any concern for sustainable development. This can lead to the problem of the "optimist's paradox" (Martínez Alier, 2002; Padilla, 2002): if we apply a high discount because we assume that future generations will be better-off, we might seriously jeopardize such an outcome by making decisions that damage their environmental resource base.

In a normative model, the use of a discount rate must be based on ethical criteria. However, the Nordhaus-type model calibrates the objective function as if the economy were effectively maximizing social welfare. So, the rate of return on market investments is considered as an indicator of social preferences between current and future consumption. However, this rate tells us nothing about the level of sacrifice people are willing to make to preserve climate stability. Moreover, a normative model would not have to share dominant current preferences (Llavador et al., 2015). Nordhaus uses the two aforementioned arguments (pure time preference and increasing consumption per capita) to discount the future, but ends up arguing that the discount rate cannot be very different from the market return on investments. For example, he criticizes both Cline (1992), arguing that his approach is philosophically satisfactory but inconsistent with actual social decisions on savings and investments, and Stern (2007), whose "unambiguous conclusions about the need for extreme immediate action will not survive the substitution of assumptions that are consistent with today's marketplace real interest rates and savings rates" (Nordhaus, 2007, p. 701). Thus, when deciding if a discount rate is "realistic" or not, the ultimate criterion for Nordhaus is what happens in the market.

28.4 Uncertainty

A problematic aspect of the DICE and similar models is the number of uncertainties associated with climate change and mitigation policies. Current evidence tends to suggest that there is no linearity in the relationship between temperature change and induced damage; rather, the latter will increase proportionately much more than the former, and there will be discontinuities in the relationship. The relationship is very complex, and after a certain heating threshold it becomes even more unpredictable. One issue that makes the modeling of the damage function difficult is the possibility of a positive feedback between the concentration rates of greenhouse gases and the temperature level. This becomes more likely as the rates increase. Furthermore, when we consider the damage that all this may cause to society, something even more uncertain appears: how will people react? (Pezzey, 2017).

The DICE model assumes that economic damage is a continuous quadratic function of the increase in temperature. It calibrates the parameters so that the damage – measured as a percentage of world GDP – is moderate, even when the changes in temperature are great. The possibility of catastrophic phenomena is forgotten or the event is given a low probability rating based on the opinion of "experts", even though these phenomena are not subject to the kind of experimentation that would allow us to assign them a probability. As a result, and also due to time discounting, possible future catastrophic events carry no significant weight in the evaluation.

Pindyck (2017) notes that "When it comes to the damage function, we know virtually nothing (…) developers of IAMs have little choice but to specify what are essentially arbitrary functional forms and corresponding parameter values" (p. 101). This is even more true given that the projected changes in temperature are well beyond the range of the last thousands of years. As Pezzey (2017) observes, the damage function of the future is not only highly unknowable but will continue to be so.

Weitzman (2012), along with other authors, takes into account the potential magnitude of climatic disasters. He states that it is not appropriate to apply conventional cost–benefit analysis to decide climate policies, and argues that the issue should be treated as a risk management problem. The main question is how much society is willing to sacrifice to insure against the risk of possible catastrophic effects.

28.5 Cost–Benefit Analysis and Commensurability

The DICE model is presented as the maximization of a sum of (discounted) utilities. Moreover, climatic damage is considered the equivalent of (negative) consumption, which requires it to be valued in monetary terms. In practical terms, the utility-maximizing model becomes simply the traditional formula of cost–benefit analysis, according to which efficiency is determined by the criterion of "potential compensation". Greenhouse gas reduction is appropriate only as long as the monetary value of mitigation "costs" is less than the (discounted) monetary "benefits" of the damage avoided. In such a setting, everything (e.g., health and environment) can be given a pecuniary value. However, there are many problems and controversies surrounding monetary valuation estimations, and these are particularly acute in the case of climate change. For instance, a 1995 IPCC report was based on an assessment that the value of life in developing countries was 15 times lower than that in developed countries, and this sparked comprehensible widespread protests (Martínez Alier & Roca Jusmet, 2013, p. 288). Martínez Alier strongly argued against the assumption of commensurability, defending a multicriteria approach and the existence of different languages of valuation (Martínez Alier et al., 1998; Martínez Alier, 2002).

The Nordhaus approach is alien to any consideration of environmental rights or justice: it is permissible to harm some people to benefit others as long as the benefit to the latter is greater than the harm to the former. Even human lives become compensable when they are given a monetary value. Climate change is characterized by deep inequalities, both in responsibilities and in impact. Nordhaus chooses functions with decreasing marginal utility with respect to the level of consumption; that is to say, he introduces theoretically the assumption of "inequality aversion." However, the DICE model compares global consumption at different moments in time, ignoring the distribution among individuals of the same generation. As we have seen, inequality aversion is used to justify discounting the future, and the discounting will be greater the greater the "inequality aversion". This aversion should, logically, be applied to the intragenerational inequality between individuals of the same generation, and greater weight should be given to the impact on the poorest (Azar & Sterner, 1996). Thus, the paradox is that concerns about inequality in the Nordhaus aggregate model are used to promote not more but less action against climate change. Intragenerational inequalities are forgotten, and aversion to inequality is introduced only to discount the negative impact on future generations caused by current generations. If we consider the great intragenerational inequalities between different countries, it is worth asking whether optimism is so high as to suppose that future generations of poor African countries will be better-off than current generations of countries such as the United States or those in the European Union. Otherwise, one of the arguments for discounting the future benefits of climate policy becomes meaningless.

28.6 Conclusions

In this chapter, we have analyzed the Nordhaus approach to climate change, which is based on the cost–benefit principle and is frequently presented as the way economics establishes the "optimal" policy. We – along with most ecological economists – do not think it possible to determine the optimal policy from economic analysis (Azar, 1998), because this depends on value judgments and an uncertain knowledge of the future. Instead, we believe that a number of useful principles should be applied.

The first is sustainability, or *inter*generational justice, which can be understood as meeting our needs without jeopardizing the welfare of future generations. The second is *intra*generational justice. It is not acceptable that the carbon-intensive consumption of certain populations should jeopardize the most basic needs of others. Ongoing uncertainties, the difficulties in modeling climate change and, most importantly, the potential magnitude of the impact of global warming on future generations have led to growing support for the view that mitigation policies should be also guided by the *precautionary principle*. We should make efforts to minimize the risk of catastrophic scenarios that threaten to last longer than the entire history of humanity up to this point (Pezzey, 2017).

In conclusion, models *à la* Nordhaus conflict with the principles of sustainability, environmental justice, and precaution. These principles do not offer *the* optimal climate change policy, but they are a good point of reference, and they each demand rapid and radical emissions reduction. In stark contrast to these principles, Nordhaus argues that "economic optimality" implies that emissions should continue to increase for several decades. Interpreting this result as the "scientific" answer of economics ignores the limitations and biases of the model. It is an example of the negative influence that a certain type of macroeconomic modeling can have on decision-making, and it also damages the prestige of economics among the disciplines that are looking for solutions.

We are not, in any way, optimistic about climate change. Unfortunately, future emissions will almost certainly be much greater than those recommended by the IPCC. However, being pessimistic about the future is very different than characterizing it as economically efficient or optimal.

References

Azar, C. (1998). Are optimal CO_2 emissions really optimal? *Environmental and Resource and Energy Economics, 11*, 301–315.

Azar, C., & Sterner, T. (1996). Discounting and distributional considerations in the context of global warming. *Ecological Economics, 19*, 169–184.

Cline, W. R. (1992). *The economics of global warming*. Institute for International Economics.

Goulder, L. H., & Williams, R. C., III. (2012). The choice of discount rate for climate change policy evaluation. *Climate Change Economics, 3*(4), 1250024.

IPCC. (2018). *Global warming of 1.5°C an IPCC special report on the impacts of global warming of 1.5°C above pre-industrial levels and related global greenhouse gas emission pathways, in the context of strengthening the global response to the threat of climate change, sustainable development, and efforts to eradicate poverty*. IPCC.

Llavador, H., Roemer, J., & Silvestre, J. (2015). *Sustainability in a warming planet*. Harvard University Press.

Martínez Alier, J. (2002). *The environmentalism of the poor. A study of ecological conflicts and valuation*. Edward Elgar Publishing.

Martínez Alier, J., & Roca Jusmet, J. (2013). *Economía Ecológica y Política Ambiental.*, tercera edición. Fondo de Cultura Económica. (Primera edición 2000).

Martínez Alier, J., Munda, G., & O'Neill, J. (1998). Weak comparability of values as a foundation of ecological economics. *Ecological Economics, 26*(3), 277–286.

Nordhaus, W. (1993). Rolling the "DICE": An optimal transition path for controlling greenhouse gases. *Resource and Energy Economics, 5*, 27–50.

Nordhaus, W. (1994). *Managing the global commons: The economics of climate change*. MIT Press.

Nordhaus, W. (2007). A review of the stern review on the economics of climate change. *Journal of Economic Literature, 45*, 686–702.

Nordhaus, W. (2018a). Evolution of modelling of the economics of global warming: changes in the DICE model, 1992–2017. *Climatic Change, 148*, 623–640.

Nordhaus, W. (2018b). Projections and uncertainties about climate change in an era of minimal climate policies. *American Economic Journal: Economic Policy, 10*(3), 333–360.

Padilla, E. (2002). Intergenerational equity and sustainability. *Ecological Economics, 41*(1), 69–83.

Padilla, E. (2004). Climate change, economic analysis and sustainable development. *Environmental Values, 13*(4), 523–544.

Pezzey, J. V. C. (2017). Why the social cost of carbon will always be disputed. *WIREs Climate Change*. https://doi.org/10.1002/wcc.558

Pindyck, R. S. (2017). The use and misuse of models for climate change. *Review of Environmental Economics and Policy, 11*, 100–114.

Ramsey, F. P. (1928). A mathematical theory of saving. *Economic Journal, 38*, 543–559.

Roca Jusmet, J. (2011). El debate sobre el crecimiento económico desde la perspectiva de la sostenibilidad y la equidad. In A. Dubois, J. L. Millán, & J. Roca (coord.), *Capitalismo, desigualdades y degradación ambiental*. Icaria.

Solow, R. (1974). Intergenerational equity and exhaustible resources. *The Review of Economic Studies, 41*, 29–45.

Stern, N. (2007). *The economics of climate change: The Stern review*. Cambridge University Press.

Strotz, R. H. (1956). Myopia and inconsistency in dynamic utility maximization. *Review of Economic Studies, 23*, 165–180.

Weitzman, M. L. (2012). GHG targets as insurance against catastrophic climate damages. *Journal of Public Economic Theory, 14*(2), 221–244.

Open Access This chapter is licensed under the terms of the Creative Commons Attribution 4.0 International License (http://creativecommons.org/licenses/by/4.0/), which permits use, sharing, adaptation, distribution and reproduction in any medium or format, as long as you give appropriate credit to the original author(s) and the source, provide a link to the Creative Commons license and indicate if changes were made.

The images or other third party material in this chapter are included in the chapter's Creative Commons license, unless indicated otherwise in a credit line to the material. If material is not included in the chapter's Creative Commons license and your intended use is not permitted by statutory regulation or exceeds the permitted use, you will need to obtain permission directly from the copyright holder.

Chapter 29
Contribution of Global Cities to Climate Change Mitigation Overrated

Jeroen C. J. M. van den Bergh

29.1 Limited Reach of Urban Policies

Work on ecological economics in Barcelona as founded by Joan Martinez-Alier tends to pay much attention to local conditions and conflicts associated with global environment challenges. This encompasses studies of cities, including policies and strategies by urban authorities. Generally, this work is characterized by considerable optimism, reflecting the fact that local policies can achieve a lot by restricting and guiding urban development or by encouraging local initiatives and bottom-up processes (e.g., Ecología Política, 2014). Incidentally, Joan himself has shown to be more skeptical, emphasizing the concern that "cities displace environmental loads to larger geographical scales" (Martinez-Alier, 2003). This links to the systemic effects discussed in Sect. 29.3 of this chapter.

Here, I consider the previous "optimistic thesis" for the case of climate change and associated mitigation (i.e., emission-reduction) policies by cities. Indeed, the message that cities can do a lot to help solving climate change is popular. It was reinforced by repeated failures of international climate negotiations, suggesting that ambitious climate mitigation policies by cities can compensate for weak national policies (Watts, 2017). This has given rise to various networks of cities combating climate change, such as the UN's Compact of Mayors or the C40 Cities Climate Leadership Group. However, despite so many cities seemingly setting ambitious targets and implementing many policies, we do not see the effect of this in terms of

---

J. C. J. M. van den Bergh (✉)
ICREA, Barcelona, Spain

Institute of Environmental Science and Technology, Universitat Autònoma de Barcelona, Barcelona, Spain

School of Business and Economics, Institute for Environmental Studies, VU University Amsterdam, Amsterdam, The Netherlands
e-mail: jeroen.bergh@uab.es

reduced global greenhouse gas (GHG) emissions (Siskova & van den Bergh, 2021). Hence, there is a need for critical and quantitative studies that depict a realistic role of cities to reduction.

The motivation in several publications for the supposed important role of cities in global GHG emissions reduction is that the largest part of the world population and emission sources are located within urban areas. This is illustrated by the following citations: "Cities are crucial to global mitigation efforts. [...] urban areas are responsible for 71% of global energy-related carbon emissions" (Rosenzweig et al., 2010); "Cities must address climate change. More than half of the world's population is urban, and cities emit 75% of all carbon dioxide from energy use" (Bai et al., 2018); and "Cities are at the heart of the decarbonisation effort. The energy landscape is shaped by cities. [...] cities account for about two-thirds of primary energy demand and 70% of total energy-related carbon dioxide (CO_2) emissions. [...] Hence, efforts aimed at fostering sustainable urban energy paths are crucial to meet national and global low-carbon ambitions." (IEA, 2016). While superficially these statements sound like reasonable points, careful scrutiny shows they raise too high expectations of what cities can contribute to global emissions reduction.

A fundamental problem is that studies tend to attribute GHG emissions associated with consumers and firms to cities, without checking if cities are really responsible, that is, can in any way control such emissions (Satterthwaite, 2008). One study reviewing 200 European cities attributes emissions reduction in urban areas to the respective cities without identifying a concrete link with feasible urban policies (Reckien et al., 2014). This evidently creates a false picture, exaggerating the contribution of cities. Even so, the study claims that if up-scaled to all cities, only 20% of required emissions reduction by the EU – i.e., to meet the 2 °C target – will be achieved. A rare study of the actual impact of urban climate policies confirms this picture. It statistically analyzes data from 478 cities in California for eight policy outputs (green building projects and standards, residential solar photovoltaic systems, street lighting, waste programs, pedestrian/bicycle infrastructure, gasoline sales, and commute vehicle share), finding little evidence for emissions reduction being causally related to urban climate plans and policies. Its conclusion is that the latter are largely formalizing outcomes that would have been realized anyway, in view of prevailing environmental preferences and national policies (Millard-Ball, 2012). Another study examines which factors explain differences in emissions among ten cities with populations varying from 432,000 to 9,519,000 (Kennedy et al., 2009). Most factors identified, including geophysical and technical ones, turn out to be not or hardly controllable by city authorities.

Studies that are optimistic about the contribution of cities tend to ignore the fact that the largest part of GHG emissions – from industry, electricity generation, tourists, consumers, and transport – cannot be controlled by urban policies. Even though arising within city borders, such emissions strongly depend on specific national policies. Indeed, actions and strategies such as levying carbon taxes on fossil fuels, creating carbon markets, providing subsidies for renewable energy or electric vehicles, regulation of industries, or setting emission standards for cars all belong to the domain of national governments. Regarding tourism, an important part of its GHG emissions is due to international and interurban travel, which is beyond the control of city governments. Moreover, while many studies suggest cities can do a lot about building-related emissions, building codes and energy efficiency standards/labels are generally set by national governments. Urban authorities merely monitor and control construction permissions on their basis.

29.2 Lack of Effective Urban Instruments

On the websites of the various city-for-climate networks, one can find attractive terminologies such as "city intelligence," "reinventing cities," "transforming urban lifestyles," "new urban agenda," and "buying local". One must not expect, though, to find much information on effective regulatory policies. It seems that – like commercial companies – cities are prone to green marketing. One review finds that "… existing initiatives are fragmented and … do not address many of the key drivers and determinants involved," concluding that "… local authorities tend to move towards rhetoric rather than meaningful responses." (Romero-Lankao, 2012). A SWOT analysis in a mid-term evaluation of the Covenant of Mayors for the European Commission is also rather critical, noting that legal constraints limit the capacity of municipalities to implement own plans, that the covenant triggers mitigation actions only to a limited extent, that country participation is very uneven (e.g., dominance of southern Europe), and that many cities do not achieve even modest targets. In addition, the report signals a focus of cities on funding local actions rather than regulating polluters (Technopolis Group, 2013). A review of 55 US cities finds that voluntary outreach programs with low participation prevail, while more effective regulatory policies are little used (Ramaswami et al., 2012). This study offers details on the city of Denver, concluding that a combination of urban voluntary and regulatory actions yields at best ~1% GHG mitigation annually in buildings and transportation.

There seems to be even confusion about what counts as urban climate mitigation policy, let alone an effective one. To illustrate this, consider a recent study that identifies 13 main urban mitigation strategies (Ürge-Vorsatz et al., 2018):

1. Urban design and form.
2. Modal shift, mobility services, traffic optimization.
3. High-efficiency, low-emissions, smaller vehicles.
4. Low-energy demanding, heat-resistant architecture.
5. High-efficiency appliances and equipment.
6. Energy efficient and low-carbon urban industries.
7. High-performance operation of buildings.
8. Reducing urban heat island.
9. Infrastructure-integrated renewable energy systems generation.
10. Fuel switch to low(er) carbon generation.
11. Affordable low-carbon, durable construction materials; timber infrastructure.

12. Carbon capture and utilization in construction materials.
13. Lifestyle, behavior, choices, sustainable consumption and production, sharing economy, circular economy.

However, this list suffers from various shortcomings. A main one is that none of these are really policies or strategies but rather desired effects. Moreover, most items lack a clear connection with effective policies set by urban authorities. To illustrate: point 2 requires partly national policies such as carbon or gasoline taxes, and vehicle taxes; point 3 is mainly driven by technology constraints (at national and EU levels) and gasoline prices; points 4 and 5 depend on building and technological standards, which are usually defined at a national level; point 6 depends largely on energy prices and national efficiency standards and labels; cities have little influence over point 7; points 9–12 depend predominantly on national innovation and industrial policies; point 13 is a very broad category of issues driven mainly by nonurban factors, notably national regulatory policies, including carbon pricing.

This is not to say that cities cannot do anything. The most concrete and effective strategies are investing in more energy-efficient public buildings and cleaner (especially non-diesel) buses or garbage trucks. Here, one should realize, though, that public buildings and transport only contribute a very small portion of all GHGs within city limits, often around 1% (UITP, 2011). A unique role of cities is altering urban form to lower emissions (strategy 1 above). This might involve restricting cars in certain zones, creating green intra-urban spaces, reducing sprawl, improving access to public transport, and creating bicycle lanes. However, emission reduction effects of these are very limited, unless bold changes in infrastructure and access regulation to cities are implemented in a majority of cities around the world – which must be judged as unlikely. And, if possible, it would require a major transition taking a very long time (Siskova & van den Bergh, 2019). In addition, major changes in urban form are a sluggish process, limited by past choices and geographical conditions.

29.3 Free Riding and Systemic Effects

A solution to climate change is difficult as solutions to climate change are hampered by free riding by countries as well as subnational actors, due to the public goods nature of the climate. An effective solution is a binding global climate agreement, committing countries to install mutually coherent, stringent policies. It has turned out to be very difficult to achieve this type of solution, and unfortunately the Paris Agreement does not count as such. It failed to solve the free rider problem by allowing for voluntary pledges. While the role of cities became a big issue in the slipstream of failures to strike an effective agreement (Hale, 2016), a focus on subnational agents does not bring a solution to the global free rider problem closer, rather the opposite. Not only is the urban level further removed from effective, systemic policies – such as nationwide carbon pricing – but also it includes many more decision-makers: compare the tens of thousands of cities worldwide with a much smaller number of countries (<200).

If stringent, a single city's climate policies will carry a significant economic and political cost while not making much of a difference in global emissions and warming. Only if a fair share of all cities worldwide reduce emissions in concert will the effects be visible. This should include the vast majority of the approximately 1700 cities worldwide with over 300,000 inhabitants, as well as the 430 cities with over a million inhabitants (UN, 2016). Current city-for-climate networks remain very far from these numbers, and anyway do not represent binding agreements. At best, city governments are then motivated to implement soft, nonbinding policies.

If cities were, hypothetically, able to implement restrictive regulations for consumers, producers, and tourists, they would quickly be punished economically due to households and firms moving out of their borders, less national and international interest in their more expensive products, and tourists opting for cheaper destinations. All these responses would furthermore contribute to carbon leakage. Other systemic effects may reduce the net climate and environmental effects of urban policies. For instance, densification aimed at reducing transport-related emissions can strengthen urban heat island effects, unless done with sound green design.

Local politicians proposing effective regulation – e.g., banning cars from the city – should, in addition, expect political repercussions during elections. This is another reason why few cities implement effective regulatory policies. It implies that in a highly urbanized country such as the United Kingdom, only two cities have implemented congestion schemes, namely London and Durham, while in a much larger country such as the United States, no such city is found.

29.4 A Tentative Quantification of Global Emissions Reduction by Urban Policies

Using the previous insights, I undertake a back-of-the-envelope calculation in Table 29.1 and the text below to derive two indicators: (1) an upper bound to maximally controllable emissions by cities (expressed as a share of global emissions), assuming that any effective urban policy is implemented in all cities around the world; and (2) emissions reduction through realistic – i.e., unilateral, voluntary, and politically feasible – urban policies, accounting for divergent ambitions and opportunities of cities worldwide. To this end, the emissions share of distinct sources is determined (column 2 in Table 29.1), followed by the maximum share of cities in these (column 3) and the realistic share (column 4).

To derive column 4, diversity of climate policy effectiveness among cities has to be considered. Few systematic studies address effectiveness of urban policies. An evaluation for Denver finds that only 2–4% of households respond to door-to-door campaigns with actions that reduce emissions and < 1% for voluntary adoption of loans for efficiency measures in homes (Kennedy et al., 2009). The authors refer to

Table 29.1 Impact of urban policies on GHG emissions: tentative upper bound estimates[a]

Sector	Share in global GHG emissions[b]	Maximally controllable emissions (share) due to limited reach of urban policies	Realistic share emissions control due to limited effectiveness urban policies
Agriculture, forestry, and other land use	24%	0.1%	0.1%
		The percentage reflects that most agriculture is outside city borders, which depends on national climate/agricultural policies, while there is a small role for urban agriculture and green space in cities. This is, however, limited by suitable plot/rooftop area and existing regulations (fire risks, fertilizer use, proximity of contaminating roads, etc.)	Cities can only stimulate urban agriculture through soft policies. It is not clear that shifting from traditional to urban agriculture will reduce emissions anyway, also in view of loss of economies of scale. Because of uncertainties, the number in column 3 is not discounted in this column
Buildings	6%	1.5%	0.105%
		Building codes and energy efficiency standards/labels are generally set by national governments. Urban authorities merely monitor and control construction permissions on their basis. Some cities, though, require roofs of residential and commercial buildings to be "solar-ready" or with actual investments in solar energy. Since new buildings make up a very small portion of total buildings, this has a limited effect, though. An optimistic estimate, accounting for a large share of buildings outside city borders and limited control by cities, is 25% on average for all cities globally	Studies indicate a very low effectiveness of voluntary programs, loans, etc. A 7% global average effectiveness for cities is applied – see the derivation in the text

(continued)

Table 29.1 (continued)

Sector	Share in global GHG emissions[b]	Maximally controllable emissions (share) due to limited reach of urban policies	Realistic share emissions control due to limited effectiveness urban policies
Electricity and heat production	25%	7.5%	0.525%
		Electricity is used by roughly 30% of services and 30% of residential.[c] With regard to heat, cities can encourage district heating or biogas/electricity production from waste. Assume optimistically that cities can maximally affect at most half of services and residential electricity use, or $0.5 \times (30\% + 30\%) \times 25\% = 7.5\%$ of emissions	Cities mostly can use persuasive, not regulatory, instruments to reduce electricity use by consumers. Studies indicate a very low effectiveness of voluntary programs and soft policies. As indicated above, an effectiveness of 7% is applied
Industry	21%	1%	0.3%
		Mainly the domain of national policies, except waste management in some cities. The percentage is reflecting global share of waste management in total activity (1.5%),[d] the share of industry in total activity (30%), ratio of household to industrial waste (1:2), and share of households worldwide living in cities (0.6). The result is $(0.015/0.3) \times 1/3 \times 0.6 = 0.83\%$, which is rounded off to 1%	Cities can reduce waste in a very limited way through voluntary programs. Pricing waste bags has proven to be very complicated and ineffective. However, cities can aim for more recycling and energy recovery from unrecyclable waste. Cities do not, however, always control what happens in the waste phase. To capture these features, we optimistically assume a 30% effectiveness as a global average

(continued)

Table 29.1 (continued)

Sector	Share in global GHG emissions[b]	Maximally controllable emissions (share) due to limited reach of urban policies	Realistic share emissions control due to limited effectiveness urban policies
Transportation	14%	5.2%	0.794%
		60% of transportation emissions is from personal cars, rest is freight transport, aircraft, rail, and water (all uncontrolled by cities). [e, f] Most use of personal cars (in terms of distance) is outside urban borders.[g] Hence, cities control roughly less than half of 60% = 30%, which applied to 14%, gives 4.2% of global emissions. Fleet of city busses and garbage trucks represents roughly 1% of all traffic, which theoretically could be converted to 100% clean energy	Few cities ban cars or limit their use with congestion charges. Cities do not control fuel prices or car emission standards. They can discourage car use through zoning or parking tariffs. Promoting public transport has been found to have a limited effect on car use. For the 4.2% of private cars we use, as above, the 7% global average policy effectiveness and for the 1% city fleet, a 50% effectiveness (highly optimistic[h]), giving 0.294 + 0.5
Other (fuel extraction, refining, processing, and transportation)	10%	1.5%	0.183%
		This is a multiplier effect of the energy industry in total fuel use. If cities can reduce fuel consumption by roughly 15.3% (sum of the above estimates), then this could additionally reduce emissions 10% of that, i.e., 1.5%	Same multiplier effect as in left cell, i.e., 10% of sum of above estimates (1.824%)
Total	100%	16.8%	2%

Notes:

[a] The second and third columns makes many optimistic assumptions which are likely contribute to an upper bound of the overall estimates in the final row. The final column of the table reflects various facts: cities lack effective regulatory instruments such as technical emission standards or serious carbon pricing; a minority of cities will implement effective measures (e.g., banning cars in cities or severely limiting their use); and cities in majority rely on soft, weak policies (information provision, voluntary programs, loans, and small subsidies), which have been shown to have a low effectiveness in terms of emissions reduction (see text). To get from column 3–4 for cells that address consumer behavior (consumption, electricity use, heating, car use), a 7% average policy effectiveness for all global cities is applied, which is motivated by calculations in the main text
[b] IPCC (2014)
[c] EEA (2013)
[d] http://siteresources.worldbank.org.vu-nl.idm.oclc.org/INTURBANDEVELOPMENT/Resources/336387-1334852610766/AnnexE.pdf
[e] IEA (2016)
[f] https://www.epa.gov/greenvehicles/fast-facts-transportation-greenhouse-gas-emissions
[g] https://www.gov.uk/government/uploads/system/uploads/attachment_data/file/676205/Transport_section_Jan_2018_Final.pdf
[h] Ramaswami et al. (2012)

various other studies to show that these participation rates are in line with those reported in national-level reviews of outreach and loan programs. In addition, it is found that local government transport fleets contribute only 1.2% to total vehicle kilometers travelled in the city, causing even aggressive programs to retrofit engines to contribute to <1% of emissions reduction. Such low effectiveness numbers are confirmed by other studies (Millard-Ball, 2012). In view of this, an effectiveness of urban climate policies of 1% may be seen as realistic, while 5–10% is very optimistic and 20% extremely high, applying at best to the most ambitious cities.

To upscale to a global level, one must distinguish between ambitious and less ambitious cities. Indeed, one has to expect a great disparity in terms of mitigation ambitions, if only because climate concerns are a luxury for many cities in developing countries that are often entombed by urgent local challenges, such as poverty, crime, slums, lack of clean water, and inadequate road or electricity systems. Surveying 55 US cities that pledged GHG mitigation, it was found that less than 30% used some regulations, while the large majority relied on weak voluntary programs (Ramaswami et al., 2012). If we upscale this to the whole world and combine it with the afore-suggested effectiveness range 1–20%, then the average emissions reduction effectiveness of urban policies can be estimated as equal to $0.7 \times 0.01 + 0.3 \times 0.20 = 6.7\%$. We apply this percentage rounded off to above, 7%, to derive results related to private consumption, heating, electricity use, and car use as we move from the third to the fourth column in Table 29.1. This results in a realistic global contribution of city policies as 2%, which must be regarded as an upper bound because of consistently adopting optimistic assumptions, as indicated above and in the table. Another reason for interpreting this value as an upper bound is that the calculation in the table excludes considerations of systemic feedbacks, such as rebound and carbon leakage, which would reduce the global effectiveness of many urban policies.

Admittedly, the estimated range of 2–16.8% in the table is tentative and merits further study, ideally by employing a disaggregate and multiregional approach. However, the order of magnitude of the estimates is sobering, given that previous studies (mentioned in "Limited Reach of Urban Policies" section) suggest that cities are responsible for 69–75% of global energy-related emissions and that this percentage will even increase as urbanization continues. Note that the range is consistent with a rough estimate by another study, which suggested that cities could contribute up to 15% of global GHG reductions required to stay within 2 °C warming (Erickson & Tempest, 2014).

29.5 Concluding Remarks

This study was motivated by a lack of evidence on the overall contribution of cities worldwide to climate change mitigation. The arguments offered here and the tentative quantification provide a consistent picture, which moreover is in line with pessimistic conclusions of the few quantitative studies and reviews in the literature (Millard-Ball, 2012; Ramaswami et al., 2012; Stone et al., 2012; Boehnke et al., 2019; van den Bergh, 2020). According to Chapter 12 in IPCC's AR5, "Thousands of cities are undertaking climate action plans, but their aggregate impact on urban emissions is uncertain" (Seto et al., 2014). Two specific assessment reports contain more hopeful statements, but without providing any serious test of the effectiveness of urban climate policies (Rosenzweig et al., 2011, 2018).

Altogether, the evidence and arguments in these and the current study suggest that we should expect cities to fulfil a modest, and in the worst case a very small, role in reducing global GHG emissions. The main reasons are summarized in Fig. 29.1. This underpins that city strategies are no excuse for serious national regulation of emissions. In other words, national politicians remain most responsible for climate solutions. This is not to deny that cities should try, through adequate urban policies, to contribute maximally to GHG emissions reduction, as well as achieve adaptation to climate change to ameliorate negative consequences for their citizens. Some argue that cities possibly can serve as experimental labs for not-yet-tried-out climate policies (Bulkeley & Castán Broto, 2013), but also here evidence of a serious impact is lacking. Surely, city mayors are well positioned to diffuse a sense of

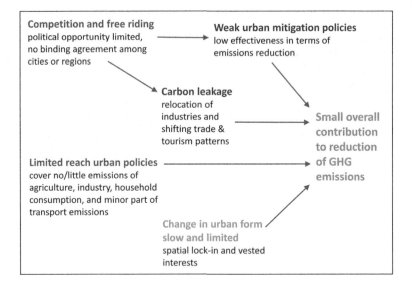

Fig. 29.1 Main reasons for a modest contribution of cities to global GHG emissions reduction

urgency about climate action. The most effective role that city councils and mayors can play, arguably, is lobby and cooperate with their national parties to achieve national climate policies that are consistently applied in all cities, thus avoiding policy competition and carbon leakage. In addition, they should encourage a binding global agreement on climate policies, not just targets (as the Paris Agreement). Only this would guarantee consistent and stringent, and hence effective, climate policies in all countries, cities, and other municipalities around the world.

Acknowledgments The author wishes to thank Isabelle Anguelovski, Wouter Botzen, Stefano Carattini, Maurie Cohen, Eric Galbraith, Jordi Roca, Gara Villalba, and Erik Verhoef for providing useful comments. The usual disclaimer applies. The research was supported by an advanced grant from the European Research Council (ERC) under the EU's Horizon 2020 research and innovation program (grant agreement no. 741087).

References

Bai, X., et al. (2018). Six priorities for cities and climate change. *Nature, 555*, 23–25.
Boehnke, R. F., Hoppe, T., Brezet, H., & Blok, K. (2019). Good practices in local climate mitigation action by small and medium-sized cities. *Journal of Cleaner Production, 207*, 630–644.
Bulkeley, H., & Castán Broto, V. (2013). Government by experiment? Global cities and the governing of climate change. *Transactions of the Institute of British Geographers, 38*, 361–375.
Ecología Política. (2014, June). "Ciudades", thematic issue of *Ecología Política* (Vol. 47).
EEA. (2013). *Final electricity consumption by sector EU-27*. https://www.eea.europa.eu/data-and-maps/figures/final-electricity-consumption-by-sector-5
Erickson, P. and Tempest, K. (2014). Advancing Climate ambition: How city-scale actions can contribute to global climate goals. SEI Working Paper No. 2014–06. Stockholm Environment Institute, . http://sei-international.org/publications?pid=2582.
Hale, T. (2016). "All hands on deck": The Paris agreement and nonstate climate action. *Global Environmental Politics, 16*(3), 12–22.
IEA. (2016). *Energy technology perspectives 2016: Towards sustainable urban energy systems*. International Energy Agency.
IPCC. (2014). *Contribution of working group III to the fifth assessment report of the intergovernmental panel on climate change*. O. Edenhofer, R. Pichs-Madruga, Y. Sokona, E. Farahani, S. Kadner, K. Seyboth, A. Adler, I. Baum, S. Brunner, P. Eickemeier, B. Kriemann, J. Savolainen, S. Schlömer, C. von Stechow, T. Zwickel, & J. C. Minx (Eds.). Cambridge University Press.
Kennedy, C., et al. (2009). Greenhouse gas emissions from global cities. *Environmental Science and Technology, 43*, 7297–7302.
Martínez-Alier, J. (2003). Scale, environmental justice, and unsustainable cities. *Capitalism Nature Socialism, 14*(4), 43–63.
Millard-Ball, A. (2012). Do city climate plans reduce emissions? *Journal of Urban Economics, 71*(3), 289–311.
Ramaswami, A., et al. (2012). Quantifying carbon mitigation wedges in U.S. cities: Near-term strategy analysis and critical review. *Environmental Science and Technology, 46*, 3629–3642.
Reckien, D., et al. (2014). Climate change response in Europe: What's the reality? Analysis of adaptation and mitigation plans from 200 urban areas in 11 countries. *Climatic Change Letters, 122*, 331–340.
Romero-Lankao, P. (2012). Governing carbon and climate in the cities: An overview of policy and planning challenges and options. *Journal of European Planning Studies, 20*, 7–26.
Rosenzweig, C., et al. (2010). Cities lead the way in climate-change action. *Nature, 467*, 909–911.

Rosenzweig, C., Solecki, W. D., Hammer, S. A., & Mehrotra, S. (Eds.). (2011). *Climate change and cities: First assessment report of the urban climate change research network*. Cambridge University Press.

Rosenzweig, C., Solecki, W. D., Romero-Lankao, P., Mehrotra, S., Dhakal, S., & Ibrahim, S. A. (Eds.). (2018). *Climate change and cities: Second assessment report of the urban climate change research network*. Cambridge University Press.

Satterthwaite, D. (2008). Cities' contribution to global warming: Notes on the allocation of greenhouse gas emissions. *Environment & Urbanization, 20*(2), 539–549.

Seto, K. C. et al. (2014). Human settlements, infrastructure and spatial planning. In *Climate change 2014 – mitigation of climate change. Contribution of working group III to the fifth assessment report of the intergovernmental panel on climate change*. Cambridge University Press. http://www.ipcc.ch/report/ar5/wg3/

Siskova, M., & van den Bergh, J. C. J. M. (2019). Optimal urban form for global and local emissions under electric vehicle and renewable energy scenarios. *Urban Climate, 29*, 100472.

Siskova, M., & van den Bergh, J. C. J. M. (2021). Are CO_2 emission targets of C40 cities realistic in view of their mayoral powers regarding climate policy? In S. Suzuki & R. Patuelli (Eds.), *A broad view of regional science: Essays in honor of Peter Nijkamp* (pp. 347–369). Springer.

Stone, B., Vargo, J., & Habeeb, D. (2012). Managing climate change in cities: Will climate action plans work? *Landscape and Urban Planning, 107*, 263–271.

Technopolis Group. (2013). Mid-term evaluation of the Covenant of Mayors, 6 February 2013, jointly with Fondazione Eni Enrico Mattei, Hinicio and Ludwig-Bölkow-Systemtechnik.

UITP. (2011). *Decarbonisation: The public transport contribution*. International Association of Public Transport. http://www.uitp.org/sites/default/files/Decarbonisation%20-%20the%20public%20transport%20contribution.pdf

UN. (2016). *The World's cities in 2016: Data booklet*. United Nations. http://www.un.org/en/development/desa/population/publications/pdf/urbanization/the_worlds_cities_in_2016_data_booklet.pdf

Ürge-Vorsatz, D., Rosenzweig, C., Dawson, R. J., Sanchez Rodriguez, R., Bai, X., Salisu Barau, A., Seto, K. C., & Dhakal, S. (2018). Locking in positive climate responses in cities. *Nature Climate Change, 8*(3), 174–185.

van den Bergh, J. (2020). Systemic assessment of urban climate policies worldwide: Decomposing effectiveness into 3 factors. *Environmental Science and Policy, 114*, 35–42.

Watts, M. (2017). Cities spearhead climate action. *Nature Climate Change, 7*, 537–538.

Open Access This chapter is licensed under the terms of the Creative Commons Attribution 4.0 International License (http://creativecommons.org/licenses/by/4.0/), which permits use, sharing, adaptation, distribution and reproduction in any medium or format, as long as you give appropriate credit to the original author(s) and the source, provide a link to the Creative Commons license and indicate if changes were made.

The images or other third party material in this chapter are included in the chapter's Creative Commons license, unless indicated otherwise in a credit line to the material. If material is not included in the chapter's Creative Commons license and your intended use is not permitted by statutory regulation or exceeds the permitted use, you will need to obtain permission directly from the copyright holder.

Chapter 30
Reconciling Waste Management and Ecological Economics

Ignasi Puig Ventosa

30.1 Introduction

Many times, I have heard Prof. Joan Martínez Alier saying: "The economy is entropic, not circular!". Thus, it has not always been easy to reconcile ecological economics, with the more conventional views coming from waste management, where the so-called "circular economy" is the new mantra.

Although waste management is not the core discipline of Prof. Martínez Alier, he has emphasised the importance of the analysis of material flows to understand how societies operate, and here the analysis of inputs (resources) and outputs (waste) are equally relevant.

He has also been involved in several conflicts related to waste management, mostly collected in the Environmental Justice Atlas. In some cases, he had firsthand involvement, as in the case of the dumping of hazardous waste in the Ebro River, in Flix, one of the areas of Catalonia he knows best for family reasons.

Besides material flow analysis and socio-economic conflicts, theories and practices on waste management have also benefited from at least two disciplines that have been central in the School of Barcelona. One is multicriteria analysis, as a way to deal with conflicting views (e.g. costs, environmental impacts, alternative location of facilities), which benefited from central contributions by Prof. Giuseppe Munda. Another one is the Multi-Scale Integrated Analysis of Societal and Ecosystem Metabolism (MuSIASEM), where the nexus between food, energy, water, land uses, urban metabolism and waste management are analysed. This approach was created by Mario Giampietro and Kozo Mayumi, and it has been applied to waste management by Giacomo D'Alisa (D'Alisa et al., 2012), Rosari Chifari, Samuele Lo Piano or Sandra Bukkens (Chifari et al., 2018), among others.

I. Puig Ventosa (✉)
ENT Environment and Management, Barcelona, Spain
e-mail: ipuig@ent.cat

These contributions have helped to understand waste management not only as a technical matter falling within the realm of engineering, as it was traditionally conceived, but also as a much more social discipline, where public participation and communication, economic analysis, public environmental policies, economic incentives and so on are fundamental.

Apart from this introduction, this chapter includes a brief discussion (Sect. 30.2) of the concept of circular economy, both globally, and in the context of the related EU policies.

Section 30.3 is the central part of the chapter and addresses a number of waste management policies, particularly based on the use of economic instruments, which proved efficient in achieving advances towards the higher tiers of the waste hierarchy (i.e. prevention, reuse and preparation for reuse and recycling). Some of the instruments presented are landfill and incineration taxes, environmental taxes on certain products, extended producer responsibility and fee-rebate schemes.

Finally, Sect. 30.4 presents the main conclusions.

30.2 A Circular Economy: Not Now, and Not Anywhere Soon

The concept of circular economy is easy to communicate, and it draws a certain parallelism with natural cycles. However, at present, the discourse on circular economy is largely disconnected from the imperatives coming from thermodynamics, and particularly from the analysis of the energy basis of the economy. In 2005, the global economy used 28 Gt/year of materials for energetic use (out of a total of 58 Gt/year of material extraction) (Haas et al., 2015). None of these materials – basically fossil fuels – are or can ever be recycled. This is, no doubt, the main criticism of the most common approach to a circular economy. A transition to a more circular economy must therefore include energy considerations as a priority and should run in parallel to a transition towards 100% renewable energy sources.

A second aspect is that current waste policies tend to focus on a rather narrow band of the material flows: especially municipal solid waste, with also some focus on industrial and construction and demolition waste. Although the targets are increasing in these areas, overall recycling remains low. At a global level, recycling was estimated to be around 6% in 2005 (Haas et al., 2015), whereas in the EU-27 it was 13%. This percentage was estimated to be around 15% in 2014 in the EU-28 when analysing only non-energy and non-food material flows (Nuss et al., 2017).

Focusing on the European Union (EU), in March 2020, the European Commission adopted the communication "A new Circular Economy Action Plan: For a cleaner and more competitive Europe" (European Commission, 2020), which succeeded the previous Circular Economy Package from 2015, which resulted in the revision of the main waste directives in 2018.

In particular, in 2018, the main objectives of the Waste Framework Directive (Directive 2008/98/EC) were revised,[1] and now a 65% recycling level of municipal waste has to be achieved by Member States by 2035, with intermediate levels by 2025 and 2030 (actual level was 47.0% in 2018; Eurostat, 2020). Although these objectives are ambitious, they are still far from circularity. If Member States comply with the objectives, in 2035, 35% of the materials will still escape from the cycle. Although we know that 100% is thermodynamically impossible, certainly, 65% still leaves a lot of room for improvement.

The Landfill Directive (Council Directive 1999/31/EC) was also revised in 2018. Its main objective, in art. 5.5, indicates that "Member States shall take the necessary measures to ensure that by 2035 the amount of municipal waste landfilled is reduced to 10% or less of the total amount of municipal waste generated (by weight)".[2] What ends up in a landfill certainly does not re-enter the economy.

Although waste incineration with (partial) energy recovery is better situated than landfills in the waste hierarchy, it is worth emphasising that in terms of the circularity of materials, it contributes exactly in the same way as landfills: what ends up in an incineration plant does not re-enter the economy either.[3]

At EU level, there are also some legal objectives regarding other important waste fractions, such as packaging waste (Directive 94/62/EC), waste from electric and electronic equipment (Directive 2012/19/EU) and construction and demolition waste (Directive 2008/98/EC), among others.

30.3 Sensible Waste Management Strategies

Traditional waste collection approaches based on environmental awareness and voluntary contributions to recycling centres and street containers for recyclables can continue to play a role, but more effective systems are needed.

Door-to-door (kerbside) separate collection schemes consistently achieve higher separate collection levels than bring schemes (Giavini et al., 2013), typically around 70–80% compared to 30–50%. This type of collection facilitates user identification and the application of pay-as-you-throw (PAYT) waste charges, thereby creating an incentive for users towards separate collection (Elia et al., 2015). Variants of pay-per-bag and pay-per-bin associated with kerbside collection are the most common PAYT schemes (Puig Ventosa et al., 2013a).

Where door-to-door is not applied, user identification is also possible using smart containers. Many possible alternatives can be applied depending on whether identification is made voluntary (open containers) or compulsory (closed containers),

---

[1] https://eur-lex.europa.eu/legal-content/EN/TXT/?uri=CELEX:02008L0098-20180705
[2] https://eur-lex.europa.eu/legal-content/EN/TXT/HTML/?uri=CELEX:01999L0031-20180704&from=EN
[3] Except for some metals recovered from the slags.

on how waste generation is measured (either volume or weight), on the waste fractions to control, on the type of technology, and so on (Saleh et al., 2019).

The question of quality should also gain importance. Separate collection is not a goal per se, it is only a means for recycling. However, truly recycling, not just *downcycling*, can only take place if the materials undertaking the recycling process are sufficiently pure, and that normally requires efficient separate collection.

In the case of biowaste (but this could also be argued for other waste fractions), the presence of impurities is statistically explained by several factors. Besides urban density and the requirement to use compostable bags, one of the factors is the type of collection system. Door-to-door collection schemes achieve the lowest average levels of impurities (Puig Ventosa et al., 2013b). This is relevant not only because the presence of impurities in biowaste causes problems during composting or anaerobic digestion (additional costs derived from the need to improve pre-treatment of biowaste and loss of treatment capacity), but specially because it has a direct impact on the quality of the output.

The quality of compost obtained in biowaste treatment facilities depends upon several variables associated with the technical specifications of the plants. However, it has been proved that the presence of improper materials (glass, plastic, etc.) leads to a negative impact on various parameters of the quality of compost, notably the concentration of heavy metals (Cu, Pb and Zn) (Rodrigues et al., 2020).

Ultimately, this is relevant because the cycle is only closed if this compost is applied to soil. Although the benefits of compost application to soil are proven (Gilbert et al., 2020), it can also cause soil pollution, if its quality is not sufficient, due to the presence of heavy metals, or due to the presence of microplastics (Lin et al., 2020).

Most European countries have landfill taxes in place for municipal waste. Some also have incineration taxes (Watkins et al., 2012; Fischer et al., 2012). These taxes are normally paid by Local Authorities in charge of waste collection, and they encourage them to adopt strategies to divert waste from landfills and incinerators, which normally include improving separate collection of biowaste and recyclables, as well as stabilising non-separately collected waste before disposal (e.g. via mechanical–biological plants).

The effectiveness of these taxes is undeniable, and they normally constitute a central piece of the waste management strategies of the countries or regions that have them in place, especially when their revenue is earmarked and dedicated to the implementation of preventive waste policies (e.g. in Catalonia; Puig Ventosa et al., 2012). However, landfill and incineration taxes do not significantly contribute to moving to the highest tiers of the waste hierarchy: advancing towards prevention and preparing for reuse implies a profound rethinking of the production and distribution strategies by the industry, and the industry is not affected by the incentives created by the landfill and incineration taxes.

Banning non-recyclable products and establishing requirements on the minimum content of recycled materials could also be positive measures in this direction, along with specific taxes on products causing significant environmental impacts (e.g. plastic bags as in Ireland (Anastasio & Nix, 2016) and disposable packaging as in Norway (Infinitum, 2020)).

Another important concept is that of extended producer responsibility (EPR). EPR "aims to make producers responsible for the environmental impacts of their products throughout the product chain, from design to the post-consumer phase" (OECD, 2016).

However, after decades of implementation, EPR has significant limitations. Public Administrations continue to sustain part of the costs that should be borne by producers even where EPR schemes exist (e.g. littering), and EPR schemes do not sufficiently incentivise recyclability and eco-design amongst individual producers (OECD, 2016; Zero Waste Europe, 2015).

Nevertheless, the main current limitation of EPR is that it is reduced to a very limited number of products. For example, within the EU, the application of EPR is only common for waste from electric and electronic equipment, batteries and accumulators, end-of-life vehicles, packaging waste, oils and tyres. For most products, it simply does not exist, and producers are allowed to put any product on the market, no matter how difficult and costly to manage when it turns into waste.

In some cases, products without EPR are a significant percentage of waste generation (such as for the case of graphic paper, furniture or textiles) or are very environmentally problematic and costly to manage (such as disposable nappies, sanitary pads, cleaning wipes, mattresses, cigarette buds, and chewing gum).

It seems unlike that extending EPR in a product-by-product approach will be able to cover a broad range of products. Thus, a new concept of generalised extended producer responsibility (GEPR) is suggested. In this case, all products put on the market would be subject to EPR, with the only exception of biowaste, which is linked to basic human metabolism.

Specific EPR schemes could continue to exist, and maybe a few more could be created, but there would also be a general scheme for all those products with no specific EPR scheme associated. GEPR could generalise incentives towards recyclability and cleaner production, particularly if some lessons are learned from past EPR experiences, and would suppose a much fairer distribution of costs, shifting them from Public Administrations to producers, and ultimately from taxpayers to consumers.

A particular form of materialising EPR is through the use of deposit-refund schemes (DRS). Although they can potentially be applied to other items, its main application so far has been on packaging, mainly beverage bottles and cans. These products are sold with a deposit, which is refunded when the empty packaging is returned. This ensures a high level of return (typically around 80–90%; Fletcher et al., 2012), sensibly higher than that achieved in street containers for packaging waste. Around 40 of these schemes are applied in different jurisdictions around the world, mainly in the EU, USA, Canada and Australia (Zhou et al., 2020). DRS can be applied to both disposable items (to ensure collection and high-quality recycling) and reusables (e.g. glass bottles are cleaned, sanitised and refilled).

There are other economic instruments with significant potential whose application has been so far very limited, such as feebate systems or Landfill Allowance Trading Schemes (LATS):

- Feebate systems make a simultaneous use of fees and rebates. Activities that take less care of the environment compared to the average are charged fees, whereas the most ecological ones receive rebates, making them more economically attractive compared to the initial situation. The more environmentally harming an activity, the higher the fee, and vice versa. Activities with the average environmental performance are neither charged nor subsidised. Globally, fees and rebates cancel each other out, and therefore beyond the administrative costs this tool is neutral for the budget of the administration that sets it up.
- Most municipalities group themselves to manage solid wastes more efficiently, sharing services and facilities. In these associations of municipalities, costs are allocated to each municipality according to some criteria (e.g. number of inhabitants or amount of waste brought to the shared facilities), which often do not provide sufficient incentives for good practices. In this context, a feebate system could be adequate to reward those municipalities making significant steps towards ecological waste management (less per capita generation, higher separate collection rates, higher biowaste quality, etc.), whilst penalising others, using the average values as a reference.
- The articulation of this instrument was proposed theoretically (Puig Ventosa, 2004), and it was successfully applied in the Metropolitan Area of Barcelona (Puig Ventosa, 2006), from 2004 to 2017.
- Landfill Allowance Trading Schemes (LATS) are useful instruments to achieve landfill diversion targets. Through LATS, allowances for landfilling of municipal solid waste (or the biodegradable part of it) are allocated to Local Authorities. The quantity of landfill allowances assigned globally is reduced annually in order to meet the landfill diversion objectives. To achieve their commitments, Local Authorities can exchange allowances among each other, or may reprofile their own allocation through banking or borrowing. Although tradable permits have been largely applied as part of climate policies, the application in the area of waste management is very limited. The main experience was the application of LATS in the United Kingdom, as an instrument to comply with Directive 1999/31/EC on the landfill of waste. Starting in 2005, it had a successful application, but it was finally abandoned in 2012 when it became redundant with the UK Landfill Tax (Calaf-Forn et al., 2014).

30.4 Conclusions

In our society, waste seems the unavoidable consequence of production and consumption. To a certain extent, waste is inevitable. Even nature produces waste. In fact, our current civilization is built entirely around one of these – fossil fuels. However, most waste could be prevented, and when not mostly recycled, as nature does.

Nevertheless, despite the advances in the political framework, especially in some regions like the EU, most of the waste is dealt with inappropriately, largely relying on incineration and landfilling, which are the lowest tiers of the waste hierarchy. Although there is a lot of room for improvement in the adoption of good practices voluntarily by citizens and industry, a main driver should come from environmental policies, and particularly from the use of economic instruments. The discipline of ecological economics has a lot to offer in terms of not only to better understand how and why waste generation is produced, but also to inform the design of such policies in the best possible way.

Waste management policies cannot only focus on the outputs of the system but also on the inputs, putting eco-design in a pre-eminent place. No recycling is possible for non-recyclable products, and still, many of them enter the market without assuming their true cost.

It is about aligning the environmental discourse with the economic. Doing things which are bad for the environment (i.e. going for the lowest tiers of the waste hierarchy) cannot come out cheaper. In this regard, advances in EPR and environmental taxation should be adopted.

Finally, the emphasis on circularity should not make us forget about the importance of the "size of the circle". Reaching higher recycling levels does not necessarily reduce environmental pressures if the system requires an increasing amount of raw materials. A broader economic vision is needed to reach social objectives, but within a framework of sufficiency.

References

Anastasio, M., & Nix, J. (2016). *Plastic bag levy in Ireland*. https://ieep.eu/uploads/articles/attachments/0817a609-f2ed-4db0-8ae0-05f1d75fbaa4/IE%20Plastic%20Bag%20Levy%20final.pdf?v=63680923242

Calaf-Forn, M., Roca, J., & Puig-Ventosa, I. (2014). "Cap and trade schemes on waste management: A case study of the Landfill Allowance Trading Scheme" (LATS) in England. *Waste Management, 34*, 919–928.

Chifari, R., Lo Piano, S., Bukkens, S. G. F., & Giampietro, M. (2018, November). A holistic framework for the integrated assessment of urban waste management systems. *Ecological Indicators, 94*, 24–36. https://doi.org/10.1016/j.ecolind.2016.03.006

D'Alisa, G., Di Nola, M. F., & Giampietro, M. (2012). A multi-scale analysis of urban waste metabolism: Density of waste disposed in Campania. *Journal of Cleaner Production, 35*, 59–70.

Elia, V., Gnoni, M. G., & Tornese, F. (2015). Designing Pay-As-You-Throw schemes in municipal waste management services: A holistic approach. *Waste Management, 44*, 188–195.

European Commission. (2020). *Communication from the Commission to the European Parliament, the Council, the European Economic and Social Committee and the Committee of the Regions. A new Circular Economy Action Plan: For a cleaner and more competitive Europe*. COM/2020/98 final https://eur-lex.europa.eu/legal-content/EN/TXT/?qid=1583933814386&uri=COM:2020:98:FIN

Eurostat. (2020). *Recycling rate of municipal waste*. https://ec.europa.eu/eurostat/databrowser/view/cei_wm011/default/table?lang=en

Fischer, C., et al. (2012). *Overview of the use of landfill taxes in Europe* (ETC/SCP Working Paper 1/2012). European Topic Centre on Sustainable Consumption and Production.

Fletcher, D., Hogg, D., von Eye, M., Elliott, T., & Bendali, L. (2012). *Evaluación de costes de introducción de un sistema de depósito, devolución y retorno en España* (Informe final para Retorna). Eunomia.

Giavini, M., Garaffa, C., Favoino, E., & Petrone, P. (2013). *Capture rates of source separated organics: A comparison across EU, with a focus on metropolitan areas*. ISWA World Congress.

Gilbert, J., Ricci-Jürgensen, M., & Ramola, A. (2020). *Benefits of compost and anaerobic digestate when applied to soil*. International Solid Waste Association.

Haas, W., Krausmann, F., Wiedenhofer, D., & Heinz, M. (2015). How circular is the global economy? An assessment of material flows, waste production, and recycling in the European Union and the World in 2005. *Journal of Industrial Ecology*. https://doi.org/10.1111/jiec.12244

Infinitum. (2020). *The environmental tax system*. https://infinitum.no/english/the-environmental-tax-system

Lin, D., Yang, G., Dou, P., Qian, S., Zhao, L., Yang, Y., & Fanin, N. (2020). Microplastics negatively affect soil fauna but stimulate microbial activity: Insights from a field-based microplastic addition experiment. *Proceedings of the Royal Society B. Biological Sciences*. https://doi.org/10.1098/rspb.2020.1268

Nuss, P., Blengini, G. A., Haas, W., Mayer, A., Nita, V., & Pennington, D. (2017). *Development of a Sankey diagram of material flows in the EU economy based on Eurostat data* (JRC Technical Reports). https://ec.europa.eu/jrc/en/publication/development-sankey-diagram-material-flows-eu-economy-based-eurostat-data

OECD. (2016). *Extended producer responsibility: Updated guidance for efficient waste management*. OECD Publishing. https://doi.org/10.1787/9789264256385-en

Puig Ventosa, I. (2004). Potential use of feebate systems to foster environmentally sound urban waste management. *International Journal of Integrated Waste Management, 24*, 3–7.

Puig Ventosa, I. (2006). Fee and rebate systems to foster ecologically sound urban waste management. In A. Cavaliere, H. Ashiabor, K. Deketelaere, L. Kreiser, & J. Milne (Eds.), *Critical issues in environmental taxation: International and comparative perspectives* (Vol. Vol. III, pp. 527–534). Richmond Law & Tax.

Puig Ventosa, I., González, A. C., & Jofra Sora, M. (2012). Landfill and waste incineration taxes in Catalonia, Spain. In L. Kreiser, A. Yábar, P. Herrera, J. E. Milne, & H. Aishabor (Eds.), *Green taxation and environmental sustainability* (Critical issues in environmental taxation) (Vol. Vol. XII, pp. 244–257). Edward Elgar.

Puig Ventosa, I., Calaf Forn, M., & Mestre Montserrat, M. (2013a). *Guide for the implementation of Pay-As-You-throw systems for municipal waste*. Agència de Residus de Catalunya. http://www20.gencat.cat/docs/arc/Home/LAgencia/Publicacions/Centre%20catala%20del%20reciclatge%20(CCR)/Guia%20PXG_EN.pdf

Puig Ventosa, I., Freire Gonzàlez, J., & Jofra-Sora, M. (2013b). Determining factors for the presence of impurities in selectively collected biowaste. *Waste Management & Research, 31*, 510–517.

Rodrigues, L. C., Puig-Ventosa, I., López, M., Martínez, F. X., Garcia Ruiz, A., & Guerrero Bertrán, T. (2020). The impact of improper materials in biowaste on the quality of compost. *Journal of Cleaner Production, 251*, 119601.

Saleh, D., Salova, M., Bulbena, B., Loderus, T., Calaf, M., & Puig, I. (2019). *User identification for municipal waste collection in high-density contexts*. ENT Environment & Management – Universitat Politècnica de Catalunya. https://ent.cat/wp-content/uploads/2019/07/User-identification-for-municipal-waste-collection_4.pdf

Watkins, E., et al. (2012). *Use of economic instruments and waste management performances*. DG Environment. European Commission. http://ec.europa.eu/environment/waste/pdf/final_report_10042012.pdf

Zero Waste Europe –Fundació per a la Prevenció de Residus i el Consum Responsable. (2015). *Redesigning producer responsibility. A new EPR is needed for a circular economy*. https://www.zerowasteeurope.eu/downloads/redesigning-producer-responsibility-a-new-epr-is-needed-for-a-circular-economy

Zhou, G., Gu, Y., Wu, Y., Gong, Y., Mu, X., Han, H., & Chang, T. (2020). A systematic review of the deposit-refund system for beverage packaging: Operating mode, key parameter and development trend. *Journal of Cleaner Production, 251*, 119660. https://doi.org/10.1016/j.jclepro.2019.119660

Open Access This chapter is licensed under the terms of the Creative Commons Attribution 4.0 International License (http://creativecommons.org/licenses/by/4.0/), which permits use, sharing, adaptation, distribution and reproduction in any medium or format, as long as you give appropriate credit to the original author(s) and the source, provide a link to the Creative Commons license and indicate if changes were made.

The images or other third party material in this chapter are included in the chapter's Creative Commons license, unless indicated otherwise in a credit line to the material. If material is not included in the chapter's Creative Commons license and your intended use is not permitted by statutory regulation or exceeds the permitted use, you will need to obtain permission directly from the copyright holder.

Chapter 31
Work and Needs in a Finite Planet: Reflections from Ecological Economics

Erik Gómez-Baggethun

31.1 Ecological Economics and Concrete Utopias

Utopia, Olin Wright (2010) notes, evokes fantasy, aspirations for a better world unconstrained by realistic considerations of human phycology and social feasibility. Political realism rejects such fantasies, arguing for accommodation to practical realities and pragmatic improvement of institutions. The ideas of 'real' and 'concrete utopias' address this tension between practice and dreams, by paying attention to feasibility and constraint while emphasizing that notions of the possible are themselves shaped by our ability to envision alternative futures (Wright, 2010: 5–6; Archer, 2019).

Drawing on the work of Joan Martínez Alier (1992; Martínez Alier & Schulpmann, 1987), this chapter draws attention to the contribution of ecological economics to the envisioning of concrete utopias (see also Ingebrigtsen & Jakobsen, 2012; Kallis & March, 2015; Gomez-Baggethun, 2020; Mair et al., 2020). To Martínez Alier (1992), concrete utopias represent radical visions to the future, but not impossible ones. They are utopian because they assume radical political change without explaining how it would come about, but they are plausible because they are elaborate and concrete, and because they acknowledge the constraints of social and ecological realities, hence having a chance of coming into being.

Specifically, this chapter explores alternative pathways to the transformation of work in the context of environmental limits to growth, a theme that has inspired ecological economists for more than a century (Popper-Lynkeus, 1912; Mair et al., 2020). It is organized in four main parts. First, I call attention to cultural, economic,

E. Gómez-Baggethun (✉)
Department of International Environment and Development Studies, Faculty of Landscape and Society, Norwegian University of Life Sciences (NMBU), Ås, Norway

Norwegian Institute for Nature Research (NINA), Oslo, Norway
e-mail: Erik.gomez@nmbu.no

technological and environmental changes that are destabilizing established conceptions of work. Second, I review alternative visions on the future of work, with an emphasis on the case for a Universal Basic Income. Next, I discuss common criticisms addressed to these visions. Finally, drawing on early ecological economics texts rescued from oblivion by Martínez Alier (1992; Martínez Alier & Schulpmann, 1987), I discuss the case for a Universal Civil Service, a variant of basic income premised in the egalitarian distribution of the volume of work required for the reproduction of society, with attention to ecological and resource limits. I call for expanding traditional emphasis in basic income debates on individual freedoms towards considerations of collective justice, and I make a case for a future of work organized around the principle of fair distribution of minimal necessary work.

31.2 The End of Work as We Know It?

Today's common understanding of work as 'paid labour' is a product of industrial capitalism that bears little in common with those that prevailed in other times and cultures (Gorz, 1988). It was introduced in the eighteenth century by the time economics took form as a discipline, and it consolidated a century later through legal codification (Komlosy, 2018). Paid labour has become not only the main means by which humans fulfil their needs but also a major pathway to social integration and a key marker of status and identity (Ackerman et al., 1998, Ehmer & Lis, 2009).

In recent decades, however, traditional understandings of work in industrialized countries are being destabilized by the compounding effects of accelerated cultural, economic, technological and environmental change (Gomez-Baggethun & Naredo, 2020). First, the relocation of industry in developing countries and the reorganization of global commodity chains with economic globalization has brought a deregulation and flexibilization of labour in the old industrialized countries. These changes have in turn changed attitudes towards work and, more generally, into ideas about work and life. For a growing amount of people in the developed countries, especially among the employed in low paid and unskilled labour, work is no longer the focus of their life, neither the main marker of their identity (Meda, 2010).

Second, digital capitalism and automation have drastically reduced the need for work (Rifkin, 1995) and are loosening the relationship between value, work and wages (Mason, 2016). This has turned much working force dispensable, raising fears of massive unemployment (Ford, 2017, OECD, 2019). Third, as current patterns of production and consumption prove incompatible with global sustainability targets (Wiedmann et al., 2020), the sustainability sciences call for reduced working hours as a key policy measure for a low carbon future (Schor, 2005; Kallis et al., 2013; Knight et al., 2013).

Finally, the COVID-19 pandemic has revolutionized worker expectations on flexibility, bringing transformations to work expected to endure beyond the pandemic (Liang, 2020).

An emerging consensus is gradually building around the notion that the current organization of work is not fit to meet the challenges emerging with current ecological and economic realities. In the face of these challenges, the literatures on ecological economics and degrowth are debating transformations pathways towards economies where the basic needs of all are met within just and safe planetary boundaries (Daly, 1996; Jackson, 2017; Raworth, 2017; Gómez-Baggethun 2022). A central question for these debates concerns the volume of work required to meet basic needs for all in a finite planet, and how work is to be organized and distributed in a post-growth future (Mair et al., 2020).

31.3 Reducing Work in the Name of Freedom

The notion that freedom starts where necessity ends and the aspiration of reducing work to enable more time for creative leisure, contemplation, self-realization and public life has been a constant in humanity from Plato to our times (Arendt, 1998; Komlosy, 2018).

Aristotle (350 B.C.E/1998) fantasized with an age where machines dispensed humans from work and slavery.[1] Marx (1894/1991: 958–959) famously wrote that the realm of freedom begins only where labour determined by necessity ends. Lafargue (2012) attacked the glorification of work by priests, economists and moralists, encouraging the working class to shift the focus of their demands from work to leisure. Keynes (2010) envisioned a future of leisure, predicting 15-h workweeks for his grandchildren. Arendt (1998) made the case for *vita activa*, a life of action where liberation from the toil of labour would enable time for broader involvement in politics and common causes. Following her steps, Gorz (1988) advocated work reduction to expand people's freedom and ability to pursue self-realization and involvement of social, community and political life.

The locus in the case for work-time reduction is that gains in work productivity from technological developments should be used to expand leisure, not consumption (Coote et al., 2010). Proposals to reduce the standard 40-h workweek and 1500–2000-h work year in developed economies range from the 6-h workday and the 32-h workweek advocated by many labour unions and green political parties, to more radical propositions like the 2–3 h workday (Lafargue, 2012; Ellul, 1954) and the 15-h workweek (Bregman, 2017; Stronge et al., 2019).

The literature on reduced working time often advocates a 'life course approach', demanding rights to flexibly reduce working time at different periods of the working life, with or without associated reductions in income (Pullinger, 2014). Gorz (1988) advocated limiting lifespan work to 20,000–30,000 h, allowing people to distribute workload over time according to need and preference. This could be

---

[1] 'If every tool […] could do the work that befits it […], if the weavers' shuttles were to weave of themselves, then there would be no need either of apprentices for the master workers, or of slaves for the lords' (Aristotle, Politics, I.4 1253 b33-1254a1).

achieved by a combination of measures, including regulations of maximum working hours and increases in minimum holiday entitlements and other statutory paid leaves (Hayden, 1999). In its maximalist strands, advocates of work reduction aspire to a post-work society, in which all tedious work is fully automated (Rifkin, 1995; Frayne, 2015).

A related body of literature shifts focus from quantitative to qualitative work transformations, envisioning futures where unpleasant, monotonous and tedious work is turned into more attractive (Fourier, 1901) and meaningful activities (Schwartz, 1982), or into some kind of pleasurable tasks where the lines between work and leisure get blurred (Black, 1985).

Proposals to increase freedom from work have long pointed to loosening or breaking the links between wages and labour (Gorz, 1988; Parijs, 1997). Most prominent among these proposals is the case for a Universal Basic Income (hereafter UBI), defined as a periodic cash payment unconditionally delivered to all on an individual basis, without means-test or work requirement (Bregman, 2017). Once a marginal vision of radical thinkers, the case for a UBI is gaining traction across the ideological spectrum, becoming one of the most influential proposal to modernize welfare states and reform capitalism (Downes & Lansley, 2018). In a juncture without precedent, financiers, Silicon Valley entrepreneurs, Nobel laureates in economics and anti-capitalists alike converge in the case for a basic income (Mason, 2016; Ford, 2017). Pilot schemes are being tested across the world, and countries such as Brazil and Spain are currently experimenting with temporary variants of basic income policies in response to threats of massive poverty and unemployment triggered by the COVID-19 pandemic (Fariza, 2020).

31.4 Work Utopias and Their Limits

Post-work futures and the case for a basic income have been criticized on the basis of various technological, psychological, economic, political and ecological considerations.

First, a long-standing line of criticism (see, e.g. Ellul, 1954) concerns an alleged excess of idealism, tacit in the assumption that time gained to work would primarily go into meaningful and laudable activities, such as arts and creative leisure or involvement in public life. Critique to UBI along these lines often revolve around fears that much free time gained from work would go to activities of no or dubious social value. This line of criticism raises some troubling questions. Is Hanna Arendt's *vita activa* a more plausible outcome from work-time reductions than a *vita pasiva*? Isn't passive life increasingly favoured by a massively expanded industry of digital entertainment (e.g. social media) and by technological gadgets that exploit human propensity to comfort (Wu, 2018)? Whether or not these fears came true, should the content of leisure be judged and administered? Who would decide what is 'good' and 'bad' leisure? In doing so, would the prevailing 'work ethic' be replaced by an equally moralizing 'leisure ethic'?

A second line of criticism concerns a long-standing reluctance to decouple work from income. Two main concerns are raised in this regard. First, the fear that doing so would erode incentives to work. Second, the widely held belief that income should bear some relation with contribution and effort. While this argument is often associated to liberal discourses on meritocracy, the idea of justice underlying this argument has footing across the ideological spectrum. In fact, this consideration lays at the basis of Marx's theory of exploitation, and in his critique of the capitalist appropriation of the surplus value produced by workers,[2] as well as in the feminist case for the paid compensation of household work (Federici, 1975).

A third line of criticism concerns an alleged excess of optimism regarding the scope for work-saving technology. As noted by Black (1985), historical records are in fact not encouraging. Despite sustained gains in productivity, industrialism's promise of work liberation through technological progress never came about. Historians and anthropologists contend that human toil increased from hunting–gathering to agrarian societies (Harari, 2014), and the industrial revolution brought the longest working weeks known to human history (Schor, 2008). Technological achievements of the digital revolution seem equally discouraging. Smartphones and other communication technologies have facilitated work's encroachment on leisure (Wajcman, 2015), de facto extending working hours in many economic sectors (Derks & Bakker, 2014). Second, digitalization has brought a new tide of 'shadow work' (Illich, 1981), shifting upon us many unpaid tasks (e.g. check-out in supermarkets, assemblage of furniture and online bookings) that were previously paid for and now parasite our time (Lambert, 2015).

The obvious but often unrecognized problem is that technology will not liberate us on its own. Left to the dictates of capital, productivity dividends will keep serving economic expansion over the contraction of work in many parts of the world. Productivity gains will be largely absorbed by robotization, while capitalist economies will keep compensating for the labour-displacing effects of productivity gains by expanding existing industries and creating new ones (Manyika et al., 2017), often of dubious social utility (Graeber, 2013). Work-time reductions through technology seem only likely if combined with concomitant changes in the institutions steering the allocation of the productivity dividends.[3] The great reductions of working time achieved in the early decades of the twentieth century did not stem from technological productivity gains alone, but from their combination with sustained pressure from organized labour.

---

[2] Sen (2009) notes that this notion of merit-based justice co-exists in Marx's work with a conflicting notion of egalitarian justice, expressed in the famous motto of *The critique of Gotha program*: 'From each according to his ability, to each according to his needs' (Marx, 1875).

[3] This view is criticized by Ellul (1954), who rejects the view that technological developments can be harnessed by extracting technology from the dictates of a capitalist economy and re-embedding into a planned economy. To Ellul, the direction and outcomes of technological change respond primarily to autonomous mechanisms of the technological system rather than to the political institutions in which it is embedded.

Critiques extend to the anti-work and post-work visions that assume that most work can be either automated or turned into pleasurable activities. First, the idea of turning work into pleasure has in fact a long history of co-option for the purpose of increasing working time and effort (Friedmann, 1963). As an example, today's tech giants in Silicon Valley such as Google and Facebook fill workplaces with amenities, persuading their employees that round-the-clock work is fun. Furthermore, they convert the leisure time tech-users spend on their screens into unpaid 'digital labour' monetized as targeted advertising (Fuchs, 2015). The work-day of digital labourers starts as we pick up our phone and begin generating data (Arrieta Ibarra et al., 2018). Second, in a low carbon economy with decreased used of fossil fuels, productivity gains can be much harder to achieve, with some projections pointing in the opposite direction (Mair et al., 2020).

Finally, and central to our discussion, there is the question of distributive justice. In her critique of UBI, Mestrum (2018: 97) notes that 'individual freedom is extremely important, but can never be dissociated from collective responsibility'. In effect, some prominent work utopias tend to downplay the fact that there will always be a share of *necessary* work that hardly can be mechanized (e.g. emotional care) or turned pleasant (e.g. sewage cleansing). The less some people do of this job, the more others will have to take up their share. If we accept this premise, a fundamental question concerns the problem of fair distribution of socially necessary work.

31.5 The Case for Sharing Minimal Necessary Work

The case for the equitable distribution of socially necessary work has a long-standing tradition of thought. We find it in the work of John Stuart Mill (1850)[4] and Bertrand Russel (1935), and it has a long footing in the feminist (Waring, 1988) and work sharing literatures (Hayden, 1999). To the reach of my knowledge, one of the most elaborated proposals along these lines is the case for a Universal Civil Service (UCS) by early ecological economist Joseph Popper-Lynkeus (1912), whose writings have been rescued from oblivion by Martínez-Alier (1992).

In the *Universal Civil Service as a Solution of a Social Problem* (1912), Popper-Lynkeus wrote about how an ecologically viable economy could cover the basic needs of all individuals based on a drastically reduced and evenly distributed *necessary* working time. Popper-Lynkeus grounded his vision on detailed accounts of available resources, with the double objective of calculating the human work required to guarantee basic needs and to investigate how consumption of exhaustible resources could be reduced. Rather than the end of work, the focus of his utopia was on how to define, distribute and organize the work required to meet basic needs

---

[4] Mill writes: 'To reduce very greatly the quantity of work required to carry on existence is as needful as to distribute it more equally' [...] 'There is a portion of work rendered necessary by the fact of each person's existence: no one could exist unless work, to a certain amount, were done either by or for him. Of this each person is bound, in justice, to perform his share'.

for all. This involved providing all individuals with goods and services of prime necessity such as food, clothes, housing, public health care, upbringing and education, a vision that is consistent with modern theories on basic needs (see, e.g. Chiappero-Martinetti, 2014).

The economic system he proposed consisted of two parts. The first part would be a collectivized economy governed by public authorities oriented to secure basic needs for all individuals. This would be achieved by means of a civil (instead of military) conscription that he calculated at a 12-year service for men and seven for women, with a 35-h week. Dividing equally by gender, this amounts to a working lifespan of less than 10 years and less than 20,000 h to all healthy members of society.[5] This civil service would entitle everyone to the right to cover necessities for their entire life course free of charge (i.e. to a basic income). After completion of the UCS, working would be an option, but no longer a necessity for survival. The second part of the economy would be market-driven, governed by supply and demand. After completing the UCS, those who wish could continue working in economic activities of the public or private sectors, either as hired employees or as free entrepreneurs. Importantly, he foresaw that with productivity increase from technological change, the duration of the UCS would gradually decrease, while the scope of the concept goods and services of primary necessity would expand.

Like any other work utopias, the case for Universal Civil Service is prone to criticism from across the ideological spectrum. Liberal and conservative status quo proponents shall dismiss it as too radical and unrealistic (as they already did in times Popper-Lynkeus), but elements of this vision can be found in the New Deal and in the universal services of Nordic welfare states. Socialists shall contend that the market-driven segment of the economy would lead to inequalities of income, wealth and power. This is true, but additional measures could be put in place to secure further redistribution of wealth, including progressive taxes on income, wealth and profits, and the enforcement of maximum–minimum income ratios (Alexander, 2014). Conspicuous consumption could be disincentivized with green taxation, while further public income could be raised from taxes on capital (including robots) and from Tech giants as social dividends for 'digital labour' (Arrieta-Ibarra et al., 2018), which could be then redistributed in the form of public goods and services. Anarchists shall contend that an army of bureaucrats would be needed to enforce and police the Universal Civil Service. This may again be true, but it is already the case in current versions of welfare states that not only rely on excessive work, resources and waste but that also fail to cover basic needs for all. A variant could be to make the public service voluntary, allowing individuals to object the service and give up the corresponding basic income. Furthermore, the case for work time reduction and the fair distribution of social necessary work (not organized by states but through free association of worker organizations) is also to be found in the anarchist literature (e.g. Kropotkin, 1892/2015).

---

[5] Consider by comparison to today's 80,000–90,000 h of work in developed countries, assuming the average 1734 annual work hours in OECD countries and a standard 50-year active working life.

Like other work utopias the UCS is no panacea, yet it could help addressing some recurrent criticisms to the UBI. First, the collectivized and planned sector of the economy could give larger for room of manoeuvre to steer the direction of technological developments, allocating productivity dividends away from growth and towards reduced working time and resource use. Second, concerns about 'inappropriate' use of leisure time would relax; upon completion of the UCS, people would have *earned* their UBI. Like today's pensioners, people could choose an active or passive life free of judgement form others. Unlike them, they would be making this choice after working, say 10 years instead of 50 in order to secure a decent living. Most importantly for our discussion, concerns about fair distribution of work would be addressed, as every healthy person would contribute a share to cover societal needs.

Acknowledgements This research received partial funding from the NMBU Sustainability Arenas 2021–2014 (grant number 1850092016AA).

References

Ackerman, F., Goodwin, N. R., Dougherty, L., & Gallagher, K. (1998). In F. Ackerman (Ed.), *The changing nature of work*. Island Press.
Alexander, S. (2014). Basic and maximum income. In G. D'Alisa, F. Demaria, & G. Kallis (Eds.), *Degrowth: A vocabulary for a new era* (pp. 146–149). Routledge.
Archer, M. S. (2019). Critical realism and concrete utopias. *Journal of Critical Realism, 18*(3), 239–257.
Arendt, H. (1998). *The human condition*. University of Chicago Press. First published in 1958.
Aristotle, D. (1998). *Politics*. Hackett Publishing Company.
Arrieta-Ibarra, I., Goff, L., Jiménez-Hernández, D., Lanier, J., & Weyl, E. G. (2018). Should we treat data as labor? Moving beyond "free". *Papers and Proceedings, 108*, 38–42.
Black, B. (1985). *The abolition of work*. Loompanics Unlimited.
Bregman, R. (2017). *Utopia for realists: And how we can get there*. Bloomsbury Publishing.
Chiappero-Martinetti, E. (2014). Basic needs. In A. C. Michalos (Ed.), *Encyclopedia of quality of life and well-being research*. Springer.
Coote, A., Franklin, J., & Simms, A. (2010). *21 hours: Why a shorter working week can help us all to flourish in the 21st century*. New Economics Foundation.
Daly, H. E. (1996). *Beyond growth: The economics of sustainable development*. Beacon Press.
Derks, D., & Bakker, A. B. (2014). Smartphone use, work–home interference, and burnout: A diary study on the role of recovery. *Applied Psychology, 63*(3), 411–440.
Downes, A., & Lansley, S. (Eds.). (2018). *It's basic income: The global debate*. Policy Press.
Ehmer, J., & Lis, C. (Eds.). (2009). *The idea of work in Europe from antiquity to modern times*. Ashgate Publishing, Ltd.
Ellul, J. (1954). *La technique ou l'enjeu du siècle*. Paris: A. Colin. Citation to the Spanish edition of 2003, *La edad de la técnica*. Octaedro.
Fariza, I. (2020). *Basic income is no longer a utopia*. https://elpais.com/economia/2020-04-06/la-renta-basica-deja-de-ser-una-utopia.html
Federici, S. (1975). *Wages against housework*. Falling Wall Press.
Ford, M. (2017). *Rise of the robots: Technology and the threat of mass unemployment*. Oneworld Publications.
Fourier, C. (1901). Attractive labor. In *Selections from the works of Fourier*. Allen & Unwin.

Frayne, D. (2015). *The refusal of work: The theory and practice of resistance to work*. Zed Books Ltd.
Friedmann, G. (1963). Où va le travail humain? *Revue Française de Sociologie, 4*(3), 359.
Fuchs, C. (2015). *Digital Labor*. The Routledge Companion to Labor and Media.
Gómez-Baggethun, E. (2020). More is more: Scaling political ecology within limits to growth. *Political Geography, 76*, 102095.
Gomez-Baggethun, E. (2022). Rethinking work for a just and sustainable future. *Ecological Economics, 200*, 107506.
Gómez-Baggethun, E., & Naredo, J. M. (2020). El mito del trabajo: Origen, evolución y perspectivas. *PAPELES de relaciones ecosociales y cambio global, 150*, 9–22.
Gorz, A. (1988). *Métamorphose du travail. Critique de la raison économique*. Éditions Galilé.
Graeber, D. (2013). On the phenomenon of bullshit jobs: A work rant. *Strike Magazine, 3*, 1–5.
Harari, Y. N. (2014). *Sapiens: A brief history of humankind*. Random House.
Hayden, A. (1999). *Sharing the work, sparing the planet: Work time, consumption, & ecology*. Zed Books.
Illich, I. (1981). *Shadow work*. Marion Boyars.
Ingebrigtsen, S., & Jakobsen, O. (2012). Utopias and realism in ecological economics—Knowledge, understanding and improvisation. *Ecological Economics, 84*, 84–90.
Jackson, T. (2017). *Prosperity without growth*. London: Earthscan.
Kallis, G., & March, H. (2015). Imaginaries of hope: the utopianism of degrowth. *Annals of the association of American geographers, 105*(2), 360–368.
Kallis, G., Kalush, M., Flynn, H., Rossiter, J., & Ashford, N. (2013). "Friday off": Reducing working hours in Europe. *Sustainability, 5*, 1545–1567.
Keynes, J. M. (2010). Economic possibilities for our grandchildren. In *Essays in persuasion* (pp. 321–332). Palgrave Macmillan. First published in 1930.
Knight, K. W., Rosa, E. A., & Schor, J. B. (2013). Could working less reduce pressures on the environment? A cross-national panel analysis of OECD countries, 1970–2007. *Global Environmental Change, 23*(4), 691–700.
Komlosy, A. (2018). *Work: The last 1,000 years*. Verso Books.
Kropotkin, P. (2015). *The conquest of bread*. Penguin UK.
Lafargue, P. (2012). *The right to be lazy*. The Floating Press. First published in 1887.
Lambert, C. (2015). *Shadow work: The unpaid, unseen jobs that fill your day*. Counterpoint.
Liang, L-H. (2020). *How Covid-19 led to a nationwide work-from-home experiment*. https://www.bbc.com/worklife/article/20200309-coronavirus-covid-19-advice-chinas-work-at-home-experiment. Retrieved 3/4/2020.
Mair, S., Druckman, A., & Jackson, T. (2020). A tale of two utopias: Work in a post-growth world. *Ecological Economics, 173*, 106653.
Manyika, J., Lund, S., Chui, M., Bughin, J., Woetzel, J., et al. (2017). *Jobs lost, jobs gained: Workforce transitions in a time of automation* (p. 150). McKinsey Global Institute.
Martinez Alier, J. (1992). Ecological economics and concrete utopias. *Utopian Studies, 3*(1), 39–52.
Martínez-Alier, J. M., & Schlüpmann, K. (1987). *Ecological economics: Energy, environment, and society*. Basil Blackwell.
Marx, K. (1991). *Capital, A critique of political economy* (Vol. 3). Penguin.
Mason, P. (2016). *Postcapitalism: A guide to our future*. Macmillan.
Méda, D. (2010). *Le Travail. une valeur en voie de disparition?* Flammarion.
Mestrum, F. (2018). Why basic income can never be a progressive solution. In A. Downes & S. Lansley (Eds.), *It's basic income: The global debate* (pp. 97-100). Policy Press.
Mill, J. S. (1850). *The negro question*. https://cruel.org/econthought/texts/carlyle/millnegro.html
OECD. (2019). *OECD employment outlook: The future of work*. OECD Publishing.
Parijs, P. V. (1997). *Real freedom for all: What (if anything) can justify capitalism?* OUP Catalogue.
Popper-Lynkeus, J. (1912). *Die Allgemeine Nährpflicht als L€osung der Sozialen Frage: Einge hend Bearbeitet und Statistische Durchgerechnet*. Reissner.

Pullinger, M. (2014). Working time reduction policy in a sustainable economy: Criteria and options for its design. *Ecological Economics, 103*, 11–19.
Raworth, K. (2017). *Doughnut economics: Seven ways to think like a 21st-century economist.* Chelsea Green Publishing.
Rifkin, J. (1995). *The end of work.* Putnam.
Russell, B. (1935). *In praise of idleness and other essays.* George Allen & Unwin.
Schor, J. B. (2005). Sustainable consumption and worktime reduction. *Journal of Industrial Ecology, 9*(1–2), 37–50.
Schor, J. (2008). *The overworked American: The unexpected decline of leisure.* Basic Books.
Schwartz, A. (1982). Meaningful work. *Ethics, 92*(4), 634–646.
Sen, A. K. (2009). *The idea of justice.* Harvard University Press.
Stronge, W., Harper, A., & Guizzo, D. (2019). *The shorter working week: A radical and pragmatic proposal.* Autonomy.
Wajcman, J. (2015). *Pressed for time: The acceleration of life in digital capitalism.* University of Chicago Press.
Waring, M. (1988). *Counting for nothing: What men value and what women are worth.* Bridget Williams Books.
Wiedmann, T., Lenzen, M., Keyßer, L. T., & Steinberger, J. K. (2020). Scientists' warning on affluence. *Nature Communications, 11*(1), 1–10.
Wright, E. O. (2010). *Envisioning real utopias* (Vol. 98). Verso.
Wu, T. (2018). The tyranny of convenience. *The New York Times, 16*.

Open Access This chapter is licensed under the terms of the Creative Commons Attribution 4.0 International License (http://creativecommons.org/licenses/by/4.0/), which permits use, sharing, adaptation, distribution and reproduction in any medium or format, as long as you give appropriate credit to the original author(s) and the source, provide a link to the Creative Commons license and indicate if changes were made.

The images or other third party material in this chapter are included in the chapter's Creative Commons license, unless indicated otherwise in a credit line to the material. If material is not included in the chapter's Creative Commons license and your intended use is not permitted by statutory regulation or exceeds the permitted use, you will need to obtain permission directly from the copyright holder.

Chapter 32
The Environmentalism of the Paid

Esteve Corbera and Santiago Izquierdo-Tort

32.1 Introduction

Since the early 1990s, deforestation and forest degradation in the tropics – home to millions of "poor" people and of a significant share of the world's biological diversity – have become one of the most pressing environmental concerns, as they contribute to global climate change and biodiversity loss (Curtis et al., 2018). Furthermore, inequalities in income, ownership of assets, and development opportunities remain dire both within and across countries (World Bank, 2016). In this context, a perplexing policy experiment has emerged over the last two decades: a wealth of actors – from governments to private companies and social organizations – offer monetary payments to landowners and rural communities in exchange for protecting forests and related ecosystem services (e.g., watershed regulation, carbon, and biodiversity sequestration). This experiment, known as Payments for Ecosystem Services (PES), is seeking to both conserve biodiversity and alleviate poverty, and it is nowadays one of the most popular conservation policy approaches worldwide.

E. Corbera (✉)
Institute of Environmental Science and Technology & Department of Geography, Universitat Autònoma de Barcelona, Barcelona, Spain

Institució Catalana de Recerca i Estudis Avançats (ICREA), Barcelona, Spain
e-mail: Esteve.Corbera@uab.cat

S. Izquierdo-Tort
Instituto de Investigaciones Económicas, Universidad Nacional Autónoma de México, Circuito Mario de La Cueva Ciudad Universitaria, Mexico City, México

Payments for Ecosystem Services are inseparable from the episteme and trajectory of the "neoliberalization of nature", i.e., the application of neoclassical economic theory and practice to nature conservation (Castree, 2008; Heynen & Robins, 2005). However, as we will argue further below, many PES embrace two contradictory societal paradigms: a neoliberal agenda endorsing market-based institutions to tackle the ecological crisis and a Keynesian vision emphasizing the role of the state in addressing such a crisis. The theoretical debates surrounding PES thus reflect this tension. Market enthusiasts, mostly within mainstream and environmental economics, advocated for PES as a promising alternative that could harness market forces and align incentives for conservation where previous policies failed (Engel et al., 2008; Ferraro & Kiss, 2002). In contrast, other scholars akin to ecological economics and political ecology strongly opposed PES as "conceits" (Fletcher & Büscher, 2017) that would inevitably cause "the poor selling cheap" (Martínez-Alier, 2002), "green grabbing" (Fairhead et al., 2012), "commodity fetishism" (Kosoy & Corbera, 2010), or "selling out on nature" (McCauley, 2006).

However, among such polarizing views, a growing body of empirical research has recently begun to shed light on the various ways in which local peasants and communities adapt, alter, resist, and respond to PES, and how such engagement leads to both expected and unintended socio-environmental consequences (Shapiro-Garza et al., 2020). This work has investigated if pro- or anti-neoliberal interpretations of PES conform with actual policy practice, and whether alternative theorizations are needed (McElwee et al., 2014; Van Hecken et al., 2018).

This chapter draws on Martínez-Alier's environmentalism of the poor (Martínez-Alier, 2002) to provide a new perspective on PES. We argue that PES have fostered the emergence of a new form of environmentalism by local peasants and communities who bear with and take advantage of an external policy agenda predicated upon conservation payments and markets, while remaining attentive to their long-term livelihood goals. Rather provocatively, we bring forward the idea of the environmentalism of the paid as a rising though unexpected consequence of "not-so-neoliberal" conservation policy, and we outline its more salient features inspired by our long-term fieldwork in Mexico. The chapter is intended to reflect on the social lives of market-based conservation.

32.2 The Environmentalism of the Poor

In 2002, Martínez-Alier coined the powerful idea of the environmentalism of the poor, a variety of environmentalism that "grows out of local, regional, national and global ecological distribution conflicts caused by economic growth and social inequalities" (Martínez-Alier, 2002: 14). Considered similar to popular environmentalism, or to the environmental justice movement, given their shared emphasis on justice as a means to achieve sustainability, the environmentalism of the poor emerged as an articulated global social movement from the 1990s onwards, inspired by two previous currents of thought. On the one hand, the civil rights and environmental justice movement in the United States (from the 1960s onwards) denounced the racialized nature of environmental degradation. On the other hand, the agrarian, rights-based, and indigenous movements in the Global South (from the 1970s onwards) emphasized self-determination and the "abuses" of development.

The environmentalism of the poor emphasizes the negative effects of environmental degradation on the well-being of the urban and rural poor, including Indigenous Peoples, and the critical role that such populations play in resource extraction and environmental dispossession conflicts. Economic growth damages nature by unsustainably exploiting nonrenewable resources and is detrimental to the global poor because these often and disproportionately bear the burden of environmental degradation and pollution. Consequently, the poor are dispossessed and alienated from their means of survival, and experience acculturation processes that damage their culture and social institutions. The environmentalism of the poor advocates for degrowth – particularly in developed countries – challenges the absolute dematerialization of the economy, warns about the impacts of new technologies, and puts justice at the center of any environmental policy through an emphasis in recognition, participation, and redistribution.

According to Martínez-Alier, the environmentalism of the poor coexists with at least two older currents of environmentalism. On the one hand, the cult of wilderness arises from an aesthetic appreciation of beautiful landscapes, not from material interests or concerns about economic growth, and seeks to preserve the remnants of "pristine" natural areas to ensure the preservation of biodiversity. As such, has mostly advocated for the strict preservation of natural habitats, through protected areas and no-take zones (or the co-management of natural resources as the last resort), and it lies beyond some recent global campaigns for biodiversity conservation, such as the 30x30 (https://30x30initiative.org/) or Half-Earth (https://www.half-earthproject.org/story/the-half-earth-project/) initiatives. On the other hand, the gospel of eco-efficiency rests on worries about the effects of economic growth, not only on "pristine" natural areas but also on the industrial, agricultural, and urban environments. However, in contrast to the environmentalism of the poor, this variety encompasses distinct currents of thought regarding the relationship between growth and nature. Ecological modernization or sustainable development advocates, for example, emphasize the role that technologies can play in minimizing resource consumption and waste, whereas market environmentalists suggest that adequate resource stewardship is entirely dependent on how well social institutions harness self-interest through individual incentives, because individual property owners are better suited than governments to manage natural resources. Policy ideas such as technological innovation, the circular economy, or sustainable certification are central to the gospel of eco-efficiency.

Evidently, these types of environmentalism are archetypes that serve analytical purposes, which have "points of contact and points of disagreement" (Martinez-Alier, 2002: 15). For instance, one environmental organization may embrace more than one type of environmentalism over its lifetime, or it may also mobilize various types of discourses and contrasting policies in a report. We argue the same may occur with policy instruments advocated by distinct environmentalisms. For example, are PES a new approach to fortress conservation inspired by the cult of wilderness? Or are they markets or subsidies for the provision of ecosystem services, inspired by market environmentalism and ecological modernization, respectively? May PES be regarded as a policy approach that facilitates an affirmative strategy of the rural poor to protect their forests, inspired by the environmentalism of the poor? We will come back to these questions later.

32.3 Payments for Ecosystem Services: Definition and Scope

Broadly defined, PES consist of voluntary (economic or in-kind) transactions between a social actor and an individual or collective landowner where the latter provides a specific environmental benefit or ecosystem service, which is enjoyed by the former actor and/or society at large. Over the years, however, this definition has been subject to reworkings and debate (Muradian et al., 2010; Tacconi, 2012; Wunder, 2005, 2015).

PES initiatives encompass different types of policy instruments, which can be classified in three broad types: (i) user-financed PES, where individuals, companies, NGOs, or public actors directly reward landowners for ecosystem services protection, enhancement, or reestablishment (e.g., voluntary payments for watershed protection, or for carbon offsets); (ii) government-financed PES, where third parties (usually governments) who act on behalf of users compensate landholders for activities that maintain or enhance ecosystem services delivery (e.g., agro-environmental measures, national programs of payments for biodiversity conservation); and (iii) compliance-based PES, where actors facing regulatory obligations compensate others for activities that maintain or enhance comparable ecosystem services or goods in exchange for a standardized credit or offset that satisfies their mitigation requirements (e.g., biodiversity offsets, water quality trading, wetlands mitigation banking, or environmental reverse auctions) (Pirard, 2012; Salzman et al., 2018). Therefore, contrary to what some often believe, PES initiatives have not always involved the establishment of markets that trade a specific unit of biodiversity or ecosystem service. They are characterized by contrasting degrees of commodification, depending on their underlying policy and regulatory framework, which has in turn influenced their design, objectives, and expected performance (Corbera, 2015).

Diversity in implementation probably explains PES appeal and expansion throughout the world. A recent global review of PES, for example, has documented over 550 active PES programs covering millions of hectares, and disbursing US\$ 36–42 billion in annual payments (Salzman et al., 2018). Ezzine et al. (2016) appear to identify 584 PES initiatives of diverse scales and goals. Bull and Strange (2018) show that 37 countries are implementing biodiversity offsetting programs – aimed at balancing environmental damage from development projects with "equivalent" gains – which occupy approximately 153,679 km^2 for an approximate total of

12,983 projects. A recent review of REDD+[1] (Maniatis et al., 2019) shows that dozens of countries in the Global South have developed pilot projects for the reduction of deforestation or the enhancement of forest stocks, as well as national programs that incentivize conservation or sustainable land management through direct payments to reduce national land-use greenhouse emissions, in order to meet their voluntary commitments under the United Nations Framework Convention on Climate Change (UNFCCC) (Dunlop & Corbera, 2016; Corbera & Schroeder, 2018). In this regard, the World Bank's Forest Carbon Partnership Facility, one of the main sources of donor funding for REDD+, has already signed emission reduction agreements with 14 countries to obtain land-use emission reductions in exchange (World Bank, 2021). These reductions will most likely be realized, at least to some extent, through direct conservation payments schemes organized at national or subnational levels.

This evidence suggests that millions of farmers and communities have already been participating in PES, while more are likely to join in the upcoming years. Unsurprisingly, this rising policy agenda has been accompanied by significant scholarly efforts that aimed to make some sense on this experiment, which quickly turned into reality. The concept of PES, their suitability to achieve certain goals, their socio-environmental effectiveness, and many other aspects have been examined and scrutinized. To provide context as to how we see the environmentalism of the paid emerging, we expand below on the academic debates surrounding PES, and we present a short vignette from our own long-standing work in Mexico to illustrate the nature of such environmentalism.

32.4 Payments for Ecosystem Services: Competing Perspectives

As a new experiment in environmental policy, PES has attracted significant scholarly attention. Though risking simplification, we argue that such interest can be explained not only because PES have opened interesting operational questions but most importantly because the apparently "simple" act of paying for conservation outcomes reflects a series of fundamental normative, political, and ideological assumptions and worldviews.

In a seminal contribution, Wunder (2005: 3) defined PES as a voluntary transaction where a well-defined environmental service (ES; or a land-use likely to secure that service) is being "bought" by an (minimum one) ES buyer from an (minimum one) ES provider, if and only if the ES provider secures ES provision (conditionality). Drawing heavily on Coase (1960) and Hardin (1968), this view frames environmental problems, such as forest loss and degradation, as a consequence of the failure

---

[1] REDD+ stands for countries' efforts to reduce emissions from deforestation and forest degradation, foster conservation, sustainable management of forests, and enhancement of forest carbon stocks.

to incorporate the environment into the market sphere. The assumption here is that those at the resource base have no mechanisms to privately capture the positive "environmental externalities" that their resource management practices provide. This "market failure" unequivocally leads to the underprovision of public environmental goods. PES should thus attempt to put into practice the "Coase theorem" (Engel et al., 2008), which states that if transaction costs are low and property rights are clearly defined, an efficient provision of environmental goods and services can be achieved through private negotiation. In doing so, PES would correct market failures by creating a market where service "buyers" and "sellers" interact with one another through conditional payments at the "right price" (Muradian et al., 2010).

Unsurprisingly, for this perspective, ensuring that PES worked as intended required well-defined ecosystem services and property rights, minimal transaction costs, and fair negotiation processes. This would involve designing contracts to elucidate opportunity costs and reduce "informational rents" (Ferraro, 2008; Schomers & Matzdorf, 2013); enhancing efficiency through spatial targeting (Alix-García et al., 2008; Wünscher et al., 2008); ensuring that adequate property rights were at place (Engel & Palmer, 2008); and creating well-defined services that could be subject to trading (Engel et al., 2008; Wunder, 2005).

The alleged "neoliberal" (and Coasean) nature of PES attracted a wealth of criticism from political ecology and political economy scholars. McCauley (2006) decried the development of markets for ecosystem services and suggested instead that "we will make more progress in the long run by appealing to people's hearts rather than to their wallets" (ibid.: 28). Others linked PES to the expansion of capitalism into new spheres of social life and they argued that PES would inevitably imply different transacting parties operating under unequal terms of exchange (McAfee, 1999; Büscher, 2012). Similarly, voices critical of environmental markets have used Harvey's (2003) concept of "accumulation by dispossession" and Martínez-Alier's (2002) notion that "the poor sell cheap" to emphasize how market-based policies might entail virtual and actual "green grabs" (Fairhead et al., 2012) and involve "conservation rents for renouncing development" (Karsenty, 2007). Others, drawing on Polanyi (1944), warned of the counterproductive ethical and practical consequences of commodifying nature. Kosoy and Corbera (2010) argued that market-based forms of PES represented both a symptom and consequence of "commodity fetishism", which disregards ecosystem complexity, reduces ecosystem values to single exchange values, and creates power asymmetries across those involved in market development. Gómez-Baggethun and Ruiz-Pérez (2011) warned about the counterproductive consequences of commodifying nature for biodiversity conservation and equity in access to benefits from environmental services.

Much to the dismay, or relief, of supporting and critical voices of PES as "neoliberal" conservation, however, a third body of mostly empirical literature began to suggest that both views had rested more on ideological assumptions and beliefs than on a careful examination of reality. Most "real world" PES initiatives did not seem to encompass a great deal of the elements associated with neoliberal policy, such as commodification, privatization and the retreat of the state (McElwee et al., 2014). Realizing simultaneously all the conditions that would guarantee the development of efficient markets for ecosystem services has proven elusive, except for compliance-based approaches with robust regulations and well-functioning governance frameworks (Wunder et al., 2020).

PES have rarely involved processes to value nature or to compartmentalize environmental services. Instead, policy makers have often used estimations of opportunity costs or arbitrary methods to determine payments and treated resource management practices as proxies for service provision (McElwee et al., 2014). On the side of "buyers", PES national programs such as those in Costa Rica, Mexico, and China involved an active role by governments as single or monopsonistic "buyers" of ecosystem services. On the side of "sellers", where land institutions involved collective property and tenure regimes, communities acted on behalf of individual landowners. Finally, many PES initiatives have been designed with an anti-poverty agenda in mind, prioritizing poverty considerations over environmental additionality in targeting approaches, accompanied by a discourse about revaluing the countryside (Shapiro-Garza, 2013).

With these empirical observations in mind, rooted in the field of institutional economics and the interface of ecological economics and political ecology, some scholars have advocated for a better understanding of social relations in PES, the intricate institutional and political arrangements in which they take place, the complexity of the ecosystems which they are intended to sustain, and the multiple and incommensurate values embedded in the nonmarket institutions in which PES operate. For example, institutional analyses of PES have shed light on how discourses and practices around PES operate at various interconnected governance levels (Corbera et al., 2007; Muradian & Rival, 2012; Muradian et al., 2010; Vatn, 2010). Others have emphasized that different actors associated with PES at various levels, from those at the local resource base to high-level officials involved in policy design, may have different and potentially conflicting notions of fairness, equity, and justice (Corbera & Pascual, 2012; Pascual et al., 2014). Others have brought forward the need to "re-politicize" PES by explicitly examining how the workings of politics and power influence multiple spheres of PES (McAfee & Shapiro, 2010; Rodríguez de Francisco et al., 2021; Shapiro-Garza, 2013; Van Hecken et al., 2015). Finally, the conceptual and empirical flaws inherent in the Coasean view of PES have also been challenged, such as the incorrect assumption of rivalry and excludability as dynamic policy variables instead of biophysical characteristics that are not dynamic at all (Farley & Costanza, 2010), the inherent problems in measuring natural capital and assigning monetary value to environmental services (Rival & Muradian, 2012), and the scientifically weak links between land use and environmental service provision (Pascual et al., 2010).

32.5 The Environmentalism of the Paid

We have argued above that PES have acquired a myriad of forms that seldom align in practice with their market-based, foundational principles. It may thus be misleading to associate PES approaches with only one sort of environmentalism and that it would instead be more appropriate to think of such approaches as serving distinct environmental discourses. Furthermore, we believe it is important to develop contextually situated, and culturally rooted understandings of PES initiatives that can reflect on the lived experiences of PES beneficiaries.

While PES are too often designed as short-term incentives for narrow conservation goals (e.g., a contract over a limited number of years, which may or may not be renewable, to conserve forests), we have observed that beneficiaries think about their livelihoods at wider temporal and spatial scales. Such broader and longer term thinking allows them to exert some degree of control on the various policies, including but not limited to PES, that they encounter along the way. Participants thus voluntarily participate in PES but do so on their own terms. They continuously adapt, respond to, and sometimes even contest PES based on their specific livelihood interests and priorities.

Strategic behavior related to environmental policies by participants and other local actors is not a new phenomenon. What is novel in the specific case of PES, however, is how local conservation actions and attitudes are increasingly predicated and dependent upon conditional monetary payments by external entities. It is within this unexpected entanglement that we see the environmentalism of the paid as a rising phenomenon that shares some similarities but is otherwise distinctive from previously identified varieties of environmentalism. The environmentalism of the paid combines certain reverence for the stewardship of nature, with a strong concern for local livelihoods and a utilitarian view of development.

Payments in Mexico's Selva Lacandona

In Mexico, there are multiple PES initiatives throughout the country, from local and mostly user-financed schemes for watershed conservation to nation-wide PES programs that focus on hydrological services and biodiversity conservation. Between 2003 and 2019, nation-wide programs have protected 6.7 million hectares of forests under conservation contracts (10.3% and 3.4% of Mexico's forestlands and total surface, respectively) (Izquierdo-Tort et al., In progress). These national PES programs have two key features: first, most contracts are signed by indigenous and agrarian communities who often manage their forests in common and, consequently, such contracts are typically signed off and managed at the community-level upon approval by landed members within each community. Second, contracts are valid for 5 years and provide annual payments per hectare (approx. US$ 50) of protected forest in exchange for the development of a series of forest management and conservation activities, and they can be renewed.

Over the years, we have investigated the role that national PES schemes have played for local livelihoods and the natural base in several communities of two contiguous municipalities (Marqués de Comillas and Benemérito de las Américas) in the Selva Lacandona of the state of Chiapas. The region is a resource and colonization frontier bordering Guatemala, populated by both mestizo and Indigenous communities (Leyva Solano & Ascencio Franco, 1996; De Vos, 2002). Selva Lacandona is Mexico's largest remaining patch of high-canopy tropical rainforest and one of the country's most biodiverse regions, but it faces high rates of deforestation for agricultural and cattle ranching expansion (Carabias et al., 2015). In this context, we have documented a high and enduring interest in participating in PES schemes among individual households and communities. Demand for PES participation has exceeded availability of public funding since PES arrived to the region in the mid-2000s, and excess demand has grown significantly from 2016 onwards due to a reduction in the country's environmental budget.

Through several publications and ongoing projects, we have observed that voluntary participation in PES in the study region is highly strategic and at times harmonious or conflicting with the longer term, and changing livelihood needs and aspirations of those involved in conservation payments. We have discussed the various manifestations of such strategic behavior, and its consequences. For example, we have shown that interest in PES and more generally in conserving nature is at constant odds with a strong "cowboy-based" livelihood aspiration, which situates PES participation at the intersection between receiving a fair compensation for environmental stewardship, the capacity of landowners to diversify livelihood activities and land uses, and the ability of local leaders and PES intermediaries to influence collective action towards conservation (Izquierdo-Tort et al., 2019, 2021). We have highlighted how people decide which lands to enroll in PES and which to leave outside for continued agricultural expansion, and made evident that households can both receive payments for conservation while continuing to deforest other areas (Izquierdo-Tort et al., 2019).

Households and communities take advantage of the lack of coordination among various governmental institutions, providing subsidies to mix-and-match the latter and maximize income (Izquierdo-Tort, 2020). They can draw upon multiple notions of justice and preexisting land-related entitlements and norms to distribute PES payments within the community (Izquierdo-Tort et al., in review), which occasionally can result in increased social conflict (Corbera et al., 2020). Specifically, conservation payments provide a form of "rent" that has seemingly raised the economic value of forestlands (previously considered "idle" and invaluable) and thus has shielded small landowners against encroachment from land speculators. Overall, we have shown that communities respond and adapt to evolving PES designs, as well as to the changing demographic and institutions of households and communities in their struggle for a more prosperous future (Izquierdo-Tort et al., 2021).

We identify at least three features in the environmentalism of the paid. First, PES beneficiaries bring forward a discourse that emphasizes the positive role that conservation payments play in maintaining and enhancing their livelihoods. PES beneficiaries should neither be considered "noble savages" nor homo economicus, but somewhere in between. Their practical engagement with conservation – and what the PES programs expect in this regard – is one that fits with existing land tenure arrangements and with both social or individual norms, regulations, and expectations regarding resource management and conservation. They engage in PES because PES rules match, without excessive tweaking, with what they are willing or are socially expected to do with their forests. In other words, they engage in PES because conservation practice does not entail excessive costs or shifts in norms and behavior.

Second, the environmentalism of the paid connects very different types of stakeholders – from landowners, communities, and local organizations and public officials in rural contexts to national and international states, organizations, and markets – through a single exchange value (i.e., money in exchange for biodiversity conservation or specific ecosystem services), yet such exchange acquires different local meanings and values. In Mexico, payments are perceived as a recognition by the State of the cost that conservation entails for peasants holding both individual and communal lands. In other countries, for example in Colombia, payments are perceived as a compensation for the economic losses incurred because of abandoning coca cultivation and avoiding the expansion of cattle grazing. In this case, the payment does not contribute to reinforce existing social norms, but to acknowledge conservation efforts in a context where land-use pressure is increasing (Moros et al., 2020).

Third, the environmentalism of the paid counts with strong allies – including donors, governments, NGOs, and companies – who promote and make PES possible, since they act as key design, funding, or implementation actors. These allies contribute to the institutionalization of PES principles at national and local levels (Lima et al., 2019), and some of them may also do so with the prospects of establishing the foundations of a "conservation basic income" (Fletcher & Büscher, 2020). However, these allies are also often unable to design and implement PES in ways that can meet the long-term expectations of the targeted landowners and communities, given the short-term funding cycles and political uncertainties they operate in. For example, the overall positive ecological and social outcomes of Mexico's national PES programs (Sims & Alix-Garcia, 2017) and in Selva Lacandona (Costedoat et al., 2015) may be soon jeopardized by a shift in government funding priorities, significant budget reductions in the environmental sector, and new policy programs that may contribute to land-use change.[2]

It is not our intention in this provocative chapter so far to overromanticize the environmentalism of the paid. PES beneficiaries, with their discourse and praxis, are only one side of local rural struggles. Much research, including our own, has shown that PES can reproduce and exacerbate preexisting inequalities in access to land and funding, exclude vulnerable actors from benefits and decision-making, and facilitate processes of "elite capture" (Corbera et al., 2007). PES can thus leave aside, either intentionally or unintentionally, some individuals or social groups, who may or may not be interested in joining PES and displaying their environmentalism.

---

[2] The "Sembrando Vida" started in 2019 and provides large subsidies to landowners for the development of agroforestry and reforestation activities. This laudable objective seems to have induced further deforestation. See, for example, (1) https://gatopardo.com/reportajes/sembrando-vida-el-proyecto-milagro-de-lopez-obrador-para-el-campo-mexicano-2021-2020/; (2) http://movilidadamable.org/WRIMexico/WRI%20M%C3%A9xico%20An%C3%A1llisis%20sobre%20los%20impactos%20ambientales%20de%20Sembrando%20Vida%20en%202019.pdf

32.6 Conclusion

This chapter has drawn on Martínez-Alier's concept of the environmentalism of the poor, and other varieties of environmentalism, to emphasize the growing importance that Payments for Ecosystem Services (PES) play in the livelihoods of the rural poor, particularly in the Global South. We reviewed the foundations of PES, related academic debates, and suggested that PES beneficiaries should be understood as living examples of a new form of environmentalism: the environmentalism of the paid. Early on, we left some questions unanswered. Does PES, as a policy approach, respond only to one type of environmentalism? We can say with confidence that it does not. The disciplinary background – which influences one's theoretical and empirical lenses – or even her own ethical and political values determine if PES represents an instrument for fortress conservation, market-based resource management, or the emancipation of the rural poor.

We argued that PES beneficiaries portray an environmentalism that combines elements of ecosystem stewardship with livelihood-focused and utilitarian perspectives on local development. It is an environmentalism that combines the pride of conservation stewardship, with a demand for economic compensation in exchange for such stewardship, which seemingly resonates with wider calls for convivial conservation (Büscher & Fletcher, 2019) and for the establishment of a "conservation basic income" (Fletcher & Büscher, 2020). This environmentalism should not probably be circumscribed only to PES beneficiaries, however. Participants in conservation-oriented programs, such as resource comanagement schemes, or integrated conservation and development projects may also be part of this growing discourse and praxis.

The environmentalism of the paid obviously owes its existence to many allies, i.e., the donors, governments, NGOs, and companies that financially support the poor's conservation efforts. These allies probably channel economic incentives to the rural poor because they believe that conservation and development are compatible, which is a far-fetched assumption if one attends to the fact that global land-use trends point towards ecological disaster (IPBES, 2019). In this regard, we suggest that the environmentalism of the paid will only stand the test of time if conservation incentives are prolonged over time, adjusted to local economic realities, and come accompanied by decisive actions that tackle the pernicious effects of contradictory land-use policies in ways that do not harm local livelihoods (Meyfroidt et al., 2022).

References

Alix-García, J., De Janvry, A., & Sadoulet, E. (2008). The role of deforestation risk and calibrated compensation in designing payments for environmental services. *Environment and Development Economics, 13*, 375–394.

Büscher, B. (2012). Payments for Ecosystem Services as neoliberal conservation: (Reinterpreting) Evidence from the Maloti-Drakensberg, South Africa. *Conservation and Society, 10*(1), 29–41.

Büscher, B., & Fletcher, R. (2019). Towards convivial conservation. *Conservation and Society, 17*(3), 283–296.
Bull, J. W., & Strange, N. (2018). The global extent of biodiversity offset implementation under no net loss policies. *Nature Sustainability, 1*, 790–798.
Carabias, J., De la Maza, J., & Cadena, R. (Eds.). (2015). *Conservación y Desarrollo Sustentable en la Selva Lacandona: 25 Años de Actividades y Experiencias*. Natura y Ecosistemas Mexicanos.
Castree, N. (2008). Neoliberalising nature: Processes, effects, and evaluations. *Environment and Planning A, 40*, 153–173.
Coase, R. (1960). The problem of social cost. *The Journal of Law and Economics, 3*, 1–44.
Corbera, E. (2015). Valuing nature, paying for ecosystem services and realizing social justice: A response to Matulis (2014). *Ecological Economics, 110*, 154–157.
Corbera, E., & Pascual, U. (2012). Ecosystem services: Heed social goals. *Science, 335*(10), 355–356.
Corbera, E., & Schroeder, H. (Eds.). (2018). *REDD+ crossroads post Paris: Politics, lessons and interplays*. MDPI. 416p.
Corbera, E., Brown, K., & Adger, N. (2007). The equity and legitimacy of markets for ecosystem services. *Development and Change, 38*(4), 587–613.
Corbera, E., Costedoat, S., Ezzine-de-Blas, D., & Van Hecken, G. (2020). Troubled encounters: Payments for Ecosystem Services in Chiapas, Mexico. *Development and Change, 51*, 167–195. https://doi.org/10.1111/dech.12540
Costedoat, S., et al. (2015). How effective are payments for biodiversity conservation in Mexico? *PLoS One, 10*(3), e0119881.
Curtis, et al. (2018). Classifying drivers of global forest loss. *Science, 361*, 1108–1111.
De Vos, J. (2002). *Una Tierra Para Sembrar Sueños. Historia Reciente de la Selva Lacandona 1950–2000*. Centro de Investigaciones y Estudios Superiores en Antropología Social (CIESAS) and Fondo de Cultura Económica (FCE).
Dunlop, T., & Corbera, E. (2016). Incentivizing REDD+: How developing countries are laying the groundwork for benefit-sharing. *Environmental Science and Policy, 63*, 44–54.
Engel, S., & Palmer, C. (2008). Payments for environmental services as an alternative to logging under weak property rights: The case of Indonesia. *Ecological Economics, 65*, 799–809.
Engel, S., Pagiola, S., & Wunder, S. (2008). Designing payments for environmental services in theory and practice: An overview of the issues. *Ecological Economics, 65*(4), 663–674.
Ezzine-de-Blas, D., Wunder, S., Ruiz-Pérez, M., & Moreno-Sanchez, R. (2016). Global patterns in the implementation of payments for environmental services. *PLoS One, 11*(3), e0149847.
Fairhead, J., Leach, M., & Scoones, I. (2012). Green grabbing: A new appropriation of nature? *Journal of Peasant Studies, 39*(2), 237–261.
Farley, J., & Costanza, R. (2010). Payments for ecosystem services: From local to global. *Ecological Economics, 69*, 2060–2068.
Ferraro, P. J. (2008). Asymmetric information and contract design for payments for environmental services. *Ecological Economics, 65*, 811–822.
Ferraro, P. J., & Kiss, A. (2002). Direct payments to conserve biodiversity. *Science, 298*, 1718–1719.
Fletcher, R., & Büscher, B. (2017). The PES Conceit: Revisiting the relationship between payments for environmental services and neoliberal conservation. *Ecological Economics, 132*, 224–231.
Fletcher, R., & Büscher, B. (2020). Conservation basic income: A non-market mechanism to support convivial conservation. *Biological Conservation, 244*, 108520.
Gómez-Baggethun, E., & Ruiz-Perez, M. (2011). Economic valuation and the commodification of ecosystem services. *Progress in Physical Geography, 35*, 613–628.
Hardin, G. (1968). The tragedy of the commons. *Science, 162*, 1243–1248.
Harvey, D. (2003). *The new imperialism*. Oxford University Press.
Heynen, N., & Robins, P. (2005). The neoliberalization of nature: Governance, privatization, enclosure and valuation. *Capitalism Nature Socialism, 16*(1), 5–8.

IPBES. (2019). *Summary for policymakers of the global assessment report on biodiversity and ecosystem services of the Intergovernmental Science-Policy Platform on Biodiversity and Ecosystem Services* (S. Daz, J. Settele, E. S. Brondizio, H. T. Ngo, M. Guze, J. Agard, A. Arneth, P. Balvanera, K. A. Brauman, S. H. M. Butchart, K. M. A. Chan, L. A. Garibaldi, K. Ichii, J. Liu, S. M. Subramanian, G. F. Midgley, P. Miloslavich, Z. Molnr, D. Obura, A. Pfaff, S. Polasky, A. Purvis, J. Razzaque, B. Reyers, R. Roy Chowdhury, Y. J. Shin, I. J. Visseren-Hamakers, K. J. Willis, and C. N. Zayas (Eds.)). IPBES Secretariat. 56p.
Izquierdo-Tort, S. (2020). Payments for Ecosystem Services and conditional cash transfers in a policy mix: Microlevel interactions in Selva Lacandona, Mexico. *Environmental Policy and Governance, 30*, 29–45. https://doi.org/10.1002/eet.1876
Izquierdo-Tort, S., Ortiz-Rosas, F., & Vázquez-Cisneros, P. A. (2019). 'Partial' participation in payments for environmental services (PES): Land enrolment and forest loss in the Mexican Lacandona Rainforest. *Land Use Policy, 87*, 103950. https://doi.org/10.1016/j.landusepol.2019.04.01
Izquierdo-Tort, S., Corbera, E., Barceinas Cruz, A., Naime, J., Angélica VázquezCisneros, P., Carabias Lillo, J., Castro-Tovar, E., Ortiz Rosas, F., Rubio, N., Torres Knoop, L., & Dupras, J. (2021). Local responses to design changes in Payments for Ecosystem Services in Chiapas, Mexico. *Ecosystem Services, 50*, 101305. https://doi.org/10.1016/j.ecoser.2021.101305
Izquierdo-Tort, S., Corbera, E., Shapiro-Garza, Alatorre, A., et al. (In progress). *How effective and equitable are Payments for Ecosystem Services (PES) in Mexico? A systematic literature review and meta-analysis.*
Izquierdo-Tort, S., Corbera, E., Martin, A., & Carabias, J. (In review). *Contradictory distributive principles and land tenure govern benefit-sharing of Payments for Ecosystem Services (PES) in Chiapas, Mexico.*
Karsenty, A. (2007). Questioning rent for development swaps: New market-based instruments for biodiversity acquisition and the land-use issue in tropical countries. *International Forestry Review, 9*(1), 503–513.
Kosoy, N., & Corbera, E. (2010). Payments for Ecosystem Services as commodity fetishism. *Ecological Economics, 69*(6), 1228–1236.
Leyva Solano, X., & Ascencio Franco, G. (1996). *Lacandonia al Filo del Agua*. Fondo de Cultura Económica (FCE).
Lima, L. S., Ramos Barón, P. A., Villamayor-Tomas, S., & Krueger, T. (2019). Will PES schemes survive in the long-term without evidence of their effectiveness? Exploring four water-related cases in Colombia. *Ecological Economics, 156*, 211–223.
Maniatis, D., Scriven, J., Jonckheere, I., Laughlin, J., & Todd, K. (2019). Toward REDD+ implementation. *Annual Review of Environment and Resources, 44*, 373–398.
Martinez-Alier, J. (2002). *The environmentalism of the poor: A study of ecological conflicts and valuation*. Edward Elgar Publishing, Inc.. 325p.
McAfee, K. (1999). Selling nature to save it? Biodiversity and green developmentalism. *Environment and Planning D: Society and Space, 17*, 133–154.
McAfee, K., & Shapiro, E. N. (2010). Payments for ecosystem Services in Mexico: Nature, neoliberalism, social movements, and the state. *Annals of the Association of American Geographers, 100*, 579–599.
McCauley, D. J. (2006). Selling out on nature. *Nature, 443*, 27–28.
McElwee, P., Nghiem, T., Le, H., Vu, H., & Tran, N. (2014). Payments for environmental services and contested neoliberalisation in developing countries: A case study from Vietnam. *Journal of Rural Studies, 36*, 423–440.
Meyfroidt, P., de Bremond, A., Ryan, C. M., Archer, E., Aspinall, R., Chhabra, A., Camara, G., Corbera, E., DeFries, R., Díaz, S., Dong, J., Ellis, E. C., Erb, K., Fisher, J. A., Garrett, R. D., Golubiewski, N. E., Grau, H. R., Grove, J. M., Haberl, H., Heinimann, A., Hostert, P., Jobbágy, E. G., Kerr, S., Kuemmerle, T., Lambin, E. F., Lavorel, S., Lele, S., Mertz, O., Messerli, P., Metternicht, G., Munroe, D. K., Nagendra, H., Nielsen, J. Ø., Ojima, D. S., Parker, D. C., Pascual, U., Porter, J. R., Ramankutty, N., Reenberg, A., Roy Chowdhury, R., Seto, K. C.,

Seufert, V., Shibata, H., Thomson, A., Turner, B. L., Urabe, J., Veldkamp, T., Verburg, P. H., Zeleke, G., & E. K. H. J. zu Ermgassen. (2022). Ten facts about land systems for sustainability. *Proceedings of the National Academy of Sciences, 119*(7), e2109217118. https://doi.org/10.1073/pnas.2109217118

Moros, L., et al. (2020). Pragmatic conservation: Discourses of Payments for Ecosystem Services in Colombia. *Geoforum, 108*, 169–183.

Muradian, R., & Rival, L. (2012). Between markets and hierarchies: The challenge of governing ecosystem services. *Ecosystem Services, 1*, 93–100.

Muradian, R., et al. (2010). Reconciling theory and practice: An alternative conceptual framework for understanding payments for environmental services. *Ecological Economics, 69*(6), 1202–1208.

Pascual, U., Muradian, R., Rodríguez, L. C., & Duraiappah, A. (2010). Exploring the links between equity and efficiency in payments for environmental services: A conceptual approach. *Ecological Economics, 69*, 1237–1244. https://doi.org/10.1016/j.ecolecon.2009.11.004

Pascual, U., et al. (2014). Social equity matters in Payments for Ecosystem Services. *BioScience, 64*(11), 1027–1036.

Pirard, R. (2012). Market-based instruments for biodiversity and ecosystem services: A lexicon. *Environmental Science and Policy, 19–20*, 59–68.

Polanyi, K. (1944). *The great transformation: The political and economic origins of our time.* Beacon Press.

Rival, L., & Muradian, R. (2012). Introduction: Governing the provision of ecosystem services. In R. Muradian & L. Rival (Eds.), *Governing the provision of ecosystem services* (pp. 1–17). Springer. ISBN: 978-94-007-5176-7.

Rodríguez-de-Francisco, J. C., et al. (2021). Post-conflict transition and REDD+ in Colombia: Challenges to reducing deforestation in the Amazon. *Forest Policy and Economics, 127*, 102450.

Salzman, J., et al. (2018). The global status and trends of Payments for Ecosystem Services. *Nature Sustainability, 1*, 136–144.

Schomers, S., & Matzdorf, B. (2013). Payments for Ecosystem Services: A review and comparison of developing and industrialized countries. *Ecosystem Services, 6*, 16–30.

Shapiro-Garza, E. (2013). Contesting the market-based nature of Mexico's national Payments for Ecosystem Services programs: Four sites of articulation and hybridization. *Geoforum, 46*, 5–15.

Shapiro-Garza, E., McElwee, P., Van Hecken, G., & Corbera, E. (2020). Beyond market logics: Payments for ecosystem services as alternative development practices in the global south. *Development and Change, 51*(1), 3–25.

Sims, K. R. E., & Alix-Garcia, J. M. (2017). Parks versus PES: Evaluating direct and incentive-based land conservation in Mexico. *Journal of Environmental Economics and Management, 86*, 8–28.

Tacconi, L. (2012). Redefining payments for environmental services. *Ecological Economics, 73*, 29–36.

Van Hecken, G., Bastiaensen, J., & Windey, C. (2015). Towards a power-sensitive and socially informed analysis of Payments for Ecosystem Services (PES): Addressing the gaps in the current debate. *Ecological Economics, 120*, 117–125.

Van Hecken, G., et al. (2018). Silencing agency in Payments for Ecosystem Services (PES) by essentializing a neoliberal "monster" into being: A response to Fletcher and Buscher's "PES Conceit". *Ecological Economics, 144*, 314–318.

Vatn, A. (2010). An institutional analysis of payments for environmental services. *Ecological Economics, 69*(6), 1245–1252.

World Bank. (2016). *Poverty and shared prosperity 2016: Taking on inequality.* World Bank. https://doi.org/10.1596/978-1-4648-0958-3

World Bank. (2021). *Forest carbon partnership facility annual report 2021.* www.forestcarbonpartnership.org

Wunder, S. (2005). *Payments for environmental services: Some nuts and bolts* (CIFOR occasional paper no. 42). Center for International Forestry Research.

Wunder, S. (2015). Revisiting the concept of payment for environmental services. *Ecological Economics, 117*, 234–243.

Wunder, S., et al. (2020). Payments for environmental services: Past performance and pending potentials. *Annual Review of Resource Economics, 12*, 209–234.

Wünscher, T., Engel, S., & Wunder, S. (2008). Spatial targeting of payments for environmental services: A tool for boosting conservation benefits. *Ecological Economics, 65*(4), 822–833.

Open Access This chapter is licensed under the terms of the Creative Commons Attribution 4.0 International License (http://creativecommons.org/licenses/by/4.0/), which permits use, sharing, adaptation, distribution and reproduction in any medium or format, as long as you give appropriate credit to the original author(s) and the source, provide a link to the Creative Commons license and indicate if changes were made.

The images or other third party material in this chapter are included in the chapter's Creative Commons license, unless indicated otherwise in a credit line to the material. If material is not included in the chapter's Creative Commons license and your intended use is not permitted by statutory regulation or exceeds the permitted use, you will need to obtain permission directly from the copyright holder.

Chapter 33
Collective Action in Ecuadorian Amazonia

Fander Falconí and Julio Oleas

This chapter aims to examine the contribution of ecological economics (EE) to the understanding of the issues of Amazonia, a key eco-region for the planet and its inhabitants, as it faces a point of no return (Lovejoy & Nobre, 2018) due to the accumulation of historical problems. The Amazon region is experiencing accelerated changes in land use, fragmentation of its ecosystems, extractive pressure (from oil and minerals), and social dispossession. Understanding all of this requires broad disciplinary approaches, consideration of uncertainties, and stakeholder participation in decision-making. The first section defines the scope of EE and its evolution toward political ecology. Next, EE is linked to the debate on sustainability and extractivism. The third section describes the contributions of the Barcelona School to the defense of Amazonia. The next section examines the connotations of the Yasuní-ITT Initiative as an opportunity for a true socio-environmental transition. Finally, several conclusions are presented.

33.1 From Ecological Economics to Political Ecology

Although not as a formal discipline, ecological economics (EE) began in the nineteenth century, studying energy and the environment in the economy (Martínez-Alier, 1990). Among its leading precursors were F. Soddy (1933) and S. Podolinsky, with their pioneering contributions on energy (Martínez-Alier & Naredo, 1982;

F. Falconí (✉)
Department of Development, Environment and Territory, Facultad Latinoamericana de Ciencias Sociales – Sede Ecuador, Quito, Ecuador
e-mail: ffalconi@flacso.edu.ec

J. Oleas
Independent Researcher, Lima, Perú

Martínez-Alier, 1998). Nowadays, EE has, as its primary research objectives, the concerns regarding the physical limits of economic growth, based on the seminal work of Georgescu-Roegen (1971), the Report to the Club of Rome by Meadows et al. (1972), and the study of energy and material flows. Another objective of EE is the revision of externalities, based on Kapp's theory of social costs (1950, 1976).

EE suggests that the economic system is part of a much wider ecological system. This basic idea – a fundamental ontological presupposition of EE – implies that reality is present at different levels that interact and coevolve in a complex way, requiring a systemic study based on an organicist ontology capable of harboring several epistemologies (Lizarazo, 2018).

Since the renaissance, Western science has advanced toward the certainty of knowledge, aspiring to control Nature, and driving technological and industrial development. However, for EE, knowledge of the interactions of biophysical reality with social reality is full of uncertainties, and it is not possible to prove absolute truths. In the post-truth era, normal science faces a good deal of uncertainty, especially in relation to the environment and to public policies (Funtowicz & Ravetz, 1993, 2000).

It is possible to understand and interpret this reality in an atmosphere of methodological pluralism, broad enough to foster open and creative dialogue between the wisdom of different bodies of knowledge and between the academy and its social surroundings. This process becomes a reality through interdisciplinary research conducted on a democratic basis, so as to include all of the interests at stake, in a scenario of multidimensional evaluation where it is understood that scientific ethics are not neutral; that consumerism and individualism are not the only human aspirations (equity, democracy in decision-making and social justice also matter); and that it is essential to make clear the interests at stake.

EE transcends the dichotomies that characterize positivism (facts versus values, knowledge versus ignorance, positive economics versus normative economics, … academic knowledge versus common knowledge) and assumes that human beings (the homo economicus of neoclassical economics) cannot control Nature, that there is more than one legitimate perspective, and that complex systems are unpredictable.

The ontological extension of EE – from market equilibrium toward a wider system that encompasses the former and is ruled by the laws of entropy – leads to the need to regard the understanding and interpretation of that reality as a social process, subject to reasoned criticisms based on multiform empirical research (Lizarazo, 2018). EE proposes an epistemological renewal that transcends the idea of scientific revolution (Kuhn, 2004), such that the change that it inspires would not sanction the passage from one state of normal science to another, but – consistent with its ontological model – evolves toward the state of post-normal science anticipated by Funtowicz and Ravetz (1993).

Presently, the EE research agenda includes the destruction of biodiversity, the sources and uses of energy, the use of land, evaluative incommensurability, the development of new methodologies for measuring and evaluating environmental goods and services, the analysis of social metabolisms and material flows, and degrowth. Martínez-Alier has focused his attention toward ecological-distributive conflicts, ecological debt, environmental justice, and poverty and its relationship to the deterioration in ecological systems.

Economic and population growth require the use of increasing amounts of natural resources and yield greater waste. This occurs in certain institutional frameworks, social relations, and power relations. This results in diverse impacts that provoke environmental justice conflicts, expressed in different valuation languages. EE rejects the possibility of reducing these languages down to a single dimension – the monetary one – and appeals to the incommensurability of values.

When studying environmental justice conflicts from the EE perspective, a question arises similar to the matrix question of political economy (*cui bono?*), but in a negative sense: who is harmed by the environmental, distributive, and financial liabilities, as well as the injustices caused by those conflicts? The likely answers are to be found in the realm of political ecology. If EE studies the relationship between the environment and the economy, political ecology studies ecological-distributive conflicts (Martínez-Alier, 2006).

These conflicts are increasingly diverse, in keeping with advances in the extraction of resources, the generation of waste, or the abusive imposition of property rights. Their broader classification takes into account not only their geographical sphere (local or global) but also more specific subjects such as bio-piracy, conflicts related to energy and materials extraction, transportation, waste generation, and contamination.

Measuring the impacts of ecological-distributive conflicts requires the use of conventional and nonconventional indicators expressed in physical and/or monetary units (Martínez-Alier, 2006). Based on the concept of social metabolism (the economy as a system that takes up useful energy and discharges waste and dissipated heat), physical indicators are necessary.

The discrepancies arising from different value systems must be added to the complexities posed by the measurement of the conflicts' impacts. The claims of those most affected can be expressed in monetary terms, but they can also begin with an argument over the valuation system to be used (Ibid.) The loss of biodiversity or cultural heritage, the damage to human livelihoods, the violations of human rights, the sanctity of the land, and the territorial rights of the Indigenous population or environmental safety are expressions of immeasurable values.

However, "who assumes the power to determine the relevant languages of valuation?" (Ibid.) This problem, which is central for EE and for political ecology, arises first, when determining the capacity to impose a decision upon others, and second, when resolving the capacity to impose a method for making decisions on ecological-distributive conflicts.

The environmentalism of the poor, another concept coined by Martínez-Alier (1994), provides an understanding of ecological-distributive conflicts and social resistance caused by productive and extractive processes, as in Amazonia. Joan Martínez-Alier has maintained a constant commitment toward Amazonia and has even put forward courses of action that would lead to a post-oil country through the NGO *Acción Ecológica*.

33.2 Ecological Economics, Sustainability, and Extractivism

EE also serves to deepen the debate on extractivism and sustainability. Orthodoxy means that public policy decisions are the result of technical processes in which models (functional expressions of a theory) are contrasted with realities or phenomena that one seeks to change. Results are drawn from these processes based on relationships of causality – generally linear. Political authorities assimilate these results, establishing objectives, targets, and instruments, for implementation in society. This method of devising public policies – economic, social, health, educational, environmental, security, etc. – is developed in a quasi-mechanical way in which it is possible to clearly distinguish the roles of political authorities, experts, and citizens in scenarios legitimized by more or less efficient democratic practices.[1]

However, if, as EE suggests, human beings cannot control Nature (although we can and do exploit it, modify it, and influence it); if other views of what is desirable (different from those of Western capitalist culture and the ideology of extractivism) are considered legitimate and therefore value incommensurability is accepted (despite the ideal of the political homogeneity of people); if it is acknowledged that complex systems are unpredictable and scientific certainty is contingent, public policy is one among several collective action options. Moreover, it is a broad area of disagreement that covers methodological topics such as national accounting, the viability of sustainable development or, in the area of social, economic, and environmental regulations, proposals such as "good living" (*buen vivir*) from the 2008 Ecuadorian Constitution.

These debates are synthesized in the way Nature is valued, the role of human-made capital, and the distribution of extractive income, which is a key aspect in the countries that comprise the Amazon region.

At the heart of these disagreements lies the discussion over the sustainability of the Amazon region and the economic growth model that exploits it. EE has extended the concept of sustainability and has contributed to the definition of the conceptual and practical limits of economic growth and development. If we accept that Nature has biophysical limits, economic growth ad infinitum, as a necessary condition for human well-being, is a myth that is leading to the collapse of humanity.

The idea of *buen vivir*, the central concept of the 2008 Constitution of the Republic of Ecuador (Asamblea Constituyente, 2008), is an alternative to the teleology of economic growth as the only possible path toward development. The regime of *buen vivir* is an integrated system of social inclusion and well-being, together with recognition of the Rights of Nature.

This political declaration, formalized as a constitutional model, is one of the challenges humanity facing in the twenty-first century. Its application would have enabled social and economic relationships to be organized in a truly sustainable

---

[1] The "recommendations" of monetary, fiscal or developmental policies coming from multilateral bodies like the International Monetary Fund or the Organization for Economic Cooperation and Development are examples of this kind of process.

manner. In addition, it would have fostered a different relationship between human beings and Nature. This is exactly what Amazonia requires to survive into the future and continue to be the home for other cultures, providing its invaluable environmental services to the entire planet. It requires aligning, without ambiguity, objectives of preservation of all forms of life, regeneration processes, development of local productive capacities, active participation of people and nationalities in decision-making, and the gradual abandonment of rentism and extractivism.

33.3 Amazonia and the Contributions of the Barcelona School

Capitalism supports its accumulation process on continuous expansion that intensifies the exploitation of Nature's goods and services. Amazonia, a region rich in biodiversity and culture, is integrated into this process on a planetary scale as a link in the chains of international trade.

It has been connected to world markets since the nineteenth century as a provider of raw materials. First came the exploitation of natural rubber in harsh conditions, then the extraction of minerals, oil, and the aggressive expansion of monoculture and livestock production. The Ecuadorian Amazonia remained almost on the margin of global trade chains until much later. With oil exploration in the 1960s and the start of oil extraction in 1972, colonization intensified, new roads were built, and a disorderly change of land use began to expand agricultural activities, such as palm oil cultivation and extensive livestock production.

Today it is at the center of worldwide discussions because of forest fires, problems generated by extractive and productive activities, and global warming and climate change. EE has played a leading role in these discussions, with its theoretical principles, analysis of socio-environmental reality, and promotion of public policies.

Many consider the involvement of Amazonia in world markets to be pernicious, in the sense of a "curse of abundance," a reference to the plentiful availability of natural resources. These are neither a necessary nor sufficient condition for development, but in the African experience during the colonial period – especially the English and Belgian colonies – natural resources were associated with ethnic cleansing, pillage, and social exclusion. The relationship between the abundance of natural resources and the so-called bad development can also be called into question. The former does not necessarily cause the latter (Acosta, 2009, examines this situation). An objective balance of the use and control of natural resources should consider the social and environmental costs of extractive processes, the distribution and redistribution of income, the quantity of stocks and environmental-distributive conflicts, and the alteration of the lives of millenary people and nationalities.

No one challenges the ideal of getting beyond extractivism. However, presently the argument against it is that, given the urgent needs of the people in poor countries or those impoverished by their governments, by their domestic public policies and by asymmetrical international relations, it is not possible for them to manage without what would be their main – if not only – source of income.

The paradoxes of extractivism are much criticized, without any viable solutions being proposed. In practice, in order to get beyond it, other sources of income need to be available. In the final analysis, given the budgetary structures of these countries, they could not maintain the social programs nor the public investment in the formation of human capabilities, without the income generated by their primary exports. Any realistic alternative proposal would need to contemplate far-reaching fiscal reforms.

The extractivism of Amazonia was intensified from the 1980s, producing several adverse outcomes:

- *Social and environmental deprivation.* Especially because of the effect on cultures and people, including those in voluntary isolation. The social indicators of this region – particularly poverty and inequality – are worse than the national averages of the Amazonian countries. The change in land use due to deforestation, the intensification of extractive activities, monoculture expansion and livestock production have resulted in permanent losses in biodiversity.
- *Unequal exchanges.* The main economic consequence of the way Amazonia has been incorporated into world trade circuits has been the growing intensity in the exploitation of its natural resources. South American countries have a structural need to constantly increase their production of raw materials in order to acquire greater income, or to maintain the income they received when there is a fall in international prices for raw materials.
- In addition to this unequal economic exchange observed by Prebisch (1950, 1959), the raw materials markets conceal an ecologically unequal exchange: a mounting extractivist intensity to export more – in physical units – with the consequent increase in the cost to the environment and to ecological-distributive conflicts, as explored by Bunker (1984, 1985) and Muradian and Martínez-Alier (2001). Exports of natural resources and primary products are undervalued, since they do not include the social and environmental damage inherent in their extractive processes.
- These exchanges are also unequal when measured in calories (Falconí et al., 2017). The deterioration in terms of trade for food (in calories) causes a loss of self-sufficiency in food and damages the quality of diets (higher rated calories in nutritional terms – such as fruit – are exported and poorly rated calories – such as oils and fats – are imported). This exchange inequality constrains product diversification and generates deficient domestic consumption as well as loss of food self-sufficiency and sovereignty.
- *Dutch disease and vulnerability in the face of crises.* Extractivism usually provokes Dutch disease: an appreciation in the exchange rate, with a loss of competitiveness in domestic sectors not connected to the global market, and little or no product diversification, due to the influx of foreign currency arising from the increase in exports of one or several primary products. This macroeconomic distortion feeds back into recurrent economic crises.

- *Little diversification.* According to ECLAC, South America has an intensive export basket in primary products (55% of the total value of exports) and manufacturing based on natural resources (23%). Latin America and the Caribbean demonstrate an "emphasis on primary export specialization in the region" and a growing tendency to specialize in the provision of minerals and raw materials, which is reflected in their minor participation in the global value chains and their physical mineral trade balance (the difference between imports and exports measured in tons).
- *Deindustrialization.* An outcome linked to the previous one, shown as an important loss in industry in total value added.

This form of insertion generates diverse environmental conflicts documented in the Environmental Justice Atlas (EJAtlas; Temper et al., 2015). In Amazonia, the possibility of leaving oil underground challenges the conventional way of approaching the debate on sustainability and the use of natural resources, and poses other options for valuing Nature and the life it harbors.

33.4 The Yasuní-ITT Initiative

The Yasuní-ITT Initiative warned the world about a reality: it is not possible to continue extracting fossil fuels at the current rate because critical thresholds of planetary stability will be exceeded. Today, it continues to be a reference for confronting the climate crisis with the logic of degrowth of rich and industrialized economies, and the need to pay ecological and environmental debts accumulated in the North at the expense of the Global South.

In 2007, the Government of Ecuador proposed not to extract oil from the Yasuní National Park fields (in Ecuadorian Amazonia), a place of extraordinary biological diversity, in exchange for financial compensation, equal to half the net income that would have been generated from the extraction of 850 million barrels of heavy crude oil, from the international community.[2] Yasuní also provides benefits through ecosystem services for the conservation and preservation of life, including of people in voluntary isolation (Vallejo et al., 2015). The multi-criteria analysis (MCA) applied to Yasuní offers alternatives with different levels of evaluation, with a broader scope than the usual monetary cost–benefit analysis. It also demonstrated the viability of the Initiative when the definition of value is expanded.[3]

---

[2] In 2016, a technical increase of 920 million barrels was declared, which fixed the reserves of the Ishpingo-Tambococha-Tiputini (ITT) fields at 1672 million (El Comercio, 2016). This increase was certified by the North American Company, Ryder Scott. However, Espinoza et al. (2019) maintain that only 8.2% of the certified reserves were proven and probable. The remainder were possible reserves (28.5%) and contingent resources (63.2%).

[3] Burbano et al. (2017) applied an MCA to find alternative scenarios to the development being followed in Ecuadorian Amazonia.

If weak comparability of values – which implies incommensurability – is essential for EE, the tool to operationalize it is multi-criteria evaluation (Martínez-Alier et al., 1998). It is not that the unparalleled biodiversity of the Yasuní, the emissions of carbon dioxide averted and the rights of the people in voluntary isolation, are worth more or less than the 850 or 1672 million barrels of heavy crude oil; they just have a "different value" (Martínez-Alier, 2010).

Rafael Correa, the former president of Ecuador, announced the withdrawal of the Initiative on August 15, 2013. The government refused to hold a referendum that would have represented an important opportunity to have a democratic debate on extractivism.

The Initiative, as well as its official abandonment, provoked strong criticisms and several assessments (Acosta et al., 2009; Martínez, 2009; Narváez et al., 2013; Pelegrini et al., 2014; Pelegrini & Arsel, 2018). The reasons for this failure have yet to be fully explored, but several facts can be mentioned that led to this outcome. "Correa […] has never been an environmentalist, he is a typical left-wing Latin American economist, he is a classical *cepalino*,"[4] said Martínez-Alier (2010). Correa's final decision had several consequences. The most obvious was that there was no way of avoiding the "production of 410 million tons of carbon dioxide which correspond to 850 million barrels from the ITT…" (Ibid.). Now we know that it would have avoided much more.

The most important consequence of the triumph of extractivism was the curtailment of what would have been an exemplary public policy decision for Latin America and for the entire world. This initiative would have marked a decisive turning point in the struggle against climate change. "The success of the Yasuní-ITT Initiative […] could lead to imitation, in other words, to more and more fossil fuels being left below ground in places that are environmentally and/or socially sensitive," wrote Martínez-Alier (Ibid.).

An initial assessment of this episode might conclude that the internal factors and the limited international commitment combined to shut down a unique project for Ecuador and the whole world. It is ironic that in the same year, the Intergovernmental Panel on Climate Change (IPCC) report opened the way for the majority of fossil fuels to be kept underground (Le Page, 2013).

Fossil fuel emissions are intensifying global warming. The effects on the climate are multiple and complex: heat waves, droughts and fires, loss of ice mass. Meanwhile, national economies are slowly reducing their carbon emissions. The capitalist economy is addicted to fossil fuels.

The Amazon region is fundamental to the planet's climate equilibrium because of its high degree of biodiversity. Its forests absorb carbon, promote water cycle, and deliver global ecosystem services. Changes in land use due to productive and extractive activities (legal and illegal) affect human cultures, including people in voluntary isolation; they reduce forests size and cause irreversible biodiversity loss.

---

[4] The gentilic for an economist of the United Nations Economic Commission for Latin America and the Caribbean, ECLAC (in Spanish, Comisión Económica para América Latina y el Caribe, CEPAL).

The idea of leaving oil underground and avoiding burning it in the atmosphere needs to be discussed in world fora, including in the context of Covid-19. According to Carbon Tracker (2015), by 2025, fossil fuel extraction companies will be investing 2 billion dollars in carbon, oil, and gas projects. These investments pose a high environmental risk, since burning these fuels would exceed the climate irreversibility threshold of 2 °C, as advised by the scientific community. If we really want to face up to climate change and promote renewable energy, these investments would be counterproductive.

Martínez-Alier asks: "Where should these fossil fuels, the product of photosynthesis from times in the distant past, be left underground?" and he answers: "It makes sense for them to be in places like Yasuní-ITT and other similar locations on account of their environmental value and the social risks" (2010). Presently, with the crisis of global capitalism made worse by the Covid-19 pandemic, the prices of oil and other raw materials have fallen (UNCTAD, 2020). This makes several oil and mining projects unviable, given their economic cost structures, not to mention environmental costs. This reinforces the idea of leaving oil underground and hastening the passage toward a post-oil economy.

An adequate assessment of the Yasuní-ITT Initiative recognizes its relevance for the mechanisms of socio-environmental transition, the need for degrowth, and social and environmental justice from a Latin American perspective. However, it also recognizes the error of proposals focused solely on financial options. Ultimately, it is necessary to broaden the universe of values implicit in the analysis.

33.5 Conclusions

Human ecology is subsumed in social and political institutions. Therefore, "the conflicts between rich and poor cannot be hidden behind the screen of an ecological pseudo-rationality," warned Joan Martínez-Alier 27 years ago. It was not possible, he said, to base human decisions on a "new technocratic ecologism" (1994), thus foreshadowing the path down which EE and environmentalism of the poor would evolve over the past quarter century.

This has been a journey toward a radical epistemic rupture, different from previous moments of paradigmatic crisis. It has not been a passage from a moment of normal science toward another such normal moment, but rather a path toward an unprecedented moment of postnormal science based on an organicist ontology, more real than that of orthodox economics and capable of harboring different epistemologies and methodologies. This blurs the boundaries between science and nonscience; it opens the doors to a broader dialogue; and it reduces the distance between the academy and society.

EE assumes that scientific ethics are not neutral. Strengthened by interdisciplinary research, it is capable of drawing attention to the heterogeneous conflicting interests that are expressions of incommensurable but equally valid values. This enables it to carry out broader analyses that include an increasing number of ecological-distributive conflicts, as economic growth requires increasing levels of resources and generates increasing amounts of wastes.

The attention given to Amazonia by EE has not happened by chance. This region is the scene of conflicts of values and interests that are both local and global in scope. It is subject to a serious process of social and environmental deterioration, which has been accelerating since the 1980s.

EE actively supported the Ecuadorian government's proposal not to extract oil from the fields under the Yasuní National Park, an area of extraordinary biological diversity occupied by communities in voluntary isolation. The initiative's failure cut short what would have been an exemplary public policy decision that would have marked a turning point in the struggle against climate crisis. The crisis of globalized capitalism, made worse by the current Covid-19 pandemic, has made several extractivist projects nonviable – including from a financial perspective.

Based on the premise of the human impossibility of controlling Nature, of value incommensurability, and uncertainty that characterizes complex systems, EE has been the main subject of several discussions concerning Amazonia. The contributions of these and other debates found their historical synthesis in the notion of *buen vivir* (*sumac kawsay*), sanctioned by Ecuadorians in the 2008 Constitution of the Republic of Ecuador. This political, cultural, and social proposal recognizes the Rights of Nature and seeks to apply an integrated system of inclusive social well-being. Its origins lie with Andean cultural traditions, not with EE. However, as this chapter has shown, *buen vivir* is an option for the achievement of prosperity, without being dependent on economic growth.

References

Acosta, A. (2009). *La maldición de la abundancia*. Comité Ecuménico de Proyectos (CEP).
Acosta, A., Gudynas, E., Martínez, E., & Vogel, J. (2009). Dejar el crudo en tierra o la búsqueda del paraíso perdido. Elementos para una propuesta política y económica para la iniciativa de no explotación del crudo del ITT. *Polis: Revista de la Universidad Bolivariana, 8*(23), 429–452.
Asamblea Constituyente. (2008, October 20). *Constitución de la República del Ecuador* (Registro Oficial N. 449). Asamblea Nacional.
Bunker, S. (1984). Modes of extraction, unequal exchange, and the progressive underdevelopment of an extreme periphery: The Brazilian Amazon, 1600–1980. *The American Journal of Sociology, 89*, 1017–1064.
Bunker, S. (1985). *Underdeveloping the Amazon: Extraction, unequal exchange and the failure of the modern state*. University of Chicago Press.
Burbano, R., Larrea, C., & Latorre, S. (2017). Análisis multicriterial sobre alternativas para el desarrollo en la Amazonia. In C. Larrea (coordinador), *¿Está agotado el período petrolero en Ecuador? Alternativas hacia una sociedad más sustentable y equitativa: Un estudio multicriterio* (pp. 419–450). Universidad Andina Simón Bolívar, Ediciones La Tierra.
Carbon Tracker. (2015). *The $2 trillion stranded assets danger zone: How fossil fuel firms risk destroying investor returns*. Carbon Tracker Initiative. https://www.carbontracker.org/reports/stranded-assets-danger-zone

El Comercio. (2016). Reservas en el ITT suben de 920 millones de barriles a 1672 millones, según Petroamazonas. *Diario El Comercio*, 14 July 2016. https://www.elcomercio.com/actualidad/reservas-itt-petroleo-petramazonas-crudo.html

Espinoza, V., Fontalvo, J., Martí-Herrero, J., Ramírez, P., & Capellán, I. (2019). Future oil extraction in Ecuador using a Hubbert approach. *Energy, 82*, 520–534.

Falconí, F., Ramos-Martin, J., & Cango, P. (2017). Caloric unequal exchange in Latin America and the Caribbean. *Ecological Economics, 134*, 140–149. https://doi.org/10.1016/j.ecolecon.2017.01.009

Funtowicz, S., & Ravetz, J. (1993, September). Science for the post-normal age. *Futures*, 25.

Funtowicz, S., & Ravetz, J. (2000). *La ciencia post normal: Ciencia con la gente*. Icaria.

Georgescu-Roegen, N. (1971). *The entropy law and economic process*. Harvard University Press.

Kapp, W. (1950). *The social costs of private enterprise*. Shocken.

Kapp, W. (1976). The open system character of the economy and its implications. In F. A. Klink (Ed.), *1995, Economía de los Recursos Naturales: Un Enfoque Institucional*. Fundación Argentaria/Visor Distribuciones.

Kuhn, T. (2004). *La estructura de las revoluciones científicas*. Fondo de Cultura Económica.

Le Page, M. (2013, September 30). IPCC digested: Just leave the fossil fuels underground. *New Scientist*. https://www.newscientist.com/article/dn24299-ipcc-digested-just-leave-the-fossil-fuels-underground/#ixzz618h3IWRN

Lizarazo, G. S. (2018). Economía ecológica y la construcción epistemológica de una ciencia revolucionaria para la sostenibilidad y la transformación del mundo. *Gestión y Ambiente, 21*(supl. 1).

Lovejoy, T., & Nobre, C. (2018). Amazon tipping point. *Science Advances, 4*(2).

Martínez, E. (2009). *Yasuní. El tortuoso camino de Kioto a Quito*. Abya-Yala.

Martínez-Alier, J. (1990). *Ecological economics: Energy, environment and society*. Basil & Blackwell.

Martínez-Alier, J. (1994). *De la economía ecológica al ecologismo popular*. Icaria.

Martínez-Alier, J. (1998). *Curso de economía ecológica*. Programa de las Naciones Unidas para el Medio Ambiente (PNUMA).

Martínez-Alier, J. (2006). Los conflictos ecológico-distributivos y los indicadores de sustentabilidad. *Polis, Revista Académica de la Universidad Bolivariana de Chile, 5*(13).

Martínez-Alier, J. (2010). En Ecuador: la Iniciativa Yasuní-ITT se encamina al triunfo. *Acta Sociológica, 54*.

Martínez-Alier, J., & Naredo, J. (1982). A Marxist precursor of energy economics: Podolinsky. *The Journal of Peasant Studies, 9*(9), 207–224.

Martínez-Alier, J., Giuseppe, M., & John O'Neill, J. (1998). Weak comparability of values as a foundation for ecological economics. *Ecological Economics, 26*(3), 277–286.

Meadows, D., et al. (1972). *Los límites del crecimiento. Informe al Club de Roma sobre el predicamento de la humanidad*. Fondo de Cultura Económica.

Muradian, R., & Martínez-Alier, J. (2001). Trade and the environment: From a 'southern' perspective. *Ecological Economics, 36*(2), 281–229.

Narváez, I., de Marchi, M., & Pappalardo, E. (2013). Yasuní: en clave de derechos y como ícono de la transición para ubicarse en la selva de proyectos. In I. Narváez, I., M. Marchi, & E. Pappalardo (Co-ordinators), *Yasuní, zona de sacrificio: análisis de la Iniciativa ITT y los derechos colectivos indígenas* (pp. 9–26). Flacso Ecuador.

Pellegrini, L., & Arsel, M. (2018). Oil and conflict in the Ecuadorian Amazon: An exploration of motives and objectives. *European Review of Latin American and Caribbean Studies/Revista Europea de Estudios Latinoamericanos y del Caribe, 106*, 209–218.

Pellegrini, L., Arsel, M., Falconí, F., & Muradian, R. (2014). The demise of a new conservation and development policy? Exploring the tensions of the Yasuní-ITT initiative. *The Extractive Industries and Society, 1*(2), 284–291.

Prebish, R. (1950). *The economic development of Latin America and its principal problems*. UN Economic Commission for Latin America.

Prebish, R. (1959). Commercial policy in the underdeveloped countries. *American Economic Review, 49*, 251–273.
Soddy, F. (1933). *Wealth, virtual wealth and debt. The solution of the economic paradox*. Britons Publishing Company.
Temper, L., Del Bene, D., & Martínez-Alier, J. (2015). Trazando las fronteras y las líneas del frente de la justicia ambiental global: las EJAtlas. *Revista de Ecología Política, 22*, 255–278.
UNCTAD. (2020). *The Covid-19 shock to developing countries: Towards a "whatever it takes" programme for the two-thirds of the world's population being left behind*. United Nations. https://unctad.org/en/PublicationsLibrary/gds_tdr2019_covid2_en.pdf
Vallejo, M. C., Burbano, R., Falconí, F., & Larrea, C. (2015). Leaving oil underground in Ecuador: The Yasuní-ITT Initiative from a multi-criteria perspective. *Ecological Economics, 109*, 175–185.

Open Access This chapter is licensed under the terms of the Creative Commons Attribution 4.0 International License (http://creativecommons.org/licenses/by/4.0/), which permits use, sharing, adaptation, distribution and reproduction in any medium or format, as long as you give appropriate credit to the original author(s) and the source, provide a link to the Creative Commons license and indicate if changes were made.

The images or other third party material in this chapter are included in the chapter's Creative Commons license, unless indicated otherwise in a credit line to the material. If material is not included in the chapter's Creative Commons license and your intended use is not permitted by statutory regulation or exceeds the permitted use, you will need to obtain permission directly from the copyright holder.

Index

A
Absolute decoupling, 167, 169–172
Activism, 4–7, 18, 37, 195, 198, 200, 206, 214, 215, 217, 229, 263, 283, 284, 286, 289, 293, 294, 297, 299, 312
Activism mobilizing science (AMS), 4, 5, 7, 22, 27, 261–268
Acuña, M., 158, 161
African countries, 205, 331
African Union, 65
Agarwal, B., xiii, 10
Agrobiodiversity, 7, 319–324
Amazon, 8, 74, 142, 147, 160, 205, 263, 265, 267, 383, 386, 390
Amazonia, 158, 383–392
Anschauung, 42, 158–162
Anthropocene, 39, 41, 127, 157
Areva, 263–265
Argentina, vii, 213
Atahualpa, 14
Ateneu de Nou Barris, 288
Ayres, R., 124

B
Babri Masjid, xiii
Baeza, xiv, 95, 96
Bagua, 14
Bangalore, ix–xi, xiv–xvii, 305
Barcelona, v, x–xv, xix–xxii, 4, 9, 10, 47–55, 83–89, 93, 95, 97–99, 133, 142, 150, 151, 198, 213, 221, 222, 224, 225, 249, 271–273, 275–278, 283–290, 295, 296, 298, 299, 301, 306, 335, 347, 352

Barcelona school, xix, 3–14, 17–28, 37–40, 42, 43, 47, 50–54, 72, 74, 76, 83–89, 110, 111, 116, 117, 124–127, 130, 131, 137, 138, 142, 147–152, 157, 167, 206, 220–229, 249, 252–254, 256, 271, 272, 277, 283–290, 293, 294, 299, 319, 383, 387–389
Beijer Institute of Ecological Economics, 10
Beltine greenbelt project, 239
Belur, xi
Berneri, L., xii
Biodiversity, 3, 8, 23, 24, 48, 71, 73–75, 99, 100, 168, 252, 310, 320–323, 367, 369, 370, 372, 374, 376, 384, 385, 387, 388, 390
Bloomington School, 19
Blut und Boden, xi
Bolivia, vii, 213, 223
Bombay, xiii
Bookchin, M., 272, 273, 275, 277
Boulding, K., 10
Breitbart, M., 272
Buen Vivir, 43, 61, 62, 65, 67, 161, 163, 386, 392
Bunker, S., 147, 388
Buryat Peoples, 311

C
Cajamarca, 14, 160, 161
Campesinos, 157, 158, 160–162, 227
Campos, P., 94, 97, 98
Can Masdeu, 277, 283, 284, 290
Capitalism, Nature, Socialism, xx
Capitalismo, Natura, Socialismo, xx

Capra, F., xviii
Care work, 290, 320
Caruajulca, Y., 162
Castro, F., 149
Catalonia, xii, xiv, xvii, xxi, 26, 272, 273, 275–277, 284, 289, 298, 347, 350
Cavalcanti, C., 10, 158, 161
Chayanov, A.V., 93, 94
China, 115, 160, 213–215, 311, 373
Chipko movement, ix, xvi
Cine Princesa, 287
Circular economy, 7, 13, 116, 117, 166, 167, 171, 172, 338, 347–349, 369
Cleveland, C.J., 10
Climate change, 7, 25, 52, 71, 72, 74, 168, 186, 252, 283, 296–297, 310, 322, 327–332, 335–345, 367, 371, 387, 390, 391
Climate crisis, 48, 165, 173, 389, 392
Collectivization, 272–273, 275, 276
Commodity extraction frontiers, 13, 14, 182, 205, 310
Commoning, 25, 62, 88, 219–230, 255
Community-based natural resource management, 6, 25, 219
Community Organizations for Water Services and Sanitation, 224
Confederación Nacional del Trabajo (CNT), 275–277
Consumption-based accounting (CBA), 168, 169
Coordination of the Indigenous Organizations of the Amazon Basin (COICA), 224
Co-production of knowledge, 7, 22, 76, 200
Córdoba, xix, 95
Costanza, B., 10
COVID-19, 165, 167, 171, 173, 293, 301, 358, 360
CRIIRAD, 263–267
Critical cartography, 197, 198, 205, 206
Cuban Revolution, vii
Cunfer, G., 98, 99

D
Daly, H., v, 10, 12
Darwin, C., xx
Decoloniality of knowledge, 55
Décroissance, 84
Degrowth, xix, 3, 6, 12, 14, 26, 27, 54, 62, 66, 83–88, 162, 163, 172, 220–222, 228–229, 249, 250, 252, 254–256, 272, 275, 277, 278, 284, 287, 359, 369, 384, 389, 391

Degrowthers, 85, 228
Deliberative ecological economics, 53, 54
Democratic Republic of the Congo, 205
Der Kreislauf des Leben, xx
Development, v, xiii, 4, 5, 7, 8, 10, 19, 22, 24, 25, 27, 38, 49, 51, 54, 55, 59–67, 74, 75, 84, 86, 94, 96, 100, 109–111, 124, 125, 139, 147, 148, 151, 163, 166, 169, 182–185, 187, 197, 198, 212–214, 216, 221, 224, 228, 237–240, 243, 254, 273, 274, 289, 295, 308–310, 320, 322–324, 327, 329, 335, 359, 361, 364, 367, 369, 370, 372, 374, 376, 377, 384, 386, 387
Distributional conflicts, 21, 23–25, 183
Domestic production, 320
Down to Earth, xx

E
Ebro River, 347
Ecofeminism, 37, 61, 65, 197
Ecología Política, xx, 96, 271, 273, 274
Ecological debt, 23, 51, 131, 149–152, 385
Ecological distribution conflicts (EDC), 6, 12, 14, 18, 21, 23–25, 27, 49, 50, 52, 54, 61, 85, 86, 88, 116, 157, 158, 162, 181, 183, 196, 219, 240, 252, 274, 309, 368
Ecological footprint, 12, 151
Écologie Politique, xx
Einstein, A., 117
El Guinardó, 276
Embodied energy analysis (EEA), 128, 133, 342
Energy Returns of Energy Investment (EROI), 10, 85, 94, 99
Environmental justice, 3, 4, 6, 7, 14, 22–25, 27, 41, 49–51, 55, 61, 89, 149, 150, 160, 195–198, 206, 211–216, 221, 223, 228, 229, 241, 261, 264, 267, 274, 277, 278, 285, 288, 289, 305–312, 332, 347, 385, 389, 391
Environmental justice movement, 6, 7, 14, 24–27, 86, 88, 185, 200, 212, 216, 219–230, 240, 305, 306, 309, 368, 369
Environmental Kuznets Curve (EKC), 166
Environmental Science, 9, 20, 126, 131, 287, 296
Environmental social sciences, 12, 47, 157, 211–217
Environmental Sociology, 213
Environmentalism of the poor, xvi, 18, 27, 63, 184, 220, 222, 241, 305–312, 368, 391

Escobar, A., 21, 59, 67, 87, 184, 189, 207, 222, 307, 308, 311
Ethnoecology, xix, 157
Europe, viii, xi, xv–xvii, xxi, 10, 86, 99, 123, 163, 213, 224, 250, 272, 288, 306, 307, 337, 348, 351
European Commission, 337, 348
European Ecosocialist Manifesto, 96
European Green Deal (EGD), 166
European Union (EU), 84, 116, 139, 166, 171, 332, 336, 338, 348, 349, 351, 353
Exergy Replacement Cost (ERC), 128, 133, 195
Extended producer responsibility (EPR), 348, 351, 353
Extractivism, 26, 66, 88, 129, 130, 162, 168, 198, 207, 214, 253, 312, 383, 386–388, 390
Extractivist projects, 183, 184, 392

F
Falconí, F., 6, 8, 10, 24, 126, 138, 151, 287, 383–392
Farley, J., 10
Federación Anarquista Ibérica (FAI), 275
Fischer-Kowalski, M., 10, 11, 18, 19, 124, 125, 127, 150
FLACSO, 9
French Royal Academy of Sciences, xix
Funtowicz, S., xviii, 9, 74, 76, 384

G
Gadgil, M., x, 21
Gandhi, x, xv–xvii, 284
García-Ramon, M.D., 272, 273, 276
Garrabou, R., 95–97
Gavaldá, M., 277, 286, 287
Geddes, P., x, xii, 274
Generalitat, 287
Georgescu-Roegen, N., 5, 10, 11, 13, 37–43, 84, 86, 109, 110, 142, 158, 159, 161, 162, 183, 384
Giampietro, M., xxi, 5, 10, 18, 26, 38, 40, 97, 109–117, 137, 142, 347
Giralt, M., xiii
Global material flows, 169
Global North, 18, 21, 26, 65, 66, 99, 148, 149, 152, 169, 188, 228, 243, 298–299, 302
Global South, 18, 21, 23, 26, 51, 66, 149–151, 169, 188, 220, 223, 236, 298–299, 302, 306, 307, 309, 369, 371, 377, 389
Global supply chain (GSC), 128, 133

Gomez-Baggethun, E., 8, 10, 24, 299, 357–364, 372
Gonsalez de Molina, M., xiv, xix, xxi, 5, 17, 26, 93–101, 124, 273
Good life, 5, 43
Gowdy, J., 10
Green Keynesianism, 63
Green New Deal, 63
Greenhouse gas (GHG), 7, 112, 168–170, 327, 331, 336, 337, 340–344
Gross domestic product (GDP), 11, 12, 27, 84, 86, 87, 139, 151, 165, 167, 168, 170, 252, 330
Guzmán, G., 97–100

H
Halebid, xi
Hall, C., 10
Hall, S., 85
Handbook of Environmental Sociology, xx
Hanley, N., 48
Harvard University, xx
Hayek, F., xii
Hayek, F.A., 12
Hernandez, M., vi, viii
Hershberg, E., xiii
Himalayas, xvi, 202
Hogben, L., xii
Homage to Catalonia, vi, xi
Homo sapiens, xvii
Howarth, R., 10, 53
Human appropriation of net primary productivity (HANPP), 12
Human-environment interactions, 19, 27

I
Iberian anarchism, 271–278
Iberian Peninsula, 274, 284, 287
India, ix–xi, xiii, xv–xvii, xix, xx, 9, 10, 13, 21, 51, 65, 75, 202, 252, 295, 305–307, 309
Indian Supreme Court, 51
Indigenous and local knowledge (ILK), 71–76
Indigenous People, xix, 14, 61, 65, 74, 157, 158, 263, 266, 284, 309–312, 369
Indigenous Peoples and local communities (IPLC), 71–76
Input-output analysis (IOA), 128, 132, 133
Institute of Environmental Science and Technology (ICTA), 4, 10, 12, 37, 83, 84, 97, 126, 222, 249, 296, 297
Institutional ecological economics, 19

Intergovernmental Panel on Climate Change (IPCC), 328, 331, 332, 342, 344, 390
International Monetary Fund, 149
International Society for Ecological Economics (ISEE), v, xix, 10, 37, 127, 162
International trade, 12, 112, 127, 129, 130, 132, 141, 148, 152, 188, 387

J
Jacinta Kerketta, 309
Jaen, viii, 96
Jaume Terradas, 9
Journal of Latin American Studies, viii
Journal of Political Ecology, 150

K
Kallis, G., 5, 7, 10, 21, 25–27, 51, 52, 83–89, 167, 170, 228, 274, 275, 277, 357, 358
Kapp, W., 384
Kerala, xvi
Khmer Rouge, viii
Kneese, A., 11, 124
Koestler, A., xviii
Kothari, A., xi, 25, 26, 55, 59, 86, 195, 309
Krausmann, F., 23, 98, 99, 170
Kuhn, T., 384
Kurian, P., ix, x

L
Landfill Allowance Trading Schemes (LATS), 351, 352
Landfill Directive, 349
Land-use change (LUC), 128, 133, 376
Languages of valuation, 5, 19, 21, 22, 25, 47–55, 85–87, 160, 162, 181, 249, 331, 385
Lankesh, G., xv
La Rábida, 95, 96
Latifundios, 94
Latin America, vii, viii, xi, xiv–xvii, 10, 23, 51, 115, 127, 139–142, 148–150, 152, 213, 220, 225, 272, 274, 295, 299, 307, 390
Latin America and the Caribbean (LAC), 6, 123–127, 129–131, 138, 140, 141, 147, 389, 390
Latouche, S., 60, 87
Leff, E., xviii
Lévi-Strauss, C., xix
Life cycle analysis (LCA), 128, 132, 133

Limits to growth, 6, 63, 85, 87, 88, 116, 166, 357
Locally unwanted land use (LULU), 25, 236
Lodi Gardens, xiii
London, vi, xv, xvi, xx, 339
López-Linage, J, 97
Los Angeles, 237
Los Huacchilleros del Perú, 220

M
Mäler, K.-G., 10
Malthus, T., 274
Malthusianism, 274
Margalef, R., 100
Martínez-Alier, J., v, vi, ix, xix, 4, 9–14, 18–24, 37–39, 47–51, 53, 55, 61, 75, 84, 109, 123, 124, 126, 130, 131, 137–139, 142, 148–151, 157–160, 181, 183, 184, 188, 189, 195, 196, 198–201, 207, 216, 220–222, 228, 229, 241, 249, 252, 253, 261, 272, 283, 297, 305, 335, 368, 369, 372, 377, 383–385, 388, 390, 391
Marull, J., xxi, 18, 100
Marx, K., x, xiii, xx, 94, 124, 183, 187, 226, 359, 361
Marxism, ix–xii, xx, 254
Masjuan, D., 276
Masjuan, E., 274, 277, 278, 284, 286
Material flow analysis (MFA), 6, 123–129, 131–133, 151, 347
May, P., 10
Mayumi, K., 38, 40, 110, 116, 158–160, 347
McNeill, J., 96
Mendes, C., 14
Mexico, xxii, 51, 97, 129, 205, 213, 223, 319–324, 368, 371, 373–376
Mientras Tanto, 95
Mill, J.S., 362
Moleschott, J., xx
Monetary-based policy tools, 51
Morelia, xxii, 97
Mujeres Libres, 273
Multi-Scale Integrated Analysis of Societal and Ecosystem Metabolism (MUSIASEM), 5, 18, 109, 132, 133, 347
Mumford, L., xii
Munda, B., 309
Munda, G., 9, 20, 21, 49, 296, 347
Muradian, R., 3–8, 10, 17–28, 126, 138, 139, 150, 151, 157, 162, 181, 183, 186, 187, 189, 370, 372, 373, 388

N

Narodnik movement, 220
Nature, xx, xxi, 5, 11, 14, 18, 21, 25, 48, 49, 52, 54, 55, 61, 63, 65–67, 72–74, 86, 87, 95, 124, 130, 149, 182, 183, 186, 200, 212, 219, 224, 225, 236–238, 241, 243, 254, 298, 308, 310, 312, 324, 338, 352, 368, 369, 371–375, 384, 386, 387, 389, 392
Neoclassical economic theory, 64, 294, 368
Neoliberal agenda, 368
Neoliberalization, 225, 227
Neoliberalization of nature, 368
Neo-Malthusian, 274
Neo-Marxist perspective, 182, 187
Neo-narodnism, xvi, 96
Nettlau, M., 274
Neurath, O., xii, 12
New Delhi, ix, xiii
NGO, x, 385
Niger, 263–267
Nobel Prize, 328
North America, xi, xvi, xvii, 99, 213, 236, 306
Not in my backyard (NIMBY), 253

O

Oaxaca, xxii, 324
Odum, H.T., 10, 85, 109, 110
OECD, 65, 138, 351, 358, 363
O'Hara, S., 10
On the Origin of Species, xx
Ortega, M., 283
Our Common Future, 63
Oxford, vii, viii, xi–xiii, 9, 93
Oxford University, vi
Özkaynak, B., 10

P

Pakistan, 205
Parc de les Aigües, 276
Paris, xi, xv, 9, 83, 84, 94, 168, 271, 274, 285, 338, 345
Participatory Action Research (PAR), 215, 262, 293–294, 296, 299–302, 320
Payments for Ecosystem Services (PES), 8, 367, 368, 370–377
Pearce, D., 10
Peet, R., 222, 272
Perrings, C., 10
Peru, vii, 9, 14, 85, 160, 162, 220, 265, 266
Pinchot, G., xii
Pizarro, 14

Pluriversal alternatives, 65, 66
Pluriversality, 62, 66, 67
Podolinsky, S., x, xi, 94, 100, 383
Policy-making, 49
Political ecology, xix, xx, 3, 5, 6, 9–14, 17–28, 37, 42, 43, 52, 72, 74, 83, 84, 96, 116, 131, 133, 157, 160, 167, 172, 181, 182, 189, 196, 198, 199, 206, 213–215, 221–223, 229, 243, 249, 271–278, 298, 299, 302, 312, 368, 372, 373, 383–385
Polochic, 299
Pontificia Universidad Católica del Perú, 162
Popper-Lynkeus, J., 357, 362, 363
Post-development, xix, 5, 54, 59–62, 65–67, 84, 86, 163
Post-Development Dictionary, 60, 63, 65
Post-normal science, xviii, 19–22, 38, 43, 49, 72, 74, 76, 137, 294, 296, 299, 384
POUM, xi
Prebisch, R., 130, 139, 148, 149, 151, 388
Primitive accumulation, 187
Proops, J., 10, 49

Q

Quito, 9

R

Rajasthan, 21
Ramos Martin, J., 6, 10, 40, 110, 116, 137–143
Ravetz, J., xviii, 20, 74, 76, 262, 294, 384
REDD+, 371
Regional Environmental Change, 98
Rentism, 387
Reproductive work, 320
Resource decoupling, 170
Royal Society, xix
Ruedo Ibérico, vi, xi
Russel, B., 362
Russi, D., 10, 21, 151, 160

S

Sacredness, 11, 21, 49–51, 86
Sacristán, M., 94, 95
Salleh, A., 59
Scheduled Tribes, 309
Schlüpmann, K., xiii, 39, 109
Selva Lacandona, 374–376
SEMARNAT, 323
Sevilla, E., xiv, xix, 294
Sevilla Guzmán, E., 95

Shiva, V., 60, 61, 295, 302
Shravanbelagola, xi
Sieferle, R.P., 96
Simon, H.S., xii
Social conflicts, 6, 18, 182, 183, 185, 186, 273, 375
Social-ecological economics, 19
Social metabolism, xviii, xx, xxi, 3–6, 11, 12, 17, 18, 23, 25–27, 49, 85, 87, 96, 97, 101, 123–125, 127, 131, 150, 157, 158, 162, 181–183, 185–189, 196, 284, 385
Social multi-criteria evaluation (SMCE), 20–22
Society and nature, xviii, xx, 125, 182
Soddy, F., x–xii, 383
South Africa, xvii, 65, 214
South America, 65, 138–140, 163, 307, 321, 389
South East Asia, 87
Soviet countries, 205
Soviet Union, 94
Spain, vi, vii, xi–xvii, xix, 9, 26, 47–55, 94–96, 98, 213, 220, 223, 225, 241, 271–273, 275–277, 290, 296, 298, 360
Spanish Civil War, vi, xii, 272, 273, 275–278
Spanish Society for Agrarian History, 96
Spash, C., 48, 49, 53
St. Anthony's College, xii
Stefan Baumgärtner, 10
Stoffwechsel, xx
Sub-Sahara Africa, 295
Subsistence, 18, 60, 61, 124, 187, 266, 320–322
Sustainable Development Goals (SDGs), 60, 62, 166
Sustainable Farm Systems (SFS), 98–101

T
Tello, E., xxi, 5, 93–101
The Ecologist, xx, 256
The Entropy Law and the Economic Process, 13, 42
The Environmentalism of the Poor, viii, xiii, xiv, xxi, 14, 50, 96, 158, 249, 284, 305, 368–370, 377, 385
The Global Atlas of Environmental Justice, xxii, 6, 131, 195, 206
Third World, x, xvi, xvii
Toledo, V., xiii
Tossa del Mar, vi, viii

Tragadero Grande, 161
Transdiscipline, 38
Trickle down, 51
Turkey, 51, 213, 223

U
UNAM, 97
UN Conference for Sustainable Development, 63
Unequal Ecological Exchange, 129, 130
United Kingdom (UK), xx, 148, 213, 339, 352
United Nations, 63, 65, 143, 371
United Nations Environmental Programme (UNEP), 63, 127, 168
United States of America (USA), v, xx, xxi, 10, 116, 139, 213, 351

V
Value incommensurability, 48, 386, 392
Value pluralism, 21, 48, 53, 54
Vernadsky, V., 94
Verona Stolcke, xiii
Via Campesina, 88, 96
Vienna Social Ecology School, 19
Vikalp Sangam, 67
Von Mises, V., 12

W
Washington, DC, xix, 10
Western Ghats, 21
Willingness-to-Pay (WTP), 48
World Bank, v, 149, 367, 371

X
Xalapa, xxii, 97

Y
Yale University, ix, xxii, 9
Yanacocha, 14, 158, 160, 161
Yasuní-ITT Initiative, 383, 389–391

Z
Zamora, Z., 162
Zapatista revolution, 220
Zaragoza, 284, 285, 289